Probabilistic Constrained Optimization

AF182563

Nonconvex Optimization and Its Applications

Volume 49

Managing Editor:

Panos Pardalos
University of Florida, U.S.A.

Advisory Board:

J.R. Birge
Northwestern University, U.S.A.

Ding-Zhu Du
University of Minnesota, U.S.A.

C. A. Floudas
Princeton University, U.S.A.

J. Mockus
Lithuanian Academy of Sciences, Lithuania

H. D. Sherali
Virginia Polytechnic Institute and State University, U.S.A.

G. Stavroulakis
Technical University Braunschweig, Germany

Probabilistic Constrained Optimization

Methodology and Applications

Edited by

Stanislav P. Uryasev
University of Florida,
Gainesville, Florida, U.S.A.

KLUWER ACADEMIC PUBLISHERS
DORDRECHT / BOSTON / LONDON

A C.I.P. Catalogue record for this book is available from the Library of Congress.

ISBN 978-1-4419-4840-3

Published by Kluwer Academic Publishers,
P.O. Box 17, 3300 AA Dordrecht, The Netherlands.

Sold and distributed in North, Central and South America
by Kluwer Academic Publishers,
101 Philip Drive, Norwell, MA 02061, U.S.A.

In all other countries, sold and distributed
by Kluwer Academic Publishers,
P.O. Box 322, 3300 AH Dordrecht, The Netherlands.

Printed on acid-free paper

All Rights Reserved
© 2001 Kluwer Academic Publishers
Softcover reprint of the hardcover 1st edition 2001
No part of the material protected by this copyright notice may be reproduced or
utilized in any form or by any means, electronic or mechanical,
including photocopying, recording or by any information storage and
retrieval system, without written permission from the copyright owner.

Contents

Preface

There has been much recent progress in the theory of probabilistic functions and related applications. Probabilistic and quantile (percentile) functions are commonly used for the analysis of models with uncertainties or variabilities in parameters. For instance, financial applications consider the probability of profitable transactions or the probability of closing of a portfolio position at a specified price (limit order), in risk and reliability, performance functions, characterizing operation of systems, are formulated as probabilities of successful or unsuccessful accomplishment of their missions, e.g., core damage probability or frequency of a nuclear power plant, in rocket engineering - probability of successful landing of a rocket or an aircraft, and percentiles of risks in public risk assessments.

In financial applications, the percentile of the losses is called Value-at-Risk (VaR). VaR, a widely used performance measure, answers the question: what is the maximum loss with a specified confidence level? Percentiles are also used for defining other relevant risk performance measures, such as Conditional Value-at-Risk (CVaR). CVaR (also called Mean Excess Loss, Mean Shortfall, or Tail VaR) is the average loss for the worst x% scenarios (e.g., 5%).

This volume discusses various theoretical aspects of sensitivity analysis and optimization of probabilistic functions, quantiles, and related issues such as, robust Monte Carlo simulation methods, and statistical characteristics of optimal solutions of stochastic programs. The volume begins with an introductory review paper, which covers several recently developed topics:

- sensitivities of probabilistic functions;
- sensitivities of percentiles (VaR);
- optimization approaches for CVaR.

The main focus of this volume is on financial applications of probabilistic functions: (1) portfolio optimization with CVaR performance functions and constraints; (2) asset and liability management; and (3) optimal trading strategies for options. However, other important applications of probabilistic functions also are well covered, including reliability analyses and optimal design of the stochastic systems, optimization of the material flow systems, and multimedia networks.

Significant attention in this book is paid to estimating, evaluating, and comparing percentile risk measures, in particular, VaR and CVaR. CVaR is a more consistent measure of risk compared to VaR since it is sub-additive, convex and has other nice mathematical properties (see paper by Pflug in this volume). Moreover, it can be optimized using linear programming optimization algorithms, which can handle applications with very large numbers of variables and scenarios (see details in the introductory review paper). Numerical experiments indicate that the minimization of CVaR also leads to near optimal solutions in VaR terms because CVaR is always greater than or equal to VaR. Similar measures as CVaR have been earlier studied in the stochastic programming literature. The conditional expectation constraints and integrated chance constraints described in the book by Prekopa (*Stochastic Programming, Kluwer, 1995*) may serve the same purpose as CVaR.

The collection of papers in this book covers a diverse number of topics related to probabilistic functions. The book will be a valuable source of information to faculty, students, researchers, and practitioners in financial engineering, operation research, optimization, computer science, and related areas. I would like to take the opportunity to thank the authors of the papers, the anonymous referees, and Kluwer Academic Publishers for helping with the publication of this volume.

Stanislav P. Uryasev
University of Florida
May 2000

Introduction to the Theory of Probabilistic Functions and Percentiles (Value-at-Risk)

S. Uryasev (uryasev@ise.ufl.edu)
University of Florida
474 Weil Hall, Gainesville, FL 32611-6595

Abstract

Probabilistic and quantile (percentile) functions are commonly used for the analysis of models with uncertainties or variabilities in parameters. In financial applications, the percentile of the losses is called Value-at-Risk (VaR). VaR, a widely used performance measure, answers the question: what is the maximum loss with a specified confidence level? Percentiles are also used for defining other relevant performance measures, such as Conditional Value-at-Risk (CVaR). CVaR (also called Mean Excess Loss, Mean Shortfall, or Tail VaR) is the average loss for the worst x% scenarios (e.g., 5%). CVaR risk measure has more attractive properties compared to VaR. This introductory paper gives basic definitions and reviews several topics:

- sensitivities of probabilistic functions;
- sensitivities of percentiles (VaR);
- optimization approaches for CVaR.

The emphasis of this paper is on issues which have been relatively recently developed.

1 Introduction

Probabilistic and quantile (percentile) functions are commonly used for the analysis of models with uncertainties or variabilities in parameters. For instance, in risk and reliability analysis, performance functions, characterizing operation of systems, are

S.P. Uryasev (ed.), Probabilistic Constrained Optimization, 1–25.
© 2000 *Kluwer Academic Publishers.*

formulated as probabilities of successful or unsuccessful accomplishment of their missions, e.g., core damage probability or frequency of a nuclear power plant, probability of successful landing of an aircraft, probability of profitable transactions in a stock market, or percentiles of the risks in public risk assessments. In financial applications, the percentile of the losses is called Value-at-Risk (VaR). VaR, a widely used performance measure, answers the question: what is the maximum loss with a specified confidence level? Percentiles are also used for defining other relevant performance measures, such as Conditional Value-at-Risk (CVaR). CVaR (also called Mean Excess Loss, Mean Shortfall, or Tail VaR) is the average loss for the worst x% scenarios (e.g., 5%). CVaR risk measure has more attractive properties compared to VaR.

This introductory paper gives basic definitions and reviews several topics:

- sensitivities of probabilistic functions;
- sensitivities of percentiles (VaR);
- optimization approaches for CVaR.

This paper is not intended as a comprehensive review. It covers only several recently developed topics and does not include many other important issues related to the analysis of the probabilistic and percentile functions. For more comprehensive discussions and analyses of probabilistic and quantile functions, see books by Prékopa [32], Kan and Kibzun [14], Pflug [29], Kall and Wallace [13], and Birge and Louveaux [3]. Also, many relevant topics and applications are subject of papers included in this volume.

2 Sensitivity Analysis of Probabilistic Functions

2.1 Background

The sensitivity analysis of probabilistic performance functions involves an evaluation of their derivatives with respect to (w.r.t.) parameters. Derivatives of the probability and quantile functions are used as well to solve stochastic optimization problems (e.g., [8, 9, 13, 48]), and to analyze the Discrete Event Dynamic Systems (DES) (e.g., [10, 11, 43]).

This chapter overviews basic results on differentiability of the probabilistic functions without going into details of formal conditions and theorems. More formal and comprehensive discussions of these topics can be found in the review paper by Kibzun and Uryasev [18]. Here, we only formulate statements and illustrate the results with examples. A probability function can be formally represented as an expectation of a discontinuous indicator function of a set, or as an integral over a domain depending upon parameters. Nevertheless, differentiability conditions of the probability function do not follow from similar conditions of the expectations of continuous (smooth or convex) functions. A differentiation formula for an expectation of continuous functions can be obtained by interchanging the gradient and expectation operators [7, 12, 40].

The derivative of the probability function has many equivalent representations: it can be represented in general form as an integral over the surface, integral over the volume, or a sum of integrals over the volume and over the surface; it can be calculated using weak derivatives of the probability measures or conditional expectations. Here, we overview only mathematical results related to the first three representations.

The first general result on sensitivity of probability functions in general form was obtained by Raik [36] who represented the gradient of the probability function with one constraint in the form of the surface integral. For volume integrals with the domain depending upon a parameter, Roenko [41] and Simon [44] obtained the gradient formula in the form of the surface integral. Uryasev [50] extended Raik's formula for probability functions with many constraints; Kibzun and Tretyakov [17] extended it to the piece-wise smooth constraint and probability density function(see, also, [14]). Special cases of the probability function with normal and gamma distributions were investigated by Prékopa [31], and Prékopa and Szántai [34]. Pflug [29] represented the gradient of the probability function in the form of an expectation using weak probability measures.

The next general result was published by Uryasev [49], in which the gradient of the probability function was written as a volume integral. Later, he [50] generalized this formula to the case of several constraints, and established relations between formulas for the gradient in the form of the surface and the volume integrals. Using a change of variables, Marti [23, 24] derived the probability function gradient in the form of the volume integral. Rubinstein [42] used a change of variables to differentiate performance functions in analyses of DESs. Marti [23] also suggested approximations of the probability function gradient by an asymptotic expansion of the integrals.

Finally, a general analytical formula for the derivative of probability functions with many constraints was obtained by Uryasev [50]; it calculates the gradient as an integral over the surface, or an integral over the volume, or the sum of integrals over the surface and the volume. The general formula calculates the gradient using the solution of a system of nonlinear equations. Special cases of this formula correspond to the Raik formula [36], Uryasev formula [49], and the change–of–variables approach [24, 42].

2.2 Notations and Definitions

Let $f(x, y)$ be a performance function, for instance losses associated with the decision vector x, to be chosen from a subset X of $\mathbb{R}^n$, and the random vector y in $\mathbb{R}^m$. In finance applications, the vector x can be interpreted as representing a portfolio, with X as the set of available portfolios (subject to various constraints), but other interpretations could be made as well. The vector y stands for the uncertainties, e.g. market parameters, that can affect the performance.

For each x, the performance function $f(x,y)$ is a random variable having a distribution in $\mathbb{R}$ induced by that of y. The underlying probability distribution of y in $\mathbb{R}^m$ will be assumed to have density, which we denote by $p(x,y)$. The density may depend upon the decision vector x. In finance applications, a two step procedure (see, for instance, RiskMetrics [37]) can be used to derive an analytical expression for $p(y)$ or construct a Monte Carlo simulation code for drawing samples from $p(x,y)$: (1) modeling of risk factors in $\mathbb{R}^{m_1}$,(with $m_1 < m$), (2) based on the characteristics of instrument i, $i = 1,\ldots,n$, the distribution $p(x,y)$ can be derived or code transforming random samples of risk factors to the random samples from density $p(x,y)$ can be constructed.

Let an integral over the volume

$$F(x) = \int\limits_{f(x,y)\leq 0} p(x,y)\,dy \tag{1}$$

is defined on the Euclidean space $\mathbb{R}^n$, where $f : \mathbb{R}^n \times \mathbb{R}^m \to \mathbb{R}^k$ and $p : \mathbb{R}^n \times \mathbb{R}^m \to \mathbb{R}$ are some functions. The inequality $f(x,y) \leq 0$ in the integral is a system of inequalities

$$f_i(x,y) \leq 0\ , \quad i = 1,\ldots,k\ .$$

Both the kernel function $p(x,y)$ and the function $f(x,y)$ defining the integration set depend upon the parameter x. For example, let

$$F(x) = P\{f(x,\zeta(\omega)) \leq 0\} \tag{2}$$

be a *probability function*, where $\zeta(\omega)$ is a random vector in $\mathbb{R}^m$. The random vector $\zeta(\omega)$ is assumed to have a probability density $p(x,y)$ that depends on a parameter $x \in \mathbb{R}^n$. The probability function can be represented as an expectation of an indicator function, which equals one on the integration set, and equals zero outside of it. For example, let

$$F(x) = \mathbb{E}\big[I_{\{f(x,\zeta)\,\leq\,0\}}\,g(x,\zeta)\big] = \int\limits_{f(x,y)\leq 0} g(x,y)\,\rho(x,y)\,dy = \int\limits_{f(x,y)\leq 0} p(x,y)\,dy\ , \tag{3}$$

where $I_{\{\}}$ is the indicator function, and random vector ζ in $\mathbb{R}^m$ has a probability density $\rho(x,y)$ that depends on the vector $x \in \mathbb{R}^n$.

2.3 Integral Over the Surface Formula.

The following formula calculates the gradient of integral (1) over the set given by nonlinear inequalities as the sum of the integral over the volume plus the integral over the surface of the integration set. We call this *the integral over the surface formula* because if density $p(x,y)$ does not depend upon x the gradient of integral (1) equals the integral over the surface. This formula, for the case of one inequality, was obtained by Raik [36] and generalized for the case with many inequalities by Uryasev [50].

Let us denote by $\mu(x)$ the integration set

$$\mu(x) = \{ y \in \mathbb{R}^m : f(x,y) \le 0 \} \stackrel{def}{=} \{ y \in \mathbb{R}^m : f_l(x,y) \le 0,\ 1 \le l \le k \},$$

and by $\partial\mu(x)$ the surface of this set $\mu(x)$. Also, let us denote by $\partial_i\mu(x)$ a part of the surface which corresponds to the function $f_i(x,y)$, i.e.,

$$\partial_i\mu(x) = \mu(x) \bigcap \{ y \in \mathbb{R}^m : f_i(x,y) = 0 \}.$$

If the constraint functions are differentiable and the following integral exists, then the gradient of integral (1) equals

$$\nabla_x F(x) = \int_{\mu(x)} \nabla_x p(x,y)\, dy - \sum_{i=1}^{k} \int_{\partial_i\mu(x)} \frac{p(x,y)}{||\nabla_y f_i(x,y)||} \nabla_x f_i(x,y)\, dS. \qquad (4)$$

A potential disadvantage of this formula is that in a multidimensional space it is difficult to calculate the integral over a nonlinear surface. Standard numerical techniques, such as Monte-Carlo algorithms, are applicable to estimate volume rather than surface integrals. Nevertheless, this formula can be quite useful in various special cases, such as the linear case.

2.3.1 Example 1. Linear Case: Integral Over the Surface Formula [50]

Let $A(\omega)$, be a random $l \times n$ matrix with the joint density $p(A)$. Suppose that $x \in \mathbb{R}^n$ and $x_j \neq 0$, $j = 1, \ldots, n$. Let us define

$$F(x) = P\{ A(\omega)x \le b,\ A(\omega) \ge 0 \}, \quad b = (b_1, \ldots, b_l) \in \mathbb{R}^l, \quad x \in \mathbb{R}^n, \qquad (5)$$

i.e. $F(x)$ is the probability that the linear constraints $A(\omega)x \le b$, $A(\omega) \ge 0$ are satisfied. The constraint, $A(\omega) \ge 0$, means that all elements $a_{ij}(\omega)$ of the matrix $A(\omega)$ are non-negativity. Let us denote by A_i and A^i the i-th row and column of the matrix A

$$A = \begin{pmatrix} A_1 \\ \vdots \\ A_l \end{pmatrix} = \left(A^1, \cdots, A^n \right),$$

then

$$f(x,A) = \begin{pmatrix} f_1(x,A) \\ \vdots \\ f_k(x,A) \end{pmatrix} = \begin{pmatrix} A_1x - b_1 \\ \vdots \\ A_lx - b_l \\ -A^1 \\ \vdots \\ -A^n \end{pmatrix},$$

$$k = l + l \times n\,.$$

The function $F(x)$ equals

$$F(x) = \int\limits_{f(x,A)\leq 0} p(A)\, dA\,. \tag{6}$$

We use formula (4) to calculate the gradient $\nabla_x F(x)$ as an integral over the surface. The function $p(A)$ does not depend upon x and $\nabla_x p(A) = 0$. Formula (4) implies that $\nabla_x F(x)$ equals

$$\nabla_x F(x) \;=\; -\sum_{i=1}^{k} \int\limits_{\partial_i \mu(x)} \frac{p(A)}{||\nabla_A f_i(x,A)||} \nabla_x f_i(x,A)\; dS\,.$$

Since $\nabla_x f_i(x,A) = 0$ for $i = l+1, \ldots, k$, the gradient $\nabla_x F(x)$ equals

$$\nabla_x F(x) \;=\; -\sum_{i=1}^{l} \int\limits_{\partial_i \mu(x)} \frac{p(A)}{||\nabla_A f_i(x,A)||} \nabla_x f_i(x,A)\; dS \;=\; -\sum_{i=1}^{l} \int\limits_{\partial_i \mu(x)} \frac{p(A)}{||x||} A_i^T\, dS$$

$$=\; -\,||x||^{-1} \sum_{i=1}^{l} \int\limits_{\substack{Ax\leq b\,,\, A\geq 0 \\ A_i x = b_i}} p(A)\, A_i^T\, dS\,.$$

2.4 Integral Over the Volume Formula.

This section represents the gradient of function (1) in the form of volume integral. Let us introduce the following shorthand notations

$$f_{1l}(x,y) = \begin{pmatrix} f_1(x,y) \\ \vdots \\ f_l(x,y) \end{pmatrix}, \quad f(x,y) = f_{1k}(x,y)\,,$$

$$\nabla_y f(x,y) = \begin{pmatrix} \frac{\partial f_1(x,y)}{\partial y_1}, & \ldots, & \frac{\partial f_k(x,y)}{\partial y_1} \\ & \vdots & \\ \frac{\partial f_1(x,y)}{\partial y_m}, & \ldots, & \frac{\partial f_k(x,y)}{\partial y_m} \end{pmatrix}.$$

Divergence for the $n \times m$ matrix H consisting of the elements h_{ji} is denoted by

$$div_y H = \begin{pmatrix} \sum_{i=1}^{m} \frac{\partial h_{1i}}{\partial y_i} \\ \vdots \\ \sum_{i=1}^{m} \frac{\partial h_{ni}}{\partial y_i} \end{pmatrix} .$$

Following [50], the derivative of function (1) is represented as the integral over the volume

$$\nabla_x F(x) = \int\limits_{\mu(x)} \nabla_x p(x,y)\, dy + \int\limits_{\mu(x)} div_y\big(p(x,y)H(x,y)\big)\, dy \,, \tag{7}$$

where a matrix function $H : \mathbb{R}^n \times \mathbb{R}^m \to \mathbb{R}^{n \times m}$ satisfies the equation

$$H(x,y)\, \nabla_y f(x,y) + \nabla_x f(x,y) = 0 \,. \tag{8}$$

The last system of equations may have many solutions. Therefore, formula (7) provides a number of equivalent expressions for the gradient. The following section gives analytical solutions of this system of equations. In some cases, this system does not have any solution, and formula (7) is not valid. The following section deals with such cases and provides a general formula where system of equations can be solved only for some of the functions defining the integration set.

2.4.1 Example 2. Linear Case: Integral Over the Volume Formula [50]

With formula (7), the gradient of probability function (5) with linear constrains considered in Example 1 can be represented as the integral over the volume. It can be shown that equation (8) does not have a solution in this case. Nevertheless, we can slightly modify the constraints, such that the integration set is not changed and equation (8) has a solution. In the vector function $f(x,A)$, we multiply column A^i by x^i if x^i is positive or multiply it by $-x^i$ if x^i is negative. Therefore, we have the following constraint function

$$f(x,A) = \begin{pmatrix} A_1 x - b_1 \\ \vdots \\ A_l x - b_l \\ -(+) x_1 A^1 \\ \vdots \\ -(+) x_n A^n \end{pmatrix} , \tag{9}$$

where $-(+)$ means that we take an appropriate sign. It can be directly checked that, the following matrix

$$H^*(x,A) = \big(h^1(x,A_1), \ldots, h^l(x,A_l)\big) \,, \quad h^i(x,A_i) = - \begin{pmatrix} a_{i1}x_1^{-1} & & 0 \\ & \ddots & \\ 0 & & a_{in}x_n^{-1} \end{pmatrix}$$

is a solution of system (8). As it will be shown in the next section, this analytical solution follows from the fact that the change of variables $Y^i = x_i A^i$, $i = 1, \ldots, n$, eliminates variables x^i, $i = 1, \ldots, n$, from constraints (9).

Since $\nabla_x p(A) = 0$ and $div_A\big(p(A)H^*(x, A)\big)$ equals

$$div_A\big(p(A)H^*(x, A)\big) = -\begin{pmatrix} x_1^{-1}\big(l\,p(A) + \sum_{i=1}^{l} a_{i1}\frac{\partial}{\partial a_{i1}}p(A)\big) \\ \vdots \\ x_n^{-1}\big(l\,p(A) + \sum_{i=1}^{l} a_{in}\frac{\partial}{\partial a_{in}}p(A)\big) \end{pmatrix}$$

formula (7) implies

$$\frac{\partial F(x)}{\partial x_j} = -x_j^{-1}\int\limits_{\substack{Ax\le b\\ A\ge 0}} \Big(l\,p(A) + \sum_{i=1}^{l} a_{ij}\frac{\partial}{\partial a_{ij}}p(A)\Big)\, dA\,. \tag{10}$$

2.5 General Formula.

Further, we give the general formula [49, 50] for the derivative of integral (1). The gradient of the integral is represented as a sum of integrals taken over the volume and over the surface. This formula can be used when system of equations (8) does not have a solution. We split the set of constraints $K \stackrel{def}{=} \{1, \ldots, k\}$ into two subsets K_1 and K_2. Without loss of generality we suppose that

$$K_1 = \{1, \ldots, l\}, \quad K_2 = \{l+1, \ldots, k\}\,.$$

The derivative of integral (1) can be represented as the sum of the volume and surface integrals

$$\nabla_x F(x) = \int\limits_{\mu(x)} \nabla_x p(x, y)\, dy + \int\limits_{\mu(x)} div_y\big(p(x, y)H_l(x, y)\big)\, dy$$

$$- \sum_{i=l+1}^{k} \int\limits_{\partial_i\mu(x)} \frac{p(x, y)}{||\nabla_y f_i(x, y)||}\big[\nabla_x f_i(x, y) + H_l(x, y)\,\nabla_y f_i(x, y)\big]\, dS\,, \tag{11}$$

where the matrix $H_l : \mathbb{R}^n \times \mathbb{R}^m \to \mathbb{R}^{n\times m}$ satisfies the equation

$$H_l(x, y)\,\nabla_y f_{1l}(x, y) + \nabla_x f_{1l}(x, y) = 0\,. \tag{12}$$

The last equation can have a lot of solutions and we can choose an arbitrary one, differentiable w.r.t. the variable y.

The general formula contains as a special case the integral over the surface formula (4) and integral over the volume formula (7). When the set K_1 is empty, the matrix

H_l is absent and the general formula is reduced to the integral over the surface. Also, when the set K_2 is empty, we have the integral over the volume formula (7). In addition to these extreme cases, the general formula provides a number of intermediate expressions for the gradient in the form of the sum of the integral over the surface and the integral over the volume. There are many equivalent representations of the gradient corresponding to the various sets K_1 and K_2 and solutions of equation (12).

Equation (12) (and equation (8) which is a partial case of equation (12)) can be solved explicitly. Usually, this equation has many solutions. For instance, the matrix

$$- \nabla_x f_{1l}(x,y) \left(\nabla_y^T f_{1l}(x,y) \times \nabla_y f_{1l}(x,y)\right)^{-1} \nabla_y^T f_{1l}(x,y) \tag{13}$$

is a solution of equation (12). Also, in a number of cases, equation (12) can be solved using a change of variables. Suppose that there is a change of variables

$$y = \gamma(x, z)$$

which eliminates the vector x from the function $f(x,y)$, defining the integration set, i.e., the function $f(x, \gamma(x,z))$ does not depend upon the variable x. Denote by $\gamma^{-1}(x,y)$ the inverse function, defined by the equation

$$\gamma^{-1}\big(x, \gamma(x,z)\big) \;=\; z\,.$$

Let us show that the following matrix

$$H(x,y) \;=\; \nabla_x \gamma(x,z)|_{z=\gamma^{-1}(x,y)} \tag{14}$$

is a solution of (12). Indeed, the gradient of the function $\gamma(x, y(x,z))$ w.r.t. x equals zero; therefore

$$0 \;=\; \nabla_x f_{1l}(x, \gamma(x,z)) = \nabla_x \gamma(x,z) \nabla_y f_{1l}(x,y)|_{y=\gamma(x,z)} + \nabla_x f_{1l}(x,y)|_{y=\gamma(x,z)}\,,$$

and the function $\nabla_x \gamma(x,z)|_{z=\gamma^{-1}(x,y)}$ is a solution of equation (12).

Formula (7) with matrix (14) gives the derivative formulas, which can be obtained with change of variables in the integration set [24].

2.5.1 Example 3.

While investigating the operational strategies for inspected components (see [33]), the following integral was considered

$$F(x) = \int\limits_{\substack{b(y)\le x\,,\\ y_i \ge \theta\,,\; i=1,\ldots,m}} p(y)\, dy\,, \tag{15}$$

where $x \in \mathbb{R}^1, y \in \mathbb{R}^m, p : \mathbb{R}^m \to \mathbb{R}^1$, $\theta > 0$, $b(y) = \sum_{i=1}^{m} y_i^\alpha$. In this case,

$$f(x,y) = \begin{pmatrix} b(y) - x \\ \theta - y_1 \\ \vdots \\ \theta - y_m \end{pmatrix},$$

and

$$F(x) = \int\limits_{f(x,y) \leq 0} p(y)\, dy \;=\; \int\limits_{\mu(x)} p(y)\, dy \,.$$

Let us consider that $l = 1$, i.e. $K_1 = \{1\}$ and $K_2 = \{2,\ldots,m+1\}$. The gradient $\nabla_x F(x)$ equals

$$\nabla_x F(x) \;=\; \int\limits_{\mu(x)} \Big[\nabla_x p(y) \;+\; div_y\big(p(y) H_1(x,y)\big) \Big]\, dy$$

$$- \sum_{i=2}^{m+1} \int\limits_{\partial_i \mu(x)} \frac{p(y)}{||\nabla_y f_i(x,y)||} \Big[\nabla_x f_i(x,y) \;+\; H_1(x,y)\, \nabla_y f_i(x,y) \Big]\, dS \,, \tag{16}$$

where the matrix $H_1(x,y)$ satisfies equation (12). In view of

$$\nabla_y f_1(x,y) = \alpha \begin{pmatrix} y_1^{\alpha-1} \\ \vdots \\ y_m^{\alpha-1} \end{pmatrix}, \quad \nabla_x f_1(x,y) = -1 \,,$$

a solution $H_1^*(x,y)$ of equation (12) equals

$$H_1^*(x,y) \;=\; h(y) \stackrel{def}{=} \big(h_1(y_1),\ldots,h_m(y_m)\big) \;=\; \frac{1}{\alpha m} \left(y_1^{1-\alpha},\ldots,y_m^{1-\alpha}\right) . \tag{17}$$

Let us denote

$$(\theta_i \mid y) \;=\; (y_1,\ldots,y_{i-1},\theta,y_{i+1},\ldots,y_m)\,,$$

$$y^{-i} \;=\; (y_1,\ldots,y_{i-1},y_{i+1},\ldots,y_m)\,, \quad b(\theta_i \mid y) \;=\; \theta^\alpha + \sum_{\substack{j=1 \\ j\neq i}}^{m} y_j^\alpha \,.$$

Also, let us denote by $y^{-i} \geq \theta$ the set of inequalities

$$y_j \geq \theta\,, \quad j = 1,\ldots,i-1,i+1,\ldots,m \,.$$

The sets $\partial_i \mu(x)$, $i = 2,\ldots,m+1$ have a simple structure

$$\partial_i \mu(x) \;=\; \mu(x) \bigcap \{y \in \mathbb{R}^m \,:\, y_i = \theta\} \;=\; \{y^{-i} \in \mathbb{R}^{m-1} \,:\, b(\theta_i|y) \leq x\,,\, y^{-i} \geq 0\} \,.$$

For $i = 2, \ldots, m+1$, we have

$$(\nabla_y f_i(y))_j = 0,\ j = 1, \ldots, m,\ j \neq i-1, \tag{18}$$

$$(\nabla_y f_i(y))_{i-1} = -1\,,\ ||\nabla_y f_i(y)|| = 1. \tag{19}$$

The function $p(y)$ and the functions $f_i(y)$, $i = 2, \ldots, m+1$ do not depend on x, consequently,

$$\nabla_x p(y) = 0\,, \tag{20}$$

$$\nabla_x f_i(y) = 0\,, \quad i = 2, \ldots, m+1\,. \tag{21}$$

Equations (16) - (21) imply

$$\nabla_x F(x) = \int\limits_{\mu(x)} div_y\big(p(y)h(y)\big)\,dy - \sum_{i=2}^{m+1} \int\limits_{\partial_i \mu(x)} \frac{p(y)}{||\nabla_y f_i(y)||} h(y) \nabla_y f_i(y)\,dS$$

$$= \int\limits_{\mu(x)} div_y\big(p(y)h(y)\big)\,dy + \sum_{i=2}^{m+1} h_{i-1}(\theta) \int\limits_{\partial_i \mu(x)} p(y)\,dS$$

$$= \int\limits_{\substack{b(y)\le x\,, \\ y_i \ge \theta\,,\, i=1,\ldots,m}} div_y\big(p(y)h(y)\big)\,dy + \sum_{i=1}^{m} \frac{\theta^{1-\alpha}}{\alpha m} \int\limits_{\substack{b(\theta_i | y)\le x\,, \\ y^{-i} \ge \theta}} p(\theta_i \mid y)\,dy^{-i}\,.$$

Since

$$div_y\big(p(y)h(y)\big) = h(y)\nabla_y p(y) + p(y) div_y h(y) = \frac{1}{\alpha m} \sum_{i=1}^{m} \frac{\partial p(y)}{\partial y_i} y_i^{1-\alpha} + p(y) \frac{1-\alpha}{\alpha m} \sum_{i=1}^{m} y_i^{-\alpha}\,,$$

we, finally, obtain that the gradient $\nabla_x F(x)$ equals

$$\nabla_x F(x) = \int\limits_{\substack{b(y)\le x\,, \\ y_i \ge \theta,\, i=1,\ldots,m}} \sum_{i=1}^{m} \frac{y_i^{-\alpha}}{\alpha m} \Big[y_i \frac{\partial p(y)}{\partial y_i} + (1-\alpha)p(y)\Big] dy + \frac{\theta^{1-\alpha}}{\alpha m} \sum_{i=1}^{m} \int\limits_{\substack{b(\theta_i | y)\le x\,, \\ y^{-i} \ge \theta}} p(\theta_i \mid y)\,dy^{-i}\,.$$

The formula for $\nabla_x F(x)$ is valid for an arbitrary sufficiently smooth function $p(y)$.

3 The Gradient of the Quantile (Value-at-Risk)

The quantile or Value-at-Risk (VaR) answers the question: what is the maximum loss (or other measure of the performance) with a specified confidence level? A description of various methodologies for the modeling of VaR can be seen, along with related resources, at www.gloriamundi.org. Mostly, approaches to calculating VaR rely on linear approximation of the risk factors and assume a joint normal (or log-normal) distribution of the underlying parameters, see, for instance, Duffie and Pan

[6], Pritsker [35], RiskMetrics [37], Simons [45], Stambaugh [46]). In financial applications, historical or Monte Carlo simulation-based tools are used when the portfolio contains nonlinear instruments such as options ([4, 26, 35, 37, 46]). Discussions of optimization problems involving VaR can be found in papers by Litterman [20, 21], Kast et al. [15], Lucas and Klaassen [22].

The gradient of the quantile function was obtained by Kibzun et al. [16]. The approach, considered in this section, for simultaneous calculating of quantiles and their gradients with Monte Carlo simulation algorithms was proposed in [53].

The probability of the performance function $f(x, y)$ not exceeding a threshold α is denoted by

$$\Psi(x, \alpha) \;=\; \int_{f(x,y) \leq \alpha} p(y)\, dy\,. \tag{22}$$

As a function of α for fixed x, $\Psi(x, \alpha)$ is the cumulative distribution function for the performance associated with x. It completely determines the behavior of this random variable and is fundamental in defining quantile (VaR). In general, $\Psi(x, \alpha)$ is nondecreasing w.r.t. α and continuous from the right, but not necessarily from the left because of the possibility of jumps. We assume however in what follows that the probability distributions are such that no jumps occur, or in other words, that $\Psi(x, \alpha)$ is everywhere continuous w.r.t. α. The required continuity follows from properties of the function $f(x, y)$ and the density $p(y)$. As it follows from derivative formula (4), the function $\Psi(x, \alpha)$ is differentiable in α if the function $f(x, y)$ is smooth, $||\nabla_y f_i(x, y)|| > 0$, and the appropriate surface integral exists.

The β-quantile (or β-VaR) for the performance random variable associated with x and any specified probability level β in $(0, 1)$ will be denoted by $\alpha_\beta(x)$. In this setting, the quantile is given by

$$\alpha_\beta(x) \;=\; \min\{\, \alpha \in \mathbb{R} : \; \Psi(x, \alpha) \geq \beta \,\}\,.$$

So, the quantile $\alpha_\beta(x)$ is the left endpoint of the nonempty interval consisting of the values α such that actually $\Psi(x, \alpha) = \beta$. This follows from $\Psi(x, \alpha)$ being continuous and nondecreasing w.r.t. α. The interval might contain more than a single point if Ψ has "flat spots."

Further, we suppose that for considered x, and β

$$\left.\frac{\partial \Psi(x, \alpha)}{\partial \alpha}\right|_{\alpha = \alpha(x, \beta)} \neq 0\,.$$

With this assumption, the quantile $\alpha(x, \beta)$ is a unique solution of the following equation

$$\Psi(x, \alpha) = \beta \tag{23}$$

w.r.t. α . The gradient of the quantile function $\alpha(x,\beta)$ in x can be obtained by differentiating the left and right parts of equation (23), i.e.,

$$\nabla_x \Psi(x,\alpha(x,\beta)) \;=\; \nabla_x \Psi(x,\alpha)\mid_{\alpha=\alpha(x,\beta)} + \nabla_\alpha \Psi(x,\alpha)\mid_{\alpha=\alpha(x,\beta)} \nabla_x \alpha(x,\beta) \;=\; 0 \,.$$

Consequently,

$$\nabla_x \alpha(x,\beta) = -\frac{\nabla_x \Psi(x,\alpha)\mid_{\alpha=\alpha(x,\beta)}}{\nabla_\alpha \Psi(x,\alpha)\mid_{\alpha=\alpha(x,\beta)}} \,. \tag{24}$$

Analogous, the derivative of quantile function $\alpha(x,\beta)$ in the parameter β can be obtained by differentiating equation (23) w.r.t. β, i.e.,

$$\nabla_\beta \Psi(x,\alpha(x,\beta)) \;=\; \nabla_\alpha \Psi(x,\alpha)\mid_{\alpha=\alpha(x,\beta)} \nabla_\beta \alpha(x,\beta) \;=\; 1 \,,$$

and

$$\nabla_\beta \alpha(x,\beta)) \;=\; (\nabla_\alpha \Psi(x,\alpha)\mid_{\alpha=\alpha(x,\beta)})^{-1} \,. \tag{25}$$

Formulas (24) and (25) involve derivatives of the probability function $\Psi(x,\alpha)$ w.r.t. x and α. These derivatives can be calculated using formulas of the previous section. We can apply various formulas to the denominator and the numerator in (24) and (25). For example, the gradient of the quantile function can be represented as the ratio of surface or volume integrals. Usually, we prefer to work with volume integrals because they can be evaluated with standard numerical methods or Monte Carlo type techniques. However, surface integrals are convenient for low dimension cases (for instance, if $x \in \mathbb{R}^1$, the surface is reduced to a few points), or when the constraint function $f(x,y)$ is linear w.r.t. y.

We finalize this section with an example showing how formulas (24) and (25) can be used to evaluate derivatives of quantile functions with Monte Carlo simulation algorithms. In this case, the derivative of the probability function is calculated using integrals over the volume.

The probability function $\Psi(x,\alpha)$, given by formula (22), is an integral over the m-dimensional volume. Suppose that with Monte Carlo simulations we evaluated the quantile $\alpha_\beta(x)$. Using formulas (7) or (11), derivatives of probability function $\Psi(x,\alpha)$ w.r.t. x and α can be represented as integrals over the same volume with the same density $p(x)$ (for instance, such representation for linear functions gives formula (10)) i.e.,

$$\nabla_x \Psi(x,\alpha) \;=\; \int\limits_{f(x,y)\le\alpha} a(x,\alpha,y)\,p(y)\,dy\,, \qquad \nabla_\alpha \Psi(x,\alpha) \;=\; \int\limits_{f(x,y)\le\alpha} b(x,\alpha,y)\,p(y)\,dy\,. \tag{26}$$

The m-dimensional kernel function $a(x,\alpha,y)$ in the first integral and the kernel function $b(x,\alpha,y)$ in the second integral can be evaluated during the same simulation run. So, using the same random samples $y(\omega)$ from density $p(y)$ we can evaluate simultaneously gradients of the probability function with formulas (26), and gradients of the quantile function with formulas (24) and (25).

Example 4. The following percentile function was studied in evaluating incremental lifetime cancer risk for children due to ingesting benzene with soil [47]. The gradient for this percentile function is calculated in [53]. The constraint function $f(x,y)$ is the ratio of two random variables

$$f(x,y) \triangleq \frac{1}{x_3 + x_4 y_2} e^{x_1 + x_2 y_1}, \tag{27}$$

where $x \in \mathbb{R}^4$, $y \in \mathbb{R}^2$, and y_1, y_2 are independent random variables having the same normal distribution $N(0,1)$. We denote densities for these random variables by $p_1(y_1)$ and $p_2(y_2)$, respectively. The numerator of the constraint function is the lognormally distributed random variable with parameters (x_1, x_2). The denominator of the constraint function is the normally distributed random variable with parameters $N(x_3, x_4)$. Thus, the probability function $\Psi(x,\alpha)$ is a two-dimensional integral with the constraint $f(x,y) \le \alpha$ and the density function $p(y) = p_1(y_1)\, p_2(y_2)$. The change of variables

$$(y_1, y_2)^T = \gamma(x, \alpha, y) \stackrel{def}{=} \left((y_1 - x_1)/x_2,\ -(\alpha\, x_3 - y_2)/(\alpha\, x_4)\right)^T$$

eliminates x and α from the function $f(x,y) - \alpha$. Therefore, formula (7) with matrix (14) gives the following expressions for the four components of the vector $a(x,\alpha,y)$ and $b(x,\alpha,y)$

$$a_1 = \frac{y_1}{x_2}, \quad a_2 = \frac{-1 + y_1^2}{x_2}, \quad a_3 = \frac{y_2}{x_4}, \quad a_4 = \frac{-1 + y_2^2}{x_4}, \quad b = \frac{-x_4 + x_3 y_2 + x_4 y_2^2}{\alpha\, x_4}.$$

Thus, in this case, we have six volume integrals, i.e., the probability function is calculated with formula (22), and the five partial derivatives are calculated with formulas (26). All integrals have the same density $p(y)$ and the same constraint function $f(x,y) \le \alpha$ defining the integration domain. Therefore, all integrals can be evaluated simultaneously with different values of α during the same simulation runs of the Monte Carlo algorithm. By solving the equation $\Psi(x,\alpha) = \beta$ w.r.t. α, we find the quantile $\alpha_\beta(x)$. Further, with formulas (24) and (25), we can evaluate derivatives of the quantile function $\alpha_\beta(x)$.

4 Optimization with Conditional Value-at-Risk Performance Function and Constraints

This chapter reviews the approach for minimization of the Conditional Value-at-Risk, which was recently developed in papers [39, 28]. In presenting this material, we follow the review paper [52]. The methodology is quite general and can be used for any application involving optimization of quantiles. However, development of this approach is mostly stimulated by finance applications. Therefore, it will be explained in the framework of the financial contents.

Value-at-Risk (VaR), which is a quantile of a loss distribution, calculates the maximum loss with a specified confidence level. Probably the most popular technique for the estimation of VaR is the RiskMetrics methodology [37]. Although VaR is a very popular measure of risk in finance applications, it has undesirable properties [2] such as lack of sub-additivity, i.e., VaR of a portfolio with two instruments may be greater than the sum of individual VaRs of these two instruments. Also, VaR is difficult to optimize when calculated using scenarios. In this case, VaR is non-convex (see definition of convexity in [38]), non-smooth as a function of positions, and it has multiple local extrema.

An alternative measure of losses, with more attractive properties, is Conditional Value-at-Risk (CVaR), which is also called Mean Excess Loss, Mean Shortfall, or Tail VaR. CVaR is a more consistent measure of risk since it is sub-additive and convex [2]. Recently, Pflug [30] proved that CVaR is a coherent risk measure having the following properties: transition-equivariant, positively homogeneous, convex, monotonic w.r.t. stochastic dominance of order 1, and monotonic w.r.t. monotonic dominance of order 2. Moreover, as it was shown in [39], it can be optimized using linear programming (LP) and nonsmooth optimization algorithms, which allows to handle portfolios with very large numbers of instruments and scenarios. Numerical experiments indicate that the minimization of CVaR also leads to near optimal solutions in VaR terms because CVaR is always greater than or equal to VaR. When the return-loss distribution is normal, these two measures are equivalent [39], i.e., they provide the same optimal portfolio.

CVaR can be used in conjunction with VaR and is applicable to the estimation of risks with non-symmetric return-loss distributions. Although CVaR has not become a standard in the finance industry, it is likely to play a major role as it currently does in the insurance industry. Similar to the Markowitz [25] mean-variance approach, CVaR can be used in return-risk analyses. For instance, we can calculate a portfolio with a specified return and minimal CVaR. Alternatively, we can constrain CVaR and find a portfolio with maximal return, see [28]. Also, rather than constraining the variance, we can specify several CVaR constraints simultaneously with various confidence levels (thereby shaping the loss distribution), which provides a flexible and powerful risk management tool.

Similar measures as CVaR have been earlier introduced in the stochastic programming literature, although not in financial mathematics context. The conditional expectation constraints and integrated chance constraints described in [32] may serve the same purpose as CVaR. The reader interested in other applications of optimization techniques in finance area can find relevant papers in [55].

Several case studies showed that risk optimization with the CVaR performance function and constraints can be done for large portfolios and a large number of scenarios with relatively small computational resources. For instance, a problem with 1,000 instruments and 20,000 scenarios can be optimized on a 300 MHz PC in less than one minute using the CPLEX LP solver. A case study on the hedging of a portfolio of options using the CVaR minimization technique is included in [39]. This problem was first studied at Algorithmics, Inc. with the minimum expected regret approach [26]. Also, the CVaR minimization approach was applied to credit risk management of a portfolio of bonds [1]. This portfolio was put together by several banks to test various credit risk modeling techniques. Earlier, the minimum expected regret optimization technique was applied to the same portfolio at Algorithmics, Inc.[27]; we have used the same set of scenarios to test the minimum CVaR technique. A case study on optimization of a portfolio of stocks with CVaR constraints is included in [28]. The reader interested in other applications of optimization techniques in the finance area can find relevant papers in [55].

4.1 Approach

This section outlines the approach suggested in [39] for simultaneous minimization of CVaR and calculation of VaR. The next section discusses how to extend this idea to problems with CVaR constraints.

Let $f(x, y)$ be a loss function depending upon the decision vector x and a random vector y. The decision vector x belongs to a feasible set of portfolios, X. For example, we may consider portfolios with non-negative positions (short positions are not allowed) and an expected return greater than 10%.

Example 5. A Two Instrument Portfolio.
A portfolio consists of two instruments (e.g., options). Let $x = (x_1, x_2)$ be a vector of positions of these two instruments, $m = (m_1, m_2)$ be a vector of initial prices, and $y = (y_1, y_2)$ be a vector of uncertain prices of these instruments in the next period. The loss function equals the difference between the current value of the portfolio, $(x_1 m_1 + x_2 m_2)$, and an uncertain value of the portfolio at the next period $(x_1 y_1 + x_2 y_2)$, i.e.,

$$f(x, y) = (x_1 m_1 + x_2 m_2) - (x_1 y_1 + x_2 y_2) = x_1(m_1 - y_1) + x_2(m_2 - y_2).$$

If we do not allow short positions, the feasible set of portfolios is a two-dimensional set of non-negative numbers

$$X = \{(x_1, x_2) : x_1 \geq 0, x_2 \geq 0\}.$$

In this case, the loss function is linear w.r.t. positions and the feasible set is defined by a set of linear inequalities.

For convenience, we assume that the random vector y has a probability density function $p(y)$. However, the existence of the density is not critical for the considered approach; this assumption can be relaxed. Denote by $\Psi(x, \alpha)$ the probability that the loss $f(x, y)$ does not exceed some threshold value α (see, (22)). The VaR function

$\alpha(x, \beta)$, which is the percentile of the loss distribution with confidence level β, is the smallest number such that $\Psi(x, \alpha(x, \beta)) = \beta$. CVaR, denoted by $\phi_\beta(x)$, which is by definition the conditional expected loss (under the condition that it exceeds VaR), is defined by

$$\phi_\beta(x) \;=\; (1-\beta)^{-1} \int_{f(x,y)\geq\alpha_\beta(x)} f(x,y)\, p(y)\, dy. \tag{28}$$

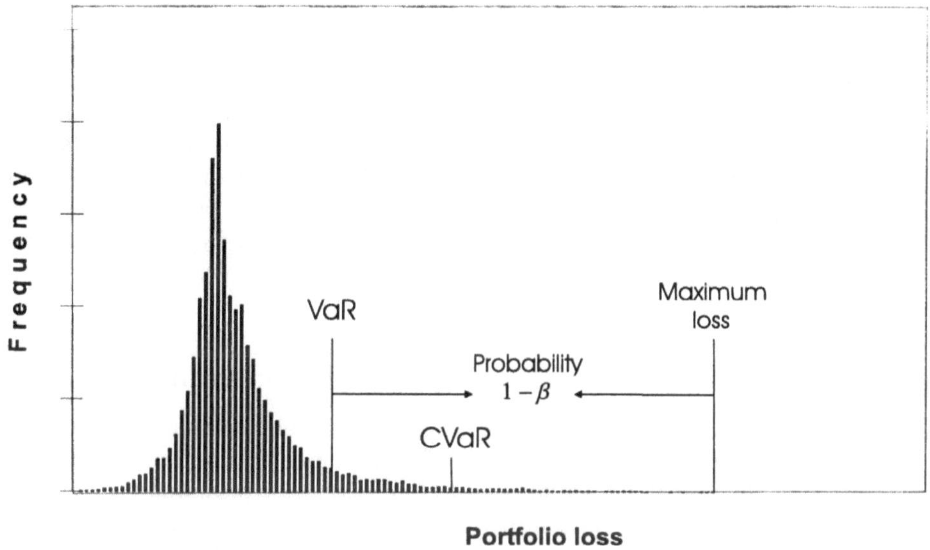

Figure 1: Portfolio Loss Distribution, VaR, and CVaR.

It is difficult to handle CVaR because of the VaR function $\alpha_\beta(x)$ involved in its definition, unless we have an analytical representation for VaR. The main idea of our approach is that we can define a much simpler function

$$F_\beta(x,\alpha) \;=\; \alpha \,+\, (1-\beta)^{-1} \int_{f(x,y)\leq\alpha} (f(x,y)-\alpha)\, p(y)\, dy\,, \tag{29}$$

which can be used instead of CVaR. It can be proved: (1) the function $F_\beta(x, \alpha)$ is convex w.r.t. α; (2) VaR is a minimum point of this function w.r.t. α; and (3) that minimizing $F_\beta(x, \alpha)$ w.r.t. α gives CVaR

$$\phi_\beta(x) \;=\; F_\beta(x, \alpha_\beta(x)) \;=\; \min_{\alpha\in\mathbb{R}} F_\beta(x,\alpha). \tag{30}$$

This follows from the fact that the derivative of the function $F_\beta(x, \alpha)$ w.r.t. α equals

$$1 + (1 - \beta)^{-1}(\Psi(x, \alpha) - 1) \,,$$

see details in [39]. By equating the derivative to zero we immediately obtain that VaR minimizes the function $F_\beta(x, \alpha)$ w.r.t. α. Furthermore, we can use the function $F_\beta(x, \alpha)$ for the simultaneous calculation of VaR and the optimization of CVaR, i.e.,

$$\min_{x \in X} \phi_\beta(x) \;=\; \min_{(x,\alpha) \in X \times \mathbb{R}} F_\beta(x, \alpha) \,. \tag{31}$$

Indeed, minimization of the function $F_\beta(x, \alpha)$ w.r.t. both variables optimizes CVaR and finds VaR in "one shot". Let (x^*, α^*) be a solution of the above minimization problem. Then, $F_\beta(x^*, \alpha^*)$ equals an optimal CVaR, the optimal portfolio equals x^*, and the corresponding VaR equals α^*. Under quite general conditions (see, Chapter 1 of this paper) the function $F_\beta(x, \alpha)$ is smooth. Moreover, if the function $f(x, y)$ is convex w.r.t. x, then the function $F_\beta(x, \alpha)$ is also convex w.r.t. x. Thus, if we want to minimize CVaR, we can use the convex smooth function $F_\beta(x, \alpha)$. Therefore, if the feasible set X is also convex, we need to solve a smooth convex optimization problem.

4.2 Optimization Problems with CVaR Constraints

Banks, investment companies, and other businesses tolerate different levels of risk, depending upon their objectives and capital. The adequate representation and management of risk is a critical task for business success. A typical approach in risk management is to estimate and control VaR with a specified confidence level, such as 0.95, 0.99, or 0.999. VaR is estimated for various periods, depending upon the risk management objectives - short term VaR is estimated usually for one day or two weeks, longer terms may include one, two, or five years. The problem of controlling VaR can be formalized as a mathematical programming problem with VaR constraints. However, such a problem is very difficult to solve using formal optimization methods because VaR is non-convex w.r.t. the portfolio positions and it has many local minima. In this section, we show that in contrast to VaR constraints, CVaR constraints can be easily handled using formal optimization approaches. Constraining CVaR also restricts VaR because CVaR $\geq$ VaR. Therefore, VaR constraints can be replaced by more conservative CVaR constraints.

Similar to CVaR minimization, we can include CVaR in constraints and replace it by the function $F_\beta(x, \alpha)$, see [28]. For instance, let us consider the problem of minimizing the mean losses, $\mu(x) \;=\; E\, f(x, y)$, subject to some balance constraints $x \in X$, and two CVaR constraints with confidence levels β and γ. In this case, the optimization problem can be stated as follows

$$\min \;\; \mu(x)$$

subject to

$$x \in X ,$$

$$\phi_\beta(x) \leq C_\beta ,$$

$$\phi_\gamma(x) \leq C_\gamma ,$$

where C_β and C_γ are some constants constraining CVaR at different confidence levels. The last two constraints can be replaced by constraints

$$F_\beta(x, \alpha_1) \leq C_\beta ,$$

$$F_\gamma(x, \alpha_1) \leq C_\gamma .$$

Indeed, if these constraints are satisfied for some α_1 and α_2, then they are satisfied for the minimal values

$$\min_{\alpha_1} F_\beta(x, \alpha_1) = \phi_\beta(x) ,$$

and

$$\min_{\alpha_2} F_\gamma(x, \alpha_2) = \phi_\gamma(x) .$$

Optimization with these constraints assures that the CVaR values are properly restricted. Moreover, if a risk constraint is active, e.g., in the first constraint, $F_\beta(x^*, \alpha_1^*) = C_\beta$, then the optimal value α_1^* equals β-VaR.

4.3 Minimizing CVaR with Finite Number of Scenarios: Linear Programming

Let us consider now the case in which an analytical representation of the density function $p(y)$ is not available, but we have J scenarios, $y_j, j = 1, ..., J$, sampled from the density $p(y)$. For instance, we may have historical observations of prices of instruments in the portfolio, or we may use Monte Carlo simulations to price the instruments. In this case, the function $F_\beta(x, \alpha)$ can be calculated approximately as follows

$$\tilde{F}(x, \alpha) \overset{def}{=} \alpha + \nu \sum_{j=1}^{J} \big(f(x, y_j) - \alpha \big)^+ ,$$

where the constant ν equals $\nu = ((1 - \beta)J)^{-1}$ and $t^+ = \max(0, t)$. If the function $f(x, y)$ is convex w.r.t. x, then the function $\tilde{F}(x, \alpha)$ is a convex nonsmooth function w.r.t. the vector (x, a). Therefore, if the feasible set X is convex, the optimization problem with the CVaR performance function can be solved using non-smooth optimization techniques. Moreover, if the function $f(x, y)$ is linear w.r.t. x, this problem can be solved using LP techniques. LP approaches are routinely used in portfolio optimization with various criteria, such as mean absolute deviation [19], maximum deviation [54], and mean regret [5].

Let us first explain how LP techniques can be used for the minimization of CVaR. Indeed, after replacing in $\tilde{F}(x,\alpha)$ the terms $\big(f(x,y_j)-\alpha\big)^+$ by auxiliary variables z_j, and imposing constraints $z_j \ge f(x,y_j)-\alpha$, $z_j \ge 0$, $j=1,\dots,J$, we can reduce the minimization of the function $\tilde{F}(x,\alpha)$ to the following optimization problem

$$\min_{x\in\mathbb{R}^n, z\in\mathbb{R}^J, \alpha\in\mathbb{R}} \quad \alpha + \nu \sum_{j=1}^{J} z_j \tag{32}$$

subject to

$$x \in X \,, \tag{33}$$

$$z_j \ge f(x,y_j) - \alpha \,, \quad z_j \ge 0 \,, \quad j=1,\dots,J \,. \tag{34}$$

Several case studies (see, [1, 28, 39]) have demonstrated that this formulation provides a very powerful and numerically stable technique which can solve problems with a large number of instruments and scenarios.

Example 6. CVaR Minimization with a Constraint on Mean Losses.
Suppose that we want to minimize CVaR of the small portfolio described in Example 5. We are interested in minimizing one day CVaR under the condition that the mean daily portfolio losses are less than or equal to $-R$ (i.e., the mean profit is bigger than or equal to R). Suppose that for two instruments in the portfolio, we have prices for J previous days. From this historical data, we can estimate J daily returns and calculate J scenarios for the next day prices, $y_j = (y_{j1}, y_{j2}), j = 1, ..., J$. The mean portfolio loss equals

$$\mu(x) = J^{-1} \sum_{j=1}^{J} f(x,y_j) = J^{-1} \sum_{j=1}^{J} (x_1(m_1 - y_{j1}) + x_2(m_2 - y_{j2})) \,.$$

The constraint on the mean losses is formulated as follows

$$J^{-1} \sum_{j=1}^{J} (x_1(m_1 - y_{j1}) + x_2(m_2 - y_{j2})) \le -R \,. \tag{35}$$

The CVaR minimization problem can be easily solved by minimizing the linear function (32) subject to linear constraints (33),(34), and (35). This problem can be solved using standard LP solvers such as CPLEX.

4.4 Linearization of CVaR Constraints with Finite Number of Scenarios

The previous section showed that the nonlinear CVaR function can be minimized using a linear objective function and linear constraints. Here, we show that a CVaR constraint in optimization problems can be approximated by a set of linear constraints. Let J scenarios, y_j, $j = 1, ..., J$, be sampled from the density $p(y)$. Suppose that a CVaR constraint, $\phi_\beta(x) \leq C_\beta$ needs to be satisfied. As it was earlier discussed, this constraint can be replaced by the constraint $F_\beta(x, \alpha) \leq C_\beta$ using the additional variable α. Further, we can approximate this constraint by the constraint $\tilde{F}_\beta(x, \alpha) \leq C_\beta$ using scenarios y_j, $j = 1, ..., J$. Finally, the last constraint can be equivalently represented by the set of constraints

$$\alpha + \nu \sum_{j=1}^{J} z_j \leq C_\beta , \tag{36}$$

$$z_j \geq f(x, y_j) - \alpha , \quad z_j \geq 0 , \quad j = 1, \ldots, J . \tag{37}$$

If constraint (36) is active, then the optimal value α^* equals VaR. A case study on the application of these techniques to the optimization of the portfolio consisting of the S&P100 stocks can be found in [28].

5 Conclusion

This introductory paper reviewed several topics related to the analysis of the probabilistic and quantile functions: (1) sensitivity analysis of probabilistic functions; (2) sensitivity analysis of quantiles (VaR); and (3) optimization approaches for the risk management problems with the CVaR performance function and constraints.

This overview is very far from being comprehensive and many important issues are beyond the scope of this paper. For instance, convexity of the probability and quantile functions is not discussed in this paper. However, this topic is well studied in the literature, see [32] and [14] and a reader interested in this material can find it in these and other publications. The emphasis of this paper is on issues which have been relatively recently developed. Also, this volume provides a lot of new results in the area of theory and applications of the probabilistic and quantile functions.

References

[1] Andersson, F., Mausser, H., Rosen, D., and S. Uryasev (2000) Credit Risk Optimization With Conditional Value-At-Risk Criterion. *Mathematical Programming*. To appear.

[2] Artzner, P., Delbaen F., Eber, J.M., and D. Heath (1999) Coherent Measures of Risk. *Mathematical Finance*, 9, 203-228.

[3] Birge J.R. and F. Louveaux (1997) Introduction to Stochastic Programming. Springer, New York.

[4] Bucay, N. and D. Rosen (1999) Credit Risk of an International Bond Portfolio: a Case Study. *ALGO Research Quarterly*. Vol.2, No. 1, 9–29.

[5] Dembo, R.S. and A.J. King (1992) Tracking Models and the Optimal Regret Distribution in Asset Allocation. *Applied Stochastic Models and Data Analysis*. Vol. 8, 151-157.

[6] Duffie, D. and J. Pan (1997) An Overview of Value-at-Risk. *Journal of Derivatives*. 4, 7–49.

[7] Ermoliev, Yu. (1976) *Stochastic Programming Methods*. Nauka, Moscow. (in Russian).

[8] Ermoliev, Yu. (1983) Stochastic Quasi-Gradient Methods and Their Applications to System Optimization. *Stochastics*, 4, 1-36.

[9] Ermoliev, Yu. and R. J-B Wets (Eds.) (1988): *Numerical Techniques for Stochastic Optimization*, Springer Series in Computational Mathematics, 10.

[10] Glasserman P. (1991) *Gradient Estimation Via Perturbation Analysis*. Kluwer, Boston.

[11] Ho, Y.C. and X.R. Cao (1991) *Perturbation Analysis of Discrete Event Dynamic Systems*. Kluwer, Boston.

[12] Ioffe, A.D. and V.L. Levin (1972) Subdifferentials of Convex Functions. *Papers of Moscow Mathematical Society* (Trudy MMO), V. 26, 3-73 (in Russian).

[13] Kall, P. and S.W. Wallace (1994) *Stochastic Programming*. Willey, Chichester.

[14] Kan, Y.S., and Kibzun, A.I. (1996). *Stochastic Programming Problems with Probability and Quantile Functions*, John Wiley & Sons, 316.

[15] Kast, R., Luciano, E., and L. Peccati (1998) *VaR and Optimization*. 2nd International Workshop on Preferences and Decisions, Trento, July 1-3 1998.

[16] Kibzun, A.I., Malyshev, V.V., and D.E. Chernov (1988) Two Approaches to Solutions of Probabilistic Optimization Problems. *Soviet Journal of Automaton and Information Silences*, V.20, No.3, 20-25.

[17] Kibzun, A.I. and G.L. Tretyakov (1996) Probability Function Differentiability. *Doklady RAN*, (in Russian).

[18] Kibzun A.I. and S. Uryasev. (1998) Differentiability of Probability Functions. Stochastic Analysis and Applications. 16(6), 1101-1128.

[19] Konno, H. and H. Yamazaki (1991): Mean Absolute Deviation Portfolio Optimization Model and Its Application to Tokyo Stock Market. *Management Science.* **37**, 519-531.

[20] Litterman, R. (1997) Hot Spots and Hedges (I). *Risk.* 10 (3), 42-45.

[21] Litterman, R. (1997) Hot Spots and Hedges (II). Risk, 10 (5), 38-42.

[22] Lucas, A., and Klaassen, P. (1998) Extreme Returns, Downside Risk, and Optimal Asset Allocation. *Journal of Portfolio Management,* Vol. 25, No. 1, 71-79.

[23] Marti,K. (1994) Approximations and Derivatives of Probability Functions. *In Approximation, Probability and Related Fields,* edited by G. Anastassiou and S.T. Rachev, Plenum Press, New York., 367-377.

[24] Marti K. (1995) Differentiation Formulas for Probability Functions: The Transformation Method. *Mathematical Programming Journal,* Series B, Vol. 75, No. 2.

[25] Markowitz, H.M. (1952) Portfolio Selection. *Journal of finance.* Vol.7, 1, 77–91.

[26] Mausser, H. and D. Rosen (1998) Beyond VaR: From Measuring Risk to Managing Risk, *ALGO Research Quarterly,* Vol. 1, No. 2, 5–20.

[27] Mausser, H. and D. Rosen (1999) Applying Scenario Optimization to Portfolio Credit Risk, *ALGO Research Quarterly,* Vol. 2, No. 2, 19–33.

[28] Palmquist, J., Uryasev, S., and P. Krokhmal (2000) Portfolio Optimization with Conditional Value-At-Risk Objective and Constraints. *The Journal of Risk.* To appear.

[29] Pflug, G.Ch. (1996) *Optimization of Stochastic Models : The Interface Between Simulation and Optimization.* Kluwer Academic Publishers, Dordrecht, Boston.

[30] Pflug, G.Ch. (2000) *Some Remarks on the Value-at-Risk and the Conditional Value-at-Risk.* In."Probabilistic Constrained Optimization: Methodology and Applications", Ed. S. Uryasev, Kluwer Academic Publishers, 2000

[31] Prékopa, A. (1970) On Probabilistic Constrained Programming, in: *Proceedings of the Princeton Symposium on Mathematical Programming,* (Princeton University Press, Princeton, N.J.), 113–138.

[32] Prekopa, A. (1995) *Stochastic Programming,* Kluwer Academic Publishers.

[33] Pulkkinen, A. and S. Uryasev (1991) *Optimal Operational Strategies for an Inspected Component – Solution Techniques.* Collaborative Paper CP-91-13, International Institute for Applied Systems Analysis, Laxenburg, Austria.

[34] Prékopa, A. and T. Szántai (1978) A New Multivariate Gamma Distribution and Its Fitting to Empirical Streamflow Data. *Water Resources Research*, 14, 19-24.

[35] Pritsker, M. (1997) Evaluating Value at Risk Methodologies, *Journal of Financial Services Research*, 12:2/3, 201-242.

[36] Raik, E. (1975) The Differentiability in the Parameter of the Probability Function and Optimization of the Probability Function via the Stochastic Pseudogradient Method. *Eesti NSV Teaduste Akadeemia Toimetised. Füüsika-Matemaatika,* 24, 1, 3-6 (in Russian).

[37] RiskMetrics^{TM} (1996) *Technical Document*, 4-th Edition, New York, NY, J.P.Morgan Inc., December.

[38] Rockafellar, R.T. (1970): *Convex Analysis.* Princeton Mathematics, Vol. 28, Princeton Univ. Press.

[39] Rockafellar, R.T. and S. Uryasev (2000) Optimization of Conditional Value-At-Risk. *The Journal of Risk*, Vol. 2, No. 3.

[40] Rockafellar, R.T. and R.J.-B. Wets (1982) On the Interchange of Subdifferentiation and Conditional Expectation for Convex Functionals. *Stochastics*, 7, 173-182.

[41] Roenko, N. (1983) *Stochastic Programming Problems with Integral Functionals over Multivalued Mappings*, (Ph.D. Thesis, Kiev, Ukraine) (in Russian).

[42] Rubinstein, R. (1992) Sensitivity Analysis of Discrete Event Systems by the "Push Out" Method. *Annals of Operations Research*, 39.

[43] Rubinstein, R. and A. Shapiro (1993) *Discrete Event Systems: Sensitivity Analysis and Stochastic Optimization via the Score Function Method.* Willey, Chichester.

[44] Simon, J. (1989) Second Variation in Domain Optimization Problems. *In International Series of Numerical Mathematics,* ed.by F.Kappel, K.Kunish and W.Schappacher, Birkhauser Verlag, 91, 361-378.

[45] Simons, K. (1996) Value-at-Risk New Approaches to Risk Management. *New England Economic Review*, Sept/Oct, 3-13.

[46] Stambaugh, F. (1996) Risk and Value-at-Risk. *European Management Journal*, Vol. 14, No. 6, 612-621.

[47] Thompson, K.M., D.E. Burmasater and E.A.C. Crouch (1992) *Monte Carlo Techniques for Quantitative Uncertainty Analysis in Public Risk Assessments.* Risk Analysis, 12.

[48] Uryasev, S. (1987) *Adaptive Algorithms for Stochastic Optimization and Game Theory.* Nauka, Moscow (in Russian).

[49] Uryasev, S. (1989) A Differentiation Formula for Integrals over Sets Given by Inclusion. *Numerical Functional Analysis and Optimization,* 10 (7 & 8), 827-841.

[50] Uryasev, S. (1994) Derivatives of Probability Functions and Integrals over Sets Given by Inequalities. *J. Computational and Applied Mathematics,* 56, 197-223.

[51] Uryasev, S. (1995) Derivatives of Probability Functions and some Applications.*Annals of Operations Research*, 56, 287-311.

[52] Uryasev, S. (2000) Conditional Value-at-Risk: Optimization Algorithms and Applications. Financial Engineering News, No. 14, February.

[53] Uryasev, S. and A. Shlyakhter (1994) *A Procedure for Simultaneous Calculation of Sensitivities in Probabilistic Risk Analysis.* In: Abstracts Society for Risk Analysis Annual Conference and Exposition, Baltimore, Maryland, December 1994.

[54] Young, M.R. (1998): A Minimax Portfolio Selection Rule with Linear Programming Solution. *Management Science.* Vol.44, No. 5, 673–683.

[55] Ziemba, W.T. and J.M. Mulvey (Eds.) (1998): *Worldwide Asset and Liability Modeling*, Cambridge Univ. Pr.

Pricing American Options by Simulation Using a Stochastic Mesh with Optimized Weights

Mark Broadie (mnb2@columbia.edu)
Columbia Business School
New York, NY 10027

Paul Glasserman (pg20@columbia.edu)
Columbia Business School
New York, NY 10027

Zachary Ha (zachary.ha@gs.com)
Goldman Sachs & Co.
New York, NY 10005

Abstract

This paper develops a simulation method for pricing path-dependent American options, and American options on a large number of underlying assets, such as basket options. Standard numerical procedures (lattice methods and finite difference methods) are generally inapplicable to such high-dimensional problems, and this has motivated research into simulation-based methods. The optimal stopping problem embedded in the pricing of American options makes this a nonstandard problem for simulation.

This paper extends the *stochastic mesh* introduced in Broadie and Glasserman [5]. In its original form, the stochastic mesh method required knowledge of the transition density of the underlying process of asset prices and other state variables. This paper extends the method to settings in which the transition density is either unknown or fails to exist. We avoid the need for a transition density by choosing mesh weights through a constrained optimization problem. If the weights are constrained to correctly price sufficiently many simple instruments, they can be expected to work well in pricing a more complex American option. We investigate two criteria for use in the optimization — maximum entropy and least squares. The methods are illustrated through numerical examples.

S.P. Uryasev (ed.), Probabilistic Constrained Optimization, 26–44.
© 2000 *Kluwer Academic Publishers.*

1 Introduction

Computational methods for pricing derivative securities can be broadly divided into deterministic methods and simulation-based methods. The first type generally involves discretizing time and discretizing the possible levels of the underlying asset prices; the discrete approximation is then solved exactly. Well-known examples of this approach include binomial and trinomial lattices, and finite difference methods. These methods are widely used, particularly in valuing relatively simple derivative securities in relatively simple models (see, e.g., [13] for background).

Deterministic methods can be very fast and effective if the dimension of the state vector representing the underlying model is 1, 2, or perhaps 3. But the time and space requirements of these methods typically grow exponentially in the dimension, rendering these methods inapplicable to high-dimensional problems.

Simulation methods are based on stochastic sampling of paths of the underlying state vector. Their space requirements generally grow linearly in the dimension of the state vector. They typically converge in proportion to the square root of the number of paths generated, a convergence rate independent of the dimension of the problem. This makes simulation-based methods attractive for valuing path-dependent and multi-asset derivatives.

A complication arises, however, with simulation techniques in pricing option contracts with American-style features—i.e., contracts in which the holder can choose the time of exercise. In this case, an optimal exercise boundary has to be determined through some type of dynamic programming procedure. The difficulty arises in combining the forward evolution of simulation with backward induction of dynamic programming. Recently, several methods have been proposed to address this issue; see [1, 3, 4, 5, 6, 7, 9, 10, 12] and references there.

In this paper we further develop the *stochastic mesh* method introduced in Broadie and Glasserman [5]. This method simulates multiple paths in parallel and uses information from all paths to estimate the continuation value (the value of holding an option rather than exercising) at each node along each path. The continuation value at each node is estimated as a discounted weighted average of the option values at the next time step across all paths. In the original mesh method, the weights were computed from the transition density of the underlying process. For complex models the transition density may be unknown, and in *singular* problems the transition density fails to exist altogether.

An example of a singular problem is an American Asian option, an option whose payoff at exercise depends on the time-average price of the underlying asset. The singularity arises from the fact that the running average is a deterministic transformation of the path of the underlying asset. Singularities also arise in any model in which the number of driving factors is smaller than the number of state variables, as is typical in term structure models. For example, a term structure model may represent 80 interest rates (quarterly rates over 20 years) and yet be driven by just a three-dimensional Brownian motion.

To address these issues, we develop a strategy for selecting weights that does not rely on the knowledge or existence of a transition density. We choose the weights through optimization subject to constraints such as matching moments of the underlying process. The goal is to choose the weights so that the mesh correctly prices simple instruments and then use those weights to price complex instruments. Because the number of constraints is typically much smaller than the number of weights, the problem is underdetermined and we have to impose an optimization criterion to choose a particular set of weights. We investigate two criteria in particular — maximum entropy and least squares.

The rest paper of this paper is organized as follow. We give a general formulation of the problem in Section 2. Section 3 reviews the stochastic mesh method and explains the crucial step of calculating the weights. Optimization problems are then formulated in Section 4 for the new approach to choosing weights. In Section 5 we give various numerical examples to illustrate the methods.

2 Problem Formulation

We denote by $S_t = (S_t^1, \ldots, S_t^n)$ the vector of underlying state variables at time t and we assume that S_t is a Markov process. The Markov property can in most cases be enforced by introducing additional variables in the state vector, if necessary. The payoff (or in some cases the discounted payoff) from exercise at time t in state S_t is given by $h(t, S_t)$ for some function h. Path-dependent payoffs can again be accommodated by introducing additional state variables, if necessary. For example, if the payoff depends on the time-average, the maximum, or the minimum of one of the state variables, we can include the running average, the running maximum, or the running minimum in the state vector.

Perhaps the simplest interesting case of a model of this type is a multivariate lognormal process. In this case, $\log S_t$ is an n-dimensional Brownian motion process with a fixed covariance matrix Σ. If the process is simulated under the risk-neutral measure, each component $\log S_t^i$ has drift r, where r is the risk-free interest rate, assumed deterministic and constant. We return to this special case later to illustrate various methods.

We assume that exercise of the option is restricted to a finite set of dates and, for simplicitly, we assume that these dates are equally spaced Δt time units apart. (Options with a finite set of exercise opportunities are sometimes called "Bermudan" and the term "American" reserved for continuous exercise opportunities. We interpret "American" to refer to either case.) Letting $T = d\Delta t$ denote the option expiration date, the problem we seek to solve is finding

$$V(0, S_0) = \max_{\tau} E[h(\tau, S_\tau)],$$

with S_0 a given initial state and τ restricted to be a stopping time (with respect to the Markov process $\{S_{k\Delta t}, k = 0, 1, \ldots\}$) taking values in $\{0, \Delta t, \ldots, d\Delta t\}$. In an important special case of this problem, the expectation is with respect to the risk-neutral measure and h denotes a payoff discounted at the risk-free rate. More generally, the expectation could be with respect to some other martingale measure and h could be discounted by the corresponding numeraire asset, which would then be one of the state variables or possibly a transformation of the state variables.

We assume that the function h is explicitly available. Thus, the value $h(T, S_T)$ from exercise at expiration is available as a function of S_T. This leads to the following backward induction for determining the value at time 0:

$$\begin{aligned} V(T,s) &= h(T,s) \\ V(k\Delta t, s) &= \max\{h(k\Delta t, s), E[V((k+1)\Delta t, S_{(k+1)\Delta t})|S_{k\Delta t} = s]\}, \\ & \quad k = d-1, d-2, \ldots, 1, 0. \end{aligned} \tag{1}$$

The first term inside the max operator is the immediate exercise value of the option and the second term (the conditional expectation) is the continuation value. Essentially all methods for pricing American options approximate this dynamic programming representation in some way. Simulation can be useful in estimating the continuation value since simulation is particularly well suited to estimating expectations.

3 Stochastic Mesh Method

The stochastic mesh method may be viewed as a particular way of choosing the points at which the immediate exercise and continuation values will be calculated and a particular way of estimating the continuation value. The method begins by generating nodes for the mesh. The next step is to construct weights for the transitions between nodes. Finally, the weights are used for estimating prices in the mesh. We give a general description of the method and illustrate it in the case of a lognormal process.

3.1 Mesh Construction

Construction of the mesh begins with simulation of multiple independent copies of the underlying process S_t, all from a common initial state S_0. A schematic representation of these paths is shown in Figure 1. In the figure, node i, for example, corresponds to the n-dimensional state vector $(S^1_{3\Delta t}, S^2_{3\Delta t}, \ldots, S^n_{3\Delta t})$. The state vector $(S^1_{4\Delta t}, S^2_{4\Delta t}, \ldots, S^n_{4\Delta t})$ at node j is generated from node i using whatever method would ordinarily be used to simulate the underlying state vector over a time period Δt. It must be stressed that the figure is purely schematic, with each path representing an

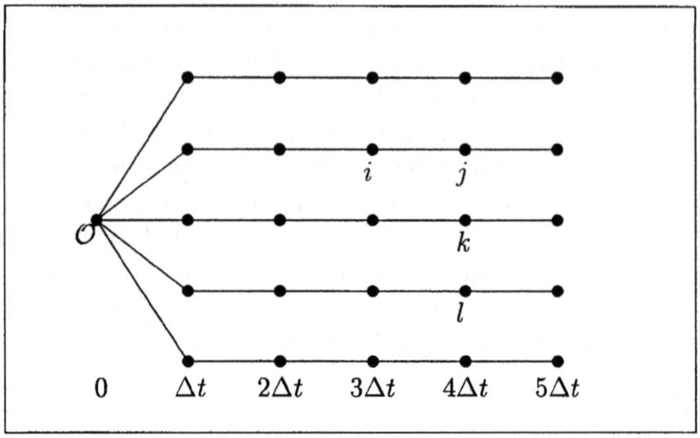

Figure 1: Stochastic mesh. Node i contains the vector $(S^1_{3\Delta t}, S^2_{3\Delta t}, \ldots, S^n_{3\Delta t})$.

independent simulated trajectory and all paths simulated from the same transition law. There is no sense in which node j is "higher" than node k, for example.

To be more concrete consider the following n-dimensional, m-factor lognormal processes

$$\frac{dS^i_t}{S^i_t} = r\,dt + \sum_{j=1}^{m} L_{ij}\,dW^j_t, \quad i = 1, \ldots, n.$$

Here, r is the risk-free interest rate, the W^j_t are independent standard Brownian motions, and L is an $n \times m$ matrix. The law of this process depends on L only through the instantaneous covariance matrix $\Sigma = LL^\top$. Paths of this process can be simulated using

$$S^i_{t+\Delta t} = S^i_t \exp\left((r - \frac{1}{2}\Sigma_{ii})\Delta t + \sqrt{\Delta t}\sum_{j=1}^{m} L_{ij} X_j \right), \quad i = 1, \ldots, n, \tag{2}$$

with the X_j's sampled from the standard normal distribution. Repeated use of the recursive relation in (2) (with independent X_j's) produces a single path for the mesh. Repeating this path generation procedure for multiple sets of independent X_j's produces a set of paths from which to construct the mesh.

3.2 Weights from a Transition Density

Once we have the paths for the mesh, we choose a set of weights w_{ij}, with w_{ij} denoting the weight attached to the transition from node i at one time slice to node j at the next time slice. The key feature of the mesh is that i and j need not be on the same path for w_{ij} to be nonzero. In fact, once we have generated the paths we deliberately "forget" which nodes were on the same paths and treat every node at time $(k+1)\Delta t$ as a potential successor of every node at time $k\Delta t$.

Given N nodes at each time slice and a set of weights w_{ij}, we approximate (1) using

$$V_i(k\Delta t) = \max\left(h_i(k\Delta t), \sum_{j=1}^{N} w_{ij}V_j((k+1)\Delta t)\right), \tag{3}$$

or

$$V_i(k\Delta t) = \max\left(h_i(k\Delta t), D_{\Delta t}^{-1}\sum_{j=1}^{N} w_{ij}V_j((k+1)\Delta t)\right),$$

with $D_{\Delta t}^{-1}$ a discount factor, depending on how the discounting is handled. Here, $V_i(k\Delta t)$ denotes the estimated value of the option at node i and time $k\Delta t$, and $h_i(k\Delta t)$ is the (explicitly available) immediate exercise value at node i and time $k\Delta t$. The weights w_{ij} will in general depend on the time index k; we suppress the time argument to simplify notation.

In the original mesh method of Broadie and Glasserman [5], the process S_t is assumed to have a known transition density and the weights are calculated from this transition density. To make this more explicit, suppose S_t satisfies

$$P(S_{t+\Delta t} \in A | S_t = x) = \int_A f(t, x, y)\, dy,$$

for some probability density $f(t, x, \cdot)$, all x, all t, and all (measurable) A. In the multivariate lognormal case (2), a transition density exists and is easily expressed in terms of the standard normal density provided the covariance matrix Σ has full rank.

For some fixed time t, let p_{ij} denote the value of the transition density from node i at time t to node j at time $t + \Delta t$. In [5], the weights are defined as

$$w_{ij} = \frac{p_{ij}}{\sum_k p_{kj}}. \tag{4}$$

This choice implies that

$$\sum_{i=1}^{N} w_{ij} = 1, \tag{5}$$

where N is the total number of mesh paths. Notice that here the destination node j is fixed and the sum is over the possible source nodes i. Because of (5), every node at time $t + \Delta t$ is assigned the same total weight; the nodes differ in how this total weight is distributed among the possible source nodes one time step earlier.

An important feature of (5) becomes evident in checking the price of a *European* option as estimated by the mesh. Of course, there is no need to use a mesh in the European case, but it is instructive to examine this case. For a European option, we replace (3) with $V_j(T) = h_j(T)$ and

$$V_i(k\Delta t) = \sum_{j=1}^{N} w_{ij}V_j((k+1)\Delta t);$$

we omit the max because in the European case the holder of the option no longer has the right to exercise early. A simple induction argument shows that in this case (5) implies

$$V(0) = \frac{1}{N}\sum_{j=1}^{N} h_j(T).$$

In other words, the mesh price telescopes to the average over the payoffs at the terminal nodes—precisely the same estimate that would be obtained from the original N paths using standard simulation rather than the mesh. The same is true if the payoff is discounted over each time step.

A natural alternative to (5) is the condition

$$\sum_{j=1}^{N} w_{ij} = 1. \tag{6}$$

Numerical experiments suggest that choosing the weights to enforce this constraint is less effective than enforcing (5), when the weights are defined from a transition density. In Section 4, where we choose weights through optimization procedures, (6) is a more convenient constraint.

3.3 High-Biased Estimator

Once the weights are determined it is straightforward to calculate the mesh price using (3). By repeating this step one obtains an estimate $V_{\mathcal{O}}(0)$ of the option value at the root node $\mathcal{O}$ of the mesh. In fact, through this procedure one also obtains an estimate of the option value and continuation value at every node in the mesh. This in turn implicitly yields an estimate of the exercise and continuation regions. At a node where the maximum in (3) is attained by the first term, the mesh estimates that it is optimal to exercise the option; wherever the maximum is attained by the second term, the mesh estimates that it is optimal to continue.

The option values estimated by the mesh through (3) tend to overestimate the true option price. This is a consequence of the convexity of the max function and Jensen's inequality. Taking conditional expectations E_i in (3) with respect to the state at node i yields

$$\begin{aligned} E_i[V_i(k\Delta t)] &= E_i\left[\max\left(h_i(k\Delta t), \sum_{j=1}^{N} w_{ij} V_j((k+1)\Delta t)\right)\right] \\ &\geq \max\left(h_i(k\Delta t), \sum_{j=1}^{N} w_{ij} E_i[V_j((k+1)\Delta t)]\right). \end{aligned}$$

This effect propagates backwards through the mesh and results in an estimate at time 0 that is biased high. For a more complete investigation of the properties of the mesh estimator, see Broadie and Glasserman [5].

3.4 Low-Biased Estimator

As in [5], the high-biased estimator can be combined with a low-biased estimator to produce an interval estimate for the option price. To generate the low-biased estimator, we generate additional independent paths through the mesh. These additional paths use the mesh only to determine *when* to exercise; the payoff assigned to each of these paths is the exact payoff upon exercise and not a value estimated from the mesh.

From the initial node $\mathcal{O}$ we begin simulation of a new path of S_t over time steps $\Delta t, 2\Delta t, \ldots, T$. At each step, to determine whether or not to stop, we calculate weights with respect to the new node on the path, use (3) to estimate the option value at that node and repeat this procedure until the estimated option value is equal to the exercise value, i.e. the estimated exercise region is reached. Upon exercise, we record a payoff by evaluating h. Figure 2 shows three paths a, b and c that are simulated and stopped when the exercise boundary (as estimated by the mesh) is reached.

The key observation here is that the original mesh determines the exercise decision at all possible states and not simply those corresponding to nodes in the mesh. Suppose for example that we have simulated a path to a state s at time $k\Delta t$. From state s (which is generally not a node in the mesh) we evaluate the transition density to each node j at time $(k+1)\Delta t$; call these weights w_{sj}. Using these weights we estimate the continuation value from state s as

$$\sum_{j=1}^{N} w_{sj} V_j((k+1)\Delta t)$$

using values V_j already calculated in the mesh. We also evaluate the immediate exercise value $h(s, k\Delta t)$ in state s. If the estimated continuation value exceeds the immediate exercise value, we continue simulating the path by generating a transition out of state s; if the immediate exercise value is greater, we stop and record a payoff of $h(s, k\Delta t)$. We repeat this procedure over many paths (all on the original mesh) and then average.

The exercise region determined by the mesh (and then used by the independent paths) is not in general the optimal exercise region. We know, however, that it cannot be better than the optimal exercise region. Thus, the average payoff generated in this way cannot be greater than the average payoff under the optimal policy. We may therefore conclude that the estimate produce by this second pass through the mesh is biased low. See [5] for a more complete investigation of this method and its properties.

The construction of the weights and the high- and low-biased estimators explained thus far apply in the case of a known transition density. For the special case of (2), this requires that the covariance matrix have full rank. To address settings in which the transition density is either unknown or fails to exist, we next develop a new approach to constructing weights.

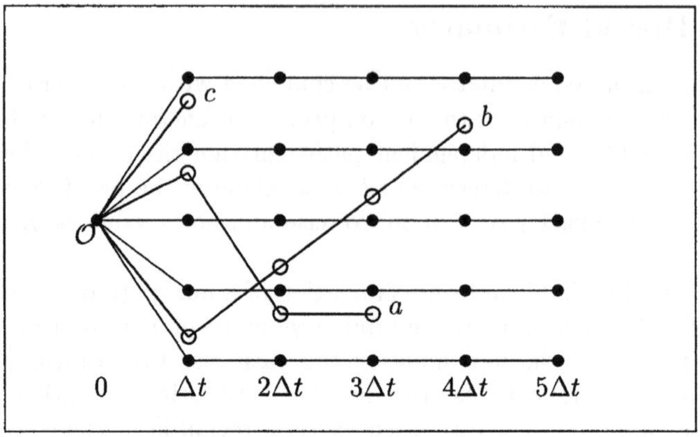

Figure 2: Schematic diagram of three randomly generated paths a, b and c for the low-biased estimator. The paths terminate upon entry into the exercise region estimated by the mesh.

4 Weights via Optimization

To obviate the need for a probability density, we instead formulate the problem of choosing "good" weights as a constrained optimization problem. The constraints we impose ensure that the mesh values of certain basic quantities — for example, low-order moments of the state variables — coincide with their theoretical values. In general, the number of constraints will be much smaller than the number of weights and the problem is underdetermined. We therefore impose an optimization criterion to choose a particular set of weights among all those satisfying the constraints. We investigate two criteria in particular: maximum entropy and least squares. Both of these criteria give preference to uniformity in the weights (subject to the constraints), which is attractive in view of the symmetry in the construction of the mesh. The maximum entropy criterion ensures nonnegativity of the weights; least squares does not but it is computationally easier to work with.

4.1 Maximum Entropy Weights

For a fixed node i, the entropy criterion is

$$L_0 = -\sum_{j=1}^{N} w_{ij} \log(w_{ij}). \tag{7}$$

This objective is maximized (subject to the sum constraint (6)) by the uniform distribution, i.e. $w_{ij} = 1/N$. However, to obtain "good" weights we impose further constraints, such as matching the first order and higher order moments for the underlying processes. The maximum entropy solution then corresponds to the "most uniform" distribution satisfying the constraints. For a different application of entropy weights in pricing derivative securities, see Avellaneda et al. [2].

As an illustration of the types of constraints we use, consider, for example, the case of a single underlying asset with value $S_t(k)$ on path k at time t. Suppose $E[S_{t+\Delta t}|S_t] = e^{r\Delta t}S_t$, as in the lognormal case (2). Then we might impose the constraint that the w_{kj} satisfy

$$S_t(k) = e^{-r\Delta t}\sum_{j=1}^{N} w_{kj}S_{t+\Delta t}(j),$$

the sum taken over all nodes at time $t + \Delta t$. This ensures that the weights w_{kj} correctly "price" the underlying asset itself at node k.

Observe that in this example the constraint is linear in the weights. Suppose more generally that at each node i there are K linear constraints given by

$$\sum_{j=1}^{N} B_{kj}w_{ij} = b_k, \quad k = 1,\ldots,K, \tag{8}$$

where B is a $K \times N$ matrix and b a K-dimensional vector. (The matrix B and vector b will in general depend on the node i.) We incorporate these constraints in the optimization criterion at node i by setting

$$L = L_0 + \lambda_0(\sum_j w_{ij} - 1) + \sum_{k,j} \lambda_k(B_{kj} - b_k)w_{ij}, \tag{9}$$

where the λ_k's are Lagrange multipliers. Notice that we have also imposed (6) as a constraint.

By explicitly solving the following equations

$$\frac{\partial L}{\partial w_{ij}} = 0, \tag{10}$$

$$\frac{\partial L}{\partial \lambda_0} = 0, \tag{11}$$

we obtain w_{ij} in terms of the Lagrange multipliers and parameters for the constraints as

$$w_{ij} = \frac{\exp(\sum_k \lambda_k(B_{kj} - b_k))}{\sum_\ell \exp(\sum_k \lambda_k(B_{k\ell} - b_k))}. \tag{12}$$

Solving the rest of the constraints given by (8), therefore, amounts to minimizing the following function with respect to λ's ([8]):

$$\log\left(\sum_\ell \exp\left(\sum_k \lambda_k(B_{k\ell} - b_k)\right)\right). \tag{13}$$

We use the Newton-Raphson method to obtain the Lagrange multipliers that minimize this function and solve for the weights using (12). The number of constraints is usually much smaller than the number of weights N and thus the numerical optimization is viable. It should be clear from the presence of the logarithm in (7) that the maximum entropy weights are always positive.

4.2 Least Squares Method

The Taylor approximation $-\log(w) \approx 1 - w$ near $w = 1$ leads to the approximation $\sum_j w_{ij}(1 - w_{ij})$ for L_0 in (9). Because we also impose the constraint that $\sum_j w_{ij} = 1$, maximizing this approximation is equivalent to choosing weights through the least squares criterion of minimizing $\sum_j w_{ij}^2$. The least squares problem has the advantage that it can be solved explicitly, without the need for numerical optimization. By solving $\partial L / \partial w_{ij} = 0$ we find that the vector $w_i = (w_{i1}, \ldots, w_{iN})$ of weights out of node i satisfy

$$w_i = \frac{1}{2} \lambda B, \tag{14}$$

with $\lambda = (\lambda_1, \ldots, \lambda_k)$. By plugging this expression into (8) we obtain the weights in terms of the known parameters as

$$w_i = B^\top (BB^\top)^{-1} b. \tag{15}$$

Intuitively, the solution for the weights is simply obtained from (8) by inverting the retangular matrix B, and $B^\top (BB^\top)^{-1}$ may be viewed as the pseudo-inverse of B.

An advantage of this method is the improvement in speed. But, there is a drawback that the weights produced are not guaranteed to be nonnegative. If we imposed nonnegativity as a constraint, solving the least squares problem would again require numerical optimization and would therefore have little or no advantage over the maximum entropy criterion.

Longstaff and Schwartz [9] also use a least squares method in pricing American options by simulation. However, they use least squares in regressing an option's continuation value against a set of basis functions rather than to price simple instruments exactly.

5 Some Numerical Examples

In this section we give various examples to illustrate the methods presented in the previous sections.

5.1 One Dimensional Examples

In order to show how the mesh method works on cases with known solutions we first price a one-dimensional American put option using the original weights from

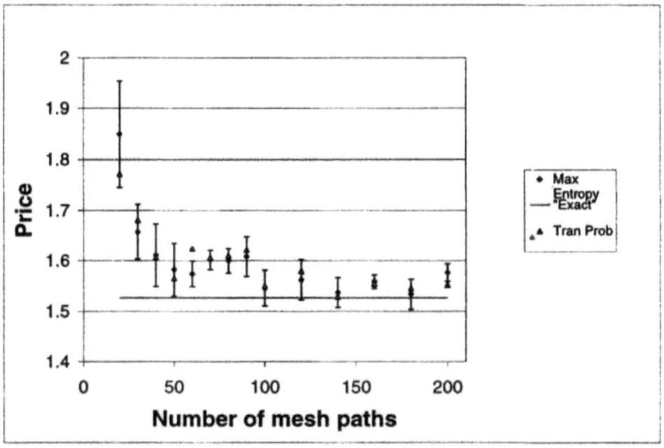

Figure 3: American put price on single asset. Exact price and the prices from weights estimated from the transition densities and the maximum entropy method are plotted. $S_0 = 40$, the strike is 40, the risk-free rate is 10%, the volatility is 20%, and the time to expiration is 5 years. There are five equally spaced exercise opportunities. Number of exercise dates = 5.

the transition probability densities and from the maximum entropy method and plot the high-biased estimators in Figure 3. The figure plots estimated prices against the number of mesh paths for an option with five exercise dates. The error bars show 95% confidence intervals around the maximum entropy estimate. The two methods are hard to distinguish from each other and they approach the true price (obtained from a binomial lattice) as the mesh size increases.

Next, we plot high-biased and low-biased estimates for the same option using the maximum entropy and the least squares methods and compare them with the exact price in Figure 4. The high-biased estimates obtained by the least squares method (LS-High) are higher than those obtained by the maximum entropy method (ME-High). The low-biased estimators are less distinguishable.

The large standand error in the two figures can be reduced if we use a larger mesh size. We use the least squares method to achieve this and plot the low-biased estimator in Figure 5. The reduction in standard error is clear. Although this method comes very close to the true value, it appears to slightly underestimate the true value. The estimated exercise region implicit in the mesh is thus slightly suboptimal.

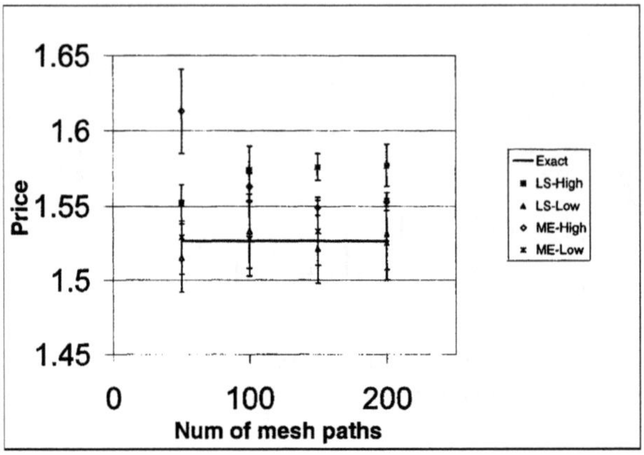

Figure 4: High- and low-biased estimators for the American put price estimated using the maximum entropy and the least squares methods.

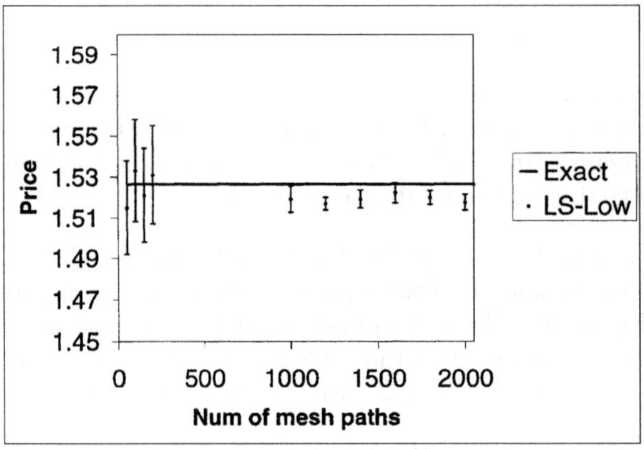

Figure 5: Low-biased estimator for the American put at large mesh sizes.

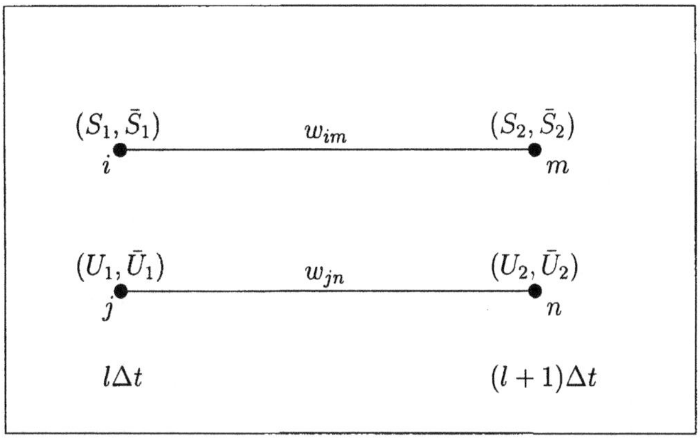

Figure 6: Illustration of singularity in American Asian option.

5.2 American Asian option

The payoff function for an American Asian option depends on the average price over the discrete exercise dates. For example, a possible payoff function for the American Asian put option is given by $\max(X - \bar{S}, 0)$ where X is the strike price and $\bar{S}$ is the average of $S_{k\Delta t}$ up to the current time. This option corresponds to a two-dimensional singular case since the underlying asset price and its average are perfectly correlated over a single time step. The problem is illustrated in Figure 6. In the mesh construction, $(S_2, \bar{S}_2)$ is simulated from $(S_1, \bar{S}_1)$ and similarly $(U_2, \bar{U}_2)$ is simulated from $(U_1, \bar{U}_1)$. When we attempt to interconnect the nodes we encounter a difficulty: if the underlying asset moves from S_1 to U_2, it is generally *not* the case that the running average moves from $\bar{S}_1$ to $\bar{U}_2$; indeed, this would happen only if $\bar{U}_2$ happens to equal $(l\bar{S}_1 + U_2)/(l+1)$, which occurs with probability zero. There is therefore no way to assign a weight w_{in} based on the probability density of moving from i to n.

Figure 7 shows prices obtained from the least-squares mesh method with two types of constraints and from a non-recombining bushy tree (used here as the benchmark). On the left the American and European prices are plotted against the number of time steps. The first pair of prices on the right denoted by a filled triangle and filled circle are Richardson extrapolated prices (see, e.g., [13] for background on Richardson extrapolation). The other two pairs on the right correspond to prices with the following two types of constraints: (i) first three moments for both the underlying and average stock prices and their first order cross moments (B3N3NB1), and (ii) first three moments for the average prices (B3). The two sets of prices are close to each other and the low-biased estimators from the mesh are approximately within 3% of the price from the Richardson extrapolated bushy tree price.

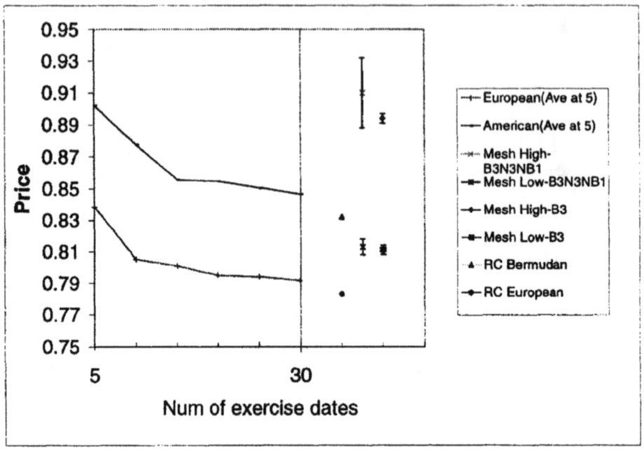

Figure 7: American Asian put with the payoff function $\max(X - \bar{S}, \cdot)$.

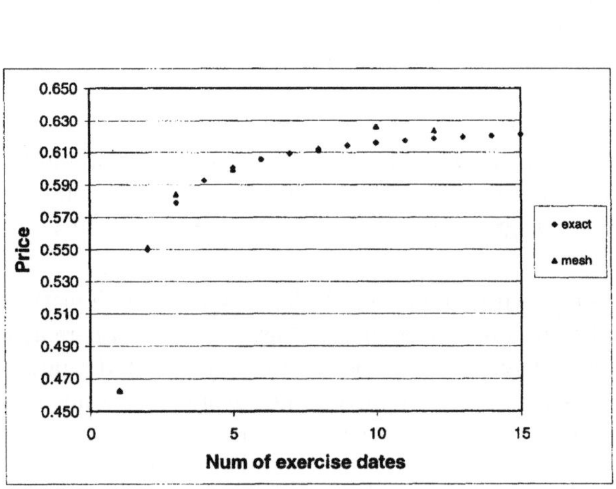

Figure 8: Four-dimensional American vs number of exercise dates

Table 1: Full-rank, multi-dimensional test cases using least squares weights. Here, r and T are the risk-free interest rate and the maturity, respectively. High estimates use replications of a 500-path mesh; Low estimates use 2000 independent paths through the mesh. High and Low interval estimates are ± 1 standard error. The covariance matrices for two cases are Σ_{2D} =((0.04,0.01),(0.01,0.04)) and Σ_{4D} = ((0.04,0.01,0.005,0.001),(0.01,0.02,0.01,0.005), (0.005,0.01,0.1,0.05),(0.001,0.005,0.05,0.08)).

Spot	Strike	r	T	High	Low	Exact	European
2D-2factor							
$(40, 40)$	40	0.1	0.5	1.176±0.007	1.126±0.009	1.137	0.982
$(38, 42)$	43	0.12	1.0	3.050±0.000	3.050±0.000	3.050	1.810
$(37, 45)$	40	0.15	1.0	0.809±0.010	0.741±0.007	0.762	0.514
4D-4factor							
$(40, 40, 40, 40)$	40	0.1	0.5	1.225±0.007	1.183±0.009	1.191	0.857
$(40, 38, 35, 45)$	42	0.12	1.0	2.669±0.004	2.603±0.001	2.665	1.496

5.3 Multi-Dimensional Options on a Geometric Average

If the payoff function depends on a vector of lognormally distributed asset prices only through their geometric average, then the option can be reduced to a one-dimensional problem, because the geometric average is again lognormal. Thus, this option provides nice test cases for our methods: we can solve it in the mesh as a multi-dimensional problem and compare with results obtained from a binomial lattice applied to the equivalent one-dimensional problem. If the multi-dimensional asset process is given by (2), then its geometric average process is given by the one-dimensional process

$$\bar{S}_t = \bar{S}_0 \exp\left((\tilde{r} - \frac{1}{2}\tilde{\sigma}^2)t + \sqrt{t}\tilde{\sigma}X\right), \tag{16}$$

where $\bar{S}_0 = (\prod_j S_0^j)^{1/n}$ and

$$\tilde{\sigma} = \frac{1}{n}\sqrt{\sum_{j,k}\Sigma_{jk}}, \tag{17}$$

$$\tilde{r} = r + \frac{1}{2n^2}\sum_{j,k}\Sigma_{j,k} - \frac{1}{2n}\sum_j \Sigma_{jj}. \tag{18}$$

In Figure 8 we plot results for an American put on the geometric average of four assets. The covariance matrix in this case has full rank and the weights are constructed from the transition probability density. Notice the rapid convergence to the continuous-exercise price as the number of exercise dates increases.

Table 1 lists American put prices for two-dimensional (2D) and four-dimensional (4D) cases having full rank. The mesh estimates (with weights constrained to match

Table 2: Multi-dimensional, singular test cases using least squares weights. All cases have a strike of 40, a risk-free rate of 10%, an expiration of 0.5 years. Under the Mesh column, the first number gives the number of paths in the mesh and the second number gives the number of independent paths simulated through the mesh to generate the Low estimate. High and Low interval estimates are ± 1 standard error.

Spot	Mesh	High	Low	Exact	European
2D-1factor					
$(40, 40)$	100/1000	1.215±0.103	0.976±0.016	1.027	0.863
	200/2000	1.065±0.128	0.982±0.013		
	200/10000	1.220±0.208	0.984±0.007		
	500/2000	1.241±0.130	1.004±0.012		
	2000/20000	1.103±0.083	1.010±0.004		
	3000/20000	1.023±0.072	1.013±0.002		
4D-2factor					
(40,40,40,40)	1000/10000	1.665±0.006	1.510±0.005	1.521	1.359
	2000/10000	1.663±0.003	1.511±0.004		
4D-1factor					
(40,40,40,40)	1000/10000	1.387±0.004	1.267±0.005	1.279	1.115
	2000/10000	1.399±0.002	1.268±0.007		
8D-3factor					
(...40...)	2000/10000	3.961±0.010	3.488±0.012	3.529	3.417
12D-3factor					
(...40...)	2000/10000	3.059±0.008	2.710±0.008	2.749	2.637
16D-3factor					
(...40...)	2000/10000	2.761±0.007	2.436±0.008	2.473	2.352
20D-3factor					
(...40...)	2000/10000	2.578±0.008	2.269±0.007	2.305	2.176

all means and covariances) are compared both with exact values (computed in a binomial lattice) and with the corresponding European prices. Including the European price helps illustrate how much of the early-exercise value is found by the mesh. Clearly, the mesh finds most of this value in these examples. As might be expected, the estimates for the longer maturity put prices are less accurate than those for the shorter maturities.

Table 2 shows results for singular cases ranging from a two-dimensional, one-factor model to a 20-dimensional three-factor model. (Details of the covariance matrices used in these examples are available from the authors.) Several of these cases are extremely singular so that constructing weights based on (nonexistent) transition densities would seem to hopeless. These are clearly harder problems, but notice that in most cases the Low estimate is accurate to the first digit at which the Exact and European prices differ.

6 Conclusion

This paper expands the scope of the stochastic mesh method for pricing multi-dimensional American options to address models in which a transition density for the underlying state variables is unknown or fails to exist. This includes multivariate lognormal processes with a singular covariance matrix. We avoid the need for a transition density by choosing mesh weights through an optimization problem. We choose weights using either a maximum entropy or least squares criterion subject to constraints that ensure the mesh correctly prices simple instruments. Numerical examples illustrate the method. Important directions for future work include improvements in speed and methods for pruning negative weights.

Acknowledgments Research support from NSF grant DMI9457189, from the Center for Applied Probability at Columbia University, from an IBM University Partnership Award, and from Goldman Sachs & Co. is gratefully acknowledged.

References

[1] Andersen, L., 1999, "A Simple Approach to the Pricing of Bermudan Swaptions in the Multi-Factor Libor Market Model," working paper, General Re Financial Products, NY.

[2] Avellaneda, M., C. Friedman, R. Holmes, D. Samperi, 1997, "Calibrating Volatility Surfaces via Relative-Entropy Minimization," *Applied Mathematical Finance*, Vol.4, No.1, 37–64.

[3] Barraquand, J., and D. Martineau, 1995, "Numerical Valuation of High Dimensional Multivariate American Securities," *Journal of Financial and Quantitative Analysis*, Vol. 30, No. 3, 383–405.

[4] Broadie, M., and P. Glasserman, 1997, "Pricing American-Style Securities Using Simulation," *Journal of Economic Dynamics and Control*, Vol. 21, Nos. 8-9, 1323–1352.

[5] Broadie, M., and P. Glasserman, 1997, "A Stochastic Mesh Method for Pricing High-Dimensional American Options," working paper, Columbia University.

[6] Broadie, M., P. Glasserman, and G. Jain, 1997, "Enhanced Monte Carlo Estimates for American Option Prices," *Journal of Derivatives*, Vol. 5, No. 1 (Fall), 25–44.

[7] Carriere, J.F., 1996, "Valuation of the Early-Exercise Price for Derivative Securities using Simulations and Splines," *Insurance: Mathematics and Economics*, Vol. 19, 19–30.

[8] Golan, A., G. Judge and D. Miller, 1996, *Maximum Entropy Econometrics: Robust Estimation with Limited Data*, Wiley, New York.

[9] Longstaff, F., and E. Schwartz, 1998, "Valuing American Options By Simulation: A Simple Least Squares Approach," working paper, UCLA Anderson Graduate School of Management.

[10] Pedersen, M., 1999, "Bermudan Swaptions in the LIBOR Market Model," working paper, SimCorp A/S, Copenhagen, Denmark.

[11] Tilley, J.A., 1993, "Valuing American Options in a Path Simulation Model," *Transactions of the Society of Actuaries*, Vol. 45, 83–104.

[12] Tsitsiklis, J., and B. Van Roy, 1997, "Optimal Stopping of Markov Processes: Hilbert Space Theory, Approximation Algorithms, and an Application to Pricing High-Dimensional Financial Derivatives," working paper, Laboratory for Information and Decision Sciences, MIT, Cambridge, MA.

[13] Wilmott, P., *Derivatives*, Wiley, Chichester, England, 1998.

On Optimization of Unreliable Material Flow Systems

Yu. Ermoliev (ermoliev@iiasa.ac.at)
International Institute for Applied Systems Analysis
A-2361 Laxenburg, Austria

S. Uryasev (uryasev@ise.ufl.edu)
University of Florida
474 Weil Hall, Gainesville, FL 32611-6595

J. Wessels (wessels@win.tue.nl)
Technical University Eindhoven
P.O. Box 513, 5600 MB Eindhoven, The Netherlands

Abstract

The paper suggests an approach for optimizing a material flow system consisting of two work-stations and an intermediate buffer. The material flow system may be a production system, a distribution system or a pollutant-deposit/removal system. The important characteristics are that one of the work-stations is unreliable (random breakdown and repair times), and that the performance function is formulated in average terms. The performance function includes random production gains and losses as well as deterministic investment and maintenance costs. Although, on average, the performance function is smooth with respect to parameters, the sample performance function is discontinuous. The performance function is evaluated analytically under general assumptions on cost function and distributions. Gradients and stochastic estimates of the gradients were calculated using Analytical Perturbation Analysis. Optimization calculations are carried out for an example system.

S.P. Uryasev (ed.), Probabilistic Constrained Optimization, 45–66.
© 2000 *Kluwer Academic Publishers.*

1 Introduction

In several types of material flow systems, there is at least one unreliable component. This feature makes it particularly important to design such a system carefully by taking into account the uncertainties introduced by the component unreliability. Such material flow systems occur in production and distribution as well in environmental remediation (removal or transformation of pollutants) systems. In production systems, work-stations may be unreliable. In distribution systems, transport mechanisms may suffer from breakdowns. In environmental systems, the removal or transformation mechanism may be unavailable (due to climatic causes, for instance). Particular, for production systems, there is extensive literature on modeling and analysis of material flows with unreliable work-stations.

Analytic approaches have primarily been developed for the case of two work-stations with an intermediate buffer. For discrete products with deterministic processing times, we may refer to Buzacott [1] and Yeralan and Muth [25]. For continuous material flows with deterministic machine speeds, important references are Wijngaard [24] and Mitra [12]. De Koster [8] gives an overview of the literature and shows how to exploit Wijngaard's approach for the construction of a numerical procedure for the analysis of larger systems. Although these approaches are very valuable for getting a better understanding of the characteristics of relevant processes, they all suffer from the fact that they are based on severe assumptions. The usual requirement is that breakdown behavior as well as repair behavior is based on a negative-exponentially distributed time length or at least something very closely related to the negative-exponential distribution like a phase-type distribution with only a few phases. Therefore, for practical system design, simulation is a frequently used tool. However, a serious drawback of simulation is that a guided search for a good design usually requires many simulations. Particular, in the case of several design parameters, this can be prohibiting.

Various modeling and optimization approaches of the manufacturing systems are discussed in [11]. General approaches for optimizing stochastic systems by using Monte Carlo simulations follow from the the techniques of stochastic optimization (see, for example, [2, 10]). For Discrete Event Dynamic Systems, algorithms for evaluating unbiased estimates of the gradients are developed in [4, 7, 13, 14]. A disadvantage of the most of these algorithms is that they are not applicable when a sample-path is discontinuous in the relevant parameter, which is the case for number of material flow systems. Rubinstein [14] proposed a technique, which is called the Push Out method, to calculate sensitivities of systems with discontinuous sample-path. In [5], Gong and Ho suggested to smooth over the sample path function (Smoothed Perturbation Analysis) by taking conditional expectations w.r.t. a σ-algebra to estimate the gradient of the performance function. This approach in application to (s,S) inventory systems was explored in [16]. Similar ideas in combination with new differentiation formulas for probability functions [21, 22] were used in Analytic Perturbation Analysis (APA) [23] to evaluate the performance function and the gradient during the same simulation run. In the framework of this approach, for systems with discontinuous sample-path, various estimates can be obtained, including the estimates of Smoothed Perturbation Analysis and the Push Out approach.

Paper [3] considered a system consisting of two machines (one is unreliable) and an intermediate buffer; based on the Monte-Carlo simulation optimization approach, it calculated gradient formulas for this system. In the present paper, for the same model, we show that using the ideas of Analytic Perturbation Analysis [23] the performance function and the gradients of this system can be calculated analytically for any distributions describing unreliability and repair characteristics of the machines. Although the sample path of this system is a discontinuous function, an average performance function is smooth with respect to control parameters. We found formulas for calculating the gradients of the performance function and their rough stochastic estimates. The paper discusses deterministic and stochastic optimization approaches for optimizing performance of the system. We evaluated with deterministic approach optimal parameters of an example system using the Variable Metric Algorithm [19] for nonsmooth optimization problems.

The following Section 2 introduces the model for a two-machine system with one unreliable machine and an intermediate buffer. Section 3 describes the optimization problem and provides analytical formulas for the discrete-continuous performance function and its gradients w.r.t. continuous variables. Section 4 considers two approaches for reducing discrete-continuous optimization problems to continuous ones. Deterministic and stochastic algorithms are described for the reduced problems. Stochastic estimates of the gradients are calculated with APA. Section 5 provides optimization results for an example system.

2 Description of the model

The system comprises two interacting processes (see Figure 1): a regular flow of material arriving at a "server" or "work-station" and a service process of this material. Each batch of material arrives at the server at equidistant points in time $t = 0,\ x_1,\ 2x_1, \dots$ The intensity of this process can be adjusted by the value $x_1 > 0$. The work-station empties the available batches one-by-one and x_2 is the time needed to process a batch by the work-station. The processing of a batch can be interrupted by the failure of the work-station; therefore, the work-station goes through alternating "operation" and "repair" intervals. The lengths of these intervals are independent random variables with density functions ν and ϱ respectively. We suppose that a batch requires a position in the storage (buffer) from the time of its arrival until the moment that processing has been finished. Denote by S the number of batches that may be stocked in the storage. If the storage is full, we suppose that a newly arriving batch is lost with cost α. The gain of each processed batch is equal to β. The cost function also includes the investments and maintenance costs and provides a tradeoff

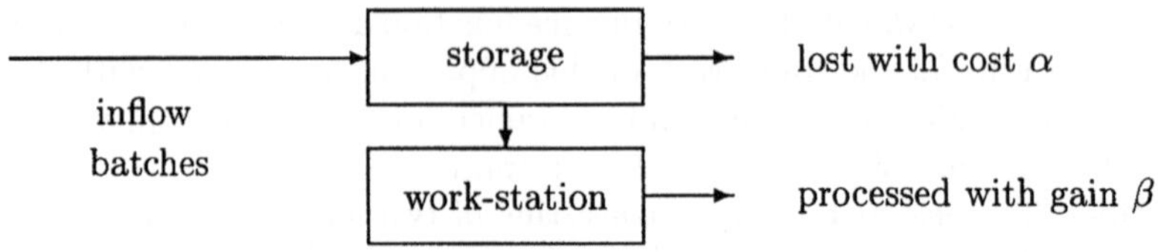

Figure 1: *System Flow Chart*

between profits and losses. In particular, in the case when $\beta \ll \alpha$, the main attention is paid to the losses due to exceeding storage capacity.

As a consequence of the interruptions, the real processing time of a batch may be essentially longer than x_2. If T is the mean time to failure and R is the average repair time, then the availability fraction is

$$\frac{T}{R+T}$$

Consequently, the real processing time will be on the average

$$\frac{R+T}{T} x_2 .$$

If x_1 would be chosen smaller than this latter value, then the work-station could not cope with the input even if the storage capacity would have infinite size.

The model sketched above was inspired by the problem of designing a production system in which the batches would be delivered by a chain oven and the work-station treats the individual products of a batch one-by-one. In this way, the platter, which bears a batch of products through the oven, occupies a position in the buffer as long as it contains some products. It is not possible to stop the oven when the buffer is full, since this would lead to the loss of several hours of production; namely all the batches which are in the oven would be lost in this case. The chain pulls the batches through the oven with a fixed speed and batches which leave the oven are mechanically delivered to the buffer. If the buffer has no position available, then the batch is set aside and lost for further processing, since outside the buffer, the products cool down too much, which is not good for the quality.

In fact, the described system is an example of the two-machine system with one unreliable machine and an intermediate buffer. The problem is to design a system which works at minimal cost. There is no constraint on the output, because it is already certain that several units will be needed. Therefore, the only goal is to find the most efficient design. We denote by $c(x_1, x_2, S)$ the cost per unit of the processed product caused by investment and maintenance costs. We suppose that this is a known function of the design parameters of the system. Also, there is a cost/gain component related to the performance of the system: each processed batch brings a gain β and lost batch a cost α. Therefore, the performance function equals

$$C(x,S) \;=\; c\,(x,S) \;+\; \alpha L(x,S) \;-\; \beta\Psi(x,S), \tag{1}$$

where $x \;=\; (x_1, x_2)$, and $L(x,S)$, $\Psi(x,S)$ are the expected numbers of lost and processed batches per supplied batch, respectively.

Let us make some useful rearrangements of the problem. Denote by $\mathbf{L}_N(x,S)$ the random number of lost batches and by $\mathbf{\Psi}_N(x,S)$ the random number of processed batches when N batches were supplied to the work-station. We use bold face style for random variables and functions. By definition

$$\mathbf{L}_N(x,S) \;+\; \mathbf{\Psi}_N(x,S) \;=\; N\,,$$

$$\lim_{N\to\infty} \frac{\mathbf{L}_N(x,S)}{N} \;=\; L(x,S)\,, \quad \lim_{N\to\infty} \frac{\mathbf{\Psi}_N(x,S)}{N} \;=\; \Psi(x,S) \quad \text{(a.s.)}$$

Therefore, a sensible estimate of $\alpha L(x,S) - \beta\Psi(x,S)$ is

$$\alpha\,\frac{\mathbf{L}_N(x,S)}{N} \;-\; \beta\,\frac{\mathbf{\Psi}_N(x,S)}{N} \;=\; \alpha\,\frac{\mathbf{L}_N(x,S)}{N} \;-\; \beta\,\frac{N-\mathbf{L}_N(x,S)}{N} \;=$$

$$(\alpha+\beta)\,\frac{\mathbf{L}_N(x,S)}{N} \;-\; \beta \;=\; \frac{(\alpha+\beta)\mathbf{L}_N(x,S)}{\mathbf{L}_N(x,S)+\mathbf{\Psi}_N(x,S)} \;-\; \beta \;=$$

$$\frac{\alpha+\beta}{\mathbf{L}_N^{-1}(x,S)\mathbf{\Psi}_N(x,S)+1} \;-\; \beta\,.$$

Let us denote

$$F(x,S) \;=\; \lim_{N\to\infty} \mathbf{L}_N(x,S)\mathbf{\Psi}_N^{-1}(x,S) \quad \text{(a.s.)},$$

then

$$\alpha L(x,S) - \beta\Psi(x,S) \;=\; \frac{\alpha+\beta}{F^{-1}(x,S)+1} \;-\; \beta\,.$$

Thus the performance function equals

$$C(x,S) \;=\; c\,(x,S) \;+\; \frac{\alpha+\beta}{F^{-1}(x,S)+1} \;-\; \beta\,. \tag{2}$$

3 Optimization problem

Usually, feasible storage sizes $S \in \{S_1, \dots, S_I\}$ are known a priori. The problem is to find values x_1, x_2, and S_i such that the function $C(x, S_i)$ is minimal. Since $x_1 > x_2$, we included the following constraint

$$x_1 \geq x_2 + \kappa ,$$

where $\kappa > 0$. As we see further, the function $C(x, S)$ involves the calculation of expectations of discontinuous functions. Although, usually, an analytical evaluation of such functions is out of the question, for this particular case, using the ideas of Analytic Perturbation Analysis [23], we found an analytical expression for the function $C(x, S)$.

The performance function $C(x, S)$ can be calculated using the function $F(x, S)$. To evaluate the function $F(x, S)$, we supposed that the number of processed batches Ψ is fixed, so the number of batches $\mathbf{N}_\Psi(x, S)$ supplied to the work-station and the number of lost batches $\mathbf{L}_\Psi(x, S)$ are random functions of the variable Ψ and the control variables x, S. As an approximation of $F(x, S)$, we consider the function

$$F_\Psi(x, S) \;=\; \Psi^{-1}\, \mathbb{E}\, \mathbf{L}_\Psi(x, S) \,. \tag{3}$$

Although the function $\mathbf{L}_\Psi(x, S)$ is discontinuous w.r.t. x, the expectation of the function $\mathbf{L}_\Psi(x, S)$, which is an integral w.r.t. random variable, is a smooth function of x.

Let us denote:

ℓ is the number of an operation interval (an interval in which the work-station is available);

$\boldsymbol{\tau}_\ell$ is the number ℓ operation interval;

$\mathbf{r}_\ell$ is the repair interval after the operation interval $\boldsymbol{\tau}_\ell$;

$\mathcal{F}_\tau$ is the σ-algebra generated by the random operation intervals $\boldsymbol{\tau}_\ell$, $\ell = 1, 2, \dots$;

$\mathbb{E}_\tau$ is the conditional expectation w.r.t. the σ-algebra $\mathcal{F}_\tau$;

$\mathbb{P}_\gamma$ is the conditional probability w.r.t. the σ-algebra $\mathcal{F}_\gamma$.

We suppose that repair intervals, $\mathbf{r}_\ell$, $\ell = 1, 2, \dots$ and operation intervals, $\boldsymbol{\tau}_\ell$, $\ell = 1, 2, \dots$ are independent random values having densities, ϱ, and, ν, respectively. Also, to simplify calculations, we suppose that operation intervals cannot be shorter than some minimal value, $\boldsymbol{\tau}_{min}$, i.e., distribution function ν equals zero for the values less then $\boldsymbol{\tau}_{min}$. For example, ν could be a normal distribution truncated at the point $\boldsymbol{\tau}_{min}$. The value $\boldsymbol{\tau}_{min}$ is chosen such that the server, working during the time $\boldsymbol{\tau}_{min}$,

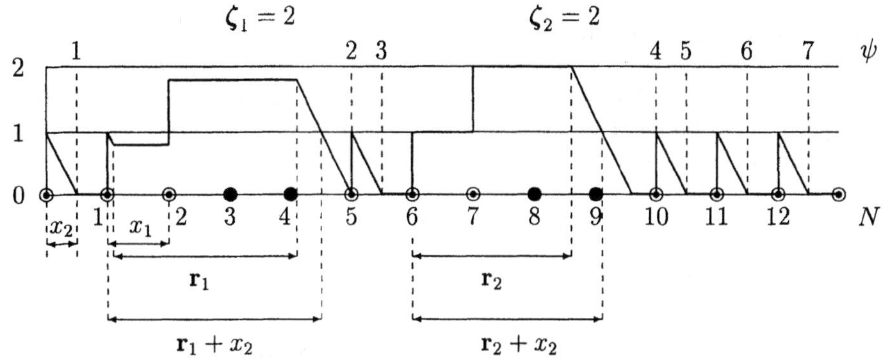

Figure 2: *Amount of material in the buffer. Buffer size equals 2. Batches with numbers 3, 4, 8, 9 are lost.*

is able to process the arriving batches and empty the buffer. Denote the maximal buffer size by

$$S_{max} = \max\{S_1, \ldots, S_I\} .$$

If the server processes only the batches in the buffer, the maximal processing time of these batches is not longer than $x_2 S_{max}$. Since the server can devote only fraction $(x_1 - x_2)/x_1$ of its time to process the batches in the buffer, $\boldsymbol{\tau}_{min}$, must be larger than $x_2 S_{max}\, x_1/(x_1 - x_2)$.

Let us denote by $\mathbf{A}(x_2)$ a random number of failures of the work-station in a run consisting of Ψ processed batches. This number $\mathbf{A}(x_2)$ does not depend upon x_1, because the random operation interval $\boldsymbol{\tau}_\ell$ does not include idle time of the work-station.

For a large Ψ, the ratio of total processing time Ψx_2 and the mean time to failure T approximately equals the expected number of failures $\mathbb{E}\,\mathbf{A}(x_2)$, i.e.,

$$\mathbb{E}\,\mathbf{A}(x_2) \;=\; \Psi x_2/T . \tag{4}$$

Let us denote by $\boldsymbol{\zeta}_\ell(x, S)$ the number of lost batches because of the repair ℓ. The expectation of the function $\mathbf{L}_\Psi(x, S)$ can be represented as

$$\mathbb{E}\,\mathbf{L}_\Psi(x,S) \;=\; \mathbb{E}\sum_{\ell=1}^{\mathbf{A}(x_2)} \boldsymbol{\zeta}_\ell(x,S) \;=\; \mathbb{E}\mathbb{E}_\tau \sum_{\ell=1}^{\mathbf{A}(x_2)} \boldsymbol{\zeta}_\ell(x,S) \;=\; \mathbb{E}\sum_{\ell=1}^{\mathbf{A}(x_2)} \mathbb{E}_\tau\, \boldsymbol{\zeta}_\ell(x,S)\,, \tag{5}$$

Denote by $\mathbf{q}_\ell(x)$ the number of batches arrived to the work-station during repair ℓ plus the number of batches arrived during processing the remaining portion of the batch after finishing the repair ℓ. Thus, $\mathbf{q}_\ell(x)$ is the total number of batches arrived to the work-station during processing the batch with repair ℓ. The number of lost batches, $\boldsymbol{\zeta}_\ell(x,S)$, can be expressed as the function of $\mathbf{q}_\ell(x)$ and S, i.e., $\boldsymbol{\zeta}_\ell(x,S) = K(\mathbf{q}_\ell(x),S)$. This number depends upon the size of the buffer: it equals zero, if $\mathbf{q}_\ell(x)$ is not larger than the size of the buffer, and equals $\mathbf{q}_\ell(x) - S$, if $\mathbf{q}_\ell(x)$ is larger than the buffer size S, i.e.,

$$K(q,S) \;=\; \max\{q - S, 0\}\,. \tag{6}$$

With the full probability formula

$$G(x) \stackrel{\text{def}}{=} \mathbb{E}_\tau\, \boldsymbol{\zeta}_\ell(x,S) \;=\; \mathbb{E}_\tau K(\mathbf{q}_\ell(x),S) \;=\; \sum_{q=1}^{\infty} K(q,S)\phi_q(x)\,, \tag{7}$$

where

$$\phi_q(x) \stackrel{\text{def}}{=} \mathbb{P}_\tau\Big[\mathbf{q}_\ell(x) = q\Big]\,. \tag{8}$$

The processing time of the batch with repair ℓ, including repair time $\mathbf{r}_\ell$, equals $x_2 + \mathbf{r}_\ell$. The constraint $\mathbf{q}_\ell(x) = q$ is equivalent to the constraints (see Fig. 2).

$$q x_1 \;\le\; x_2 + \mathbf{r}_\ell \;\le\; (q+1)\,x_1\,. \tag{9}$$

Therefore, the function $\phi_q(x)$ can be calculated using a cumulative distribution function $D(r)$ of the random repair times $\mathbf{r}_\ell$

$$\phi_q(x) \;=\; D\big((q+1)x_1 - x_2\big) \;-\; D\big(q x_1 - x_2\big)\,. \tag{10}$$

Equations (4), (5), and (7) imply

$$\mathbb{E}\,\mathbf{L}_\Psi(x,S) \;=\; \mathbb{E}\,[\mathbf{A}(x_2)]\,G(x) \;=\; \frac{\Psi x_2}{T}\,G(x)\,.$$

Therefore, with (3) and (6)

$$F_\Psi(x,S) \;=\; T^{-1} x_2\, G(x) \;=\; T^{-1} x_2 \sum_{q=S+1}^{\infty} (q - S)\,\phi_q(x)\,. \tag{11}$$

The function $\phi_q(x)$ tends to zero when q tends to infinity. Although formula (11) contains infinite summation, for practical purposes, it is sufficient to include only a finite number Q of terms. This number depends on the distribution function D and the values x_1 and x_2. Therefore, the function $F_\Psi(x,S)$ approximately equals

$$F(x,S) \;=\; T^{-1} x_2 \sum_{q=S+1}^{Q} (q - S)\,\phi_q(x)$$

$$= \sum_{q=S+1}^{Q} (q-S)\left[D\big((q+1)x_1 - x_2\big) \; - \; D\big(qx_1 - x_2\big)\right]$$

$$= \sum_{q=S+1}^{Q} \left[(q-S)\, D\big((q+1)x_1 - x_2\big) \; - \; (q-1-S)\, D\big(qx_1 - x_2\big) \; - \; D\big(qx_1 - x_2\big)\right]$$

$$= T^{-1}x_2 \left((Q-S)D\big((Q+1)x_1 - x_2\big) - \sum_{q=S+1}^{Q} D\big(qx_1 - x_2\big) \right) . \qquad (12)$$

Thus, finally, we came to the following minimization problem:

Minimization problem

$$C(x, S) \;\; \rightarrow \min_{\substack{x \in X, \\ S \in \{S_1, \ldots, S_I\}}} \qquad (13)$$

subject to

$$X \; = \; \{ \, x \in \mathbb{R}^2 \; : \; x_1 \; \geq \; x_2 + \kappa \, , \; x_2 \; \geq \; 0 \, \} \, , \qquad (14)$$

where the performance function equals

$$C(x, S) \; = \; c\,(x, S) \; + \; \frac{\alpha + \beta}{F^{-1}(x, S) + 1} \; - \; \beta \, , \qquad (15)$$

and the function $F(x, S)$ is given by equation (12).

4 Approaches to solve the optimization problem

This section discusses optimization approaches for solving the mixed discrete-continuous optimization problem (13) with constraints (14). Further, we consider two approaches for reducing this problem to a continuous one. Also, we described deterministic and stochastic algorithms for solving this reduced continuous optimization problem.

4.1 Decomposition approach

The optimization problem (13) is continuous w.r.t. x and is discrete w.r.t. S. If the buffer size S_i is fixed, this is a typical nonlinear optimization problem with linear constraints with respect to x. For each buffer size S_i, we can solve the problem

$$C(x, S_i) \;\; \rightarrow \min_{x \in X} \qquad (16)$$

with respect to x and find an optimum vector x_i^*. Then, we can find optimal buffer size S_i minimizing $C(x_i^*, S_i)$ w.r.t. S_i. Because for each buffer size we need to run a nonlinear programming algorithm, this brute-force approach is applicable for the problems with a relatively small number of feasible buffer sizes S_i. The function $C(x, S)$ is smooth w.r.t. the variable x for the fixed buffer size S; moreover the gradient w.r.t. x can be calculated in analytical form. Indeed, we supposed that the first term $c(x, S)$ of the performance function $C(x, S)$ is smooth; also, the second term is smooth because it is expressed (see (12)) through the smooth function $F(x, S)$. Since the function $D\big(qx_1 - x_2\big)$ can be analytically differentiated w.r.t. x_1 and x_2, i.e.,

$$\frac{\partial}{\partial x_1} D\big(qx_1 - x_2\big) \;=\; q\varrho\left(qx_1 - x_2\right), \quad \frac{\partial}{\partial x_2} D\big(qx_1 - x_2\big) \;=\; -\varrho\left(qx_1 - x_2\right), \tag{17}$$

the function $F(x, S)$ and, consequently, the performance function $C(x, S)$ can be analytically differentiated w.r.t. x_1 and x_2. So, we can use efficient nonlinear gradient algorithms to solve subproblem (16).

4.2 Artificial variables approach

An alternative way to reduce problem (13) to a continuous optimization problem is using artificial variables. It can be proved that problem (13) is equivalent to the minimization problem

$$\Phi(x, y) \stackrel{\text{def}}{=} \sum_{i=1}^{I} C(x, S_i)\, y_i \;\rightarrow\; \min_{(x,y)\in \mathbb{R}^2 \times \mathbb{R}^I}, \tag{18}$$

subject to constraints

$$\sum_{i=1}^{I} y_i \;=\; 1\,; \quad y_i \geq 0\,,\; i = 1, \ldots, I\,; \tag{19}$$

$$x \in X\,. \tag{20}$$

Denoting

$$Y \;=\; \{y \in \mathbb{R}^I : \sum_{i=1}^{I} y_i \;=\; 1\,,\; y_i \geq 0\,,\; i = 1, \ldots, I\}\,,$$

the problem (13) can be reformulated as

$$\Phi(x, y) \;\rightarrow\; \min_{(x,y)\in X \times Y}\,. \tag{21}$$

This is a continuous optimization problem with linear constraints. Despite the fact that the original problem is a mixed discrete-continuous optimization problem, we reduced it to a problem with continuous variables by using additional variables $y_1, \ldots, y_I$. To calculate the performance function $\Phi(x, y)$, the function $C(x, S)$ should be calculated I times. A potential problem with this approach is that for the case with large I, the exact evaluation of the function $\Phi(x, y)$ may involve a tremendous amount of calculations. The next section shows that these numerical difficulties can be overcome with stochastic quasigradient algorithms, which use only rough stochastic estimates of the gradients of the performance function.

4.3 Stochastic quasigradient algorithm

As we mentioned in the previous section, for problem (21) nonlinear programming methods may be a poor choice, because of prohibitively large amount of calculations involved in evaluating the performance function and the gradients. This section considers an alternative approach which is called *stochastic quasigradient* algorithms (on background of stochastic quasigradient algorithms see [2]). One of the most simple stochastic quasi-gradient algorithms for problem (21) can be represented in the following form

$$(x^{s+1}, y^{s+1}) = \Pi_{X \times Y}\Big((x^s, y^s) - \rho_s \xi^s\Big), \tag{22}$$

where s is a number of algorithm iterations; (x^s, y^s) is the approximation point of the extremum on the sth iteration; $\Pi_{X \times Y}(\cdot)$ is the orthoprojection operation on the convex set $X \times Y$; $\rho_s > 0$ is a step size; and ξ^s is a stochastic quasi-gradient satisfying the following property

$$E\left[\xi^s \mid (x^0, y^0), \ldots, (x^s, y^s)\right] = \nabla_x \Phi(x^s, y^s).$$

The conditional expectation of the vector ξ^s is equal to the gradient of the function $\Phi(x, y)$ at the point (x^s, y^s). This algorithm is quite efficient for non ill-conditioned performance functions, i.e., for non-"ravine" functions. In case when the function $\Phi(x, y)$ is "ravine", algorithm (22) may get stuck "at the bottom of the ravine". In such a case, more complicated stochastic quasigradient algorithms, such as algorithms with averaging (see, for example [6], [9],[17]) or variable metrics algorithm [20] may be used. Also, practical convergence rate of algorithm (22) may be improved using adaptively controlled step sizes [18] and the scaling procedure, suggested by Saridis [15].

Further, we provide formulas for calculating stochastic quasigradients of the function $\Phi(x, y)$. An advantage of the stochastic quasigradient algorithms is that they may use rough stochastic estimates of the gradient, which can be obtained with very little computational effort compared to the effort needed for calculating the exact gradient of the performance function $\Phi(x, y)$. Formula (15) implies that

$$C(x, S) \;=\; \Lambda\big(x, S, F(x, S)\big)$$

and

$$\Phi(x,y) \;=\; \sum_{i=1}^{I} \Lambda\big(x, S_i, F(x,S_i)\big)\, y_i \tag{23}$$

Hence,

$$\nabla_x \Phi(x,y) \;=\; \sum_{i=1}^{I} y_i \left[\nabla_x \Lambda(x,S_i,z) + \nabla_z \Lambda(x,S_i,z) \nabla_x F(x,S_i)\right]_{z=F(x,S_i)} , \tag{24}$$

$$\nabla_y \Phi(x,y) \;=\; \begin{pmatrix} \Lambda\big(x,S_1,F(x,S_1)\big) \\ \vdots \\ \Lambda\big(x,S_I,F(x,S_I)\big) \end{pmatrix} . \tag{25}$$

To calculate stochastic quasigradients, instead of exact values $F(x,S_i)$, $\nabla_x F(x,S_i)$, $i = 1,\dots,I$ in formulas (24) and (25), we can use rough stochastic estimates. In this case, estimates of the gradients $\nabla_x \Phi(x,y)$, $\nabla_y \Phi(x,y)$ are biased, because the function $\Lambda\big(x,S_i,z)\big)$ is nonlinear w.r.t. z. However, this bias is relatively small because the function $\Lambda\big(x,S_i,z)\big)$ is close to linear w.r.t. z. Indeed, by definition, the function $F(x,S)$ is a ratio of the number of lost and processed batches. From engineering considerations, this number is much less than 1. Therefore, the function $\Lambda\big(x,S_i,F(x,S)\big)$ approximately equals the following linear function $F(x,S)$

$$\Lambda\big(x,S,F(x,S)\big) \;=\; C(x,S) \;=\; c\,(x,S) \;+\; \frac{\alpha+\beta}{F^{-1}(x,S)+1} \;-\; \beta$$

$$\approx\; c\,(x,S) \;+\; (\alpha+\beta)F(x,S) \;-\; \beta\,.$$

As follows from (7) and (11) the function $F(x,S)$ equals

$$F_\Psi(x,S) \;=\; T^{-1}x_2\, G(x) \;=\; T^{-1}x_2\, \mathbb{E}_\tau\, \zeta_\ell(x,S)\,. \tag{26}$$

Consequently, $T^{-1}x_2\, \zeta_\ell(x,S)$, is an unbiased estimate of the function $F(x,S)$ which can be obtained by sampling the lost number of batches $\zeta_\ell(x,S)$. Formula (26) implies that the gradient of the function $F_\Psi(x,S)$ equals

$$\nabla_x F_\Psi(x,S) \;=\; T^{-1} \begin{pmatrix} 0 \\ 1 \end{pmatrix} \;+\; T^{-1}x_2\, \nabla_x G(x)\,. \tag{27}$$

Gradient of the function $G(x)$ can be calculated using APA [23]; see brief description of APA in Appendix A. Similar to (35), equation (7) represents the function $G(x)$ as a sum of indicator functions, where (see (8),(9))

$$\phi_q(x) \;=\; \mathbb{E}_\tau\Big[I_{\{q x_1 \,\le\, x_2 + r_\ell \,\le\, (q+1)\, x_1\}}\Big] \;=\; \int_{q x_1 \,\le\, x_2 + r \,\le\, (q+1)\, x_1} \varrho(r)\, \mathrm{d}r$$

$$= \int\limits_{q \le (x_2+r)/x_1 \le q+1} \varrho(r)\,\mathrm{d}r \,. \tag{28}$$

To use APA, the gradient $\nabla_x \phi_q(x)$ should be represented in a form similar to the integral (28). Since analytical expression (10) for $\phi_q(x)$ and its derivative (see (17)) is available, we can write

$$\nabla_x \phi_q(x) \;=\; \frac{\nabla_x \phi_q(x)}{\phi_q(x)}\,\phi_q(x) \;= b^q(x)\,\phi_q(x) \,,$$

where

$$b^q(x) \;=\; \frac{\begin{pmatrix} (q+1)\varrho\Big((q+1)x_1 - x_2\Big) - q\,\varrho\Big(qx_1 - x_2\Big) \\ -\varrho\Big((q+1)x_1 - x_2\Big) + \varrho\Big(qx_1 - x_2\Big) \end{pmatrix}}{\Big(D\Big((q+1)x_1 - x_2\Big) \;-\; D\Big(qx_1 - x_2\Big)\Big)} \,.$$

As it follows from formula (39) in Appendix A, a unbiased estimate of the gradient $\nabla_x G(x)$ equals

$$K(\mathbf{q}_\ell(x), S)\, b^{\mathbf{q}_\ell}(x) \,, \tag{29}$$

and

$$\nabla_x G(x) \;=\; \mathbb{E}_\tau[K(\mathbf{q}_\ell(x), S)\, b^{\mathbf{q}_\ell}(x)] \,. \tag{30}$$

Also, this formula can be obtained with Smoothed Infinitesimal Perturbation Analysis [5]. APA provides one more expression for the estimate of the gradient of the function $G(x)$, which is based on integral over volume formula (44). Since the change of variables $z = (x_2+r)/x_1$ eliminates variables x_1, x_2 from constraints in integral (28), the matrix $H(x,r)$ can be calculated with formula (46)

$$H(x,r) \;=\; \nabla_x (x_1 z \;-\; x_2)|_{z=(x_2+r)/x_1} \;=\; \begin{pmatrix} z \\ -1 \end{pmatrix}\Bigg|_{z=(x_2+r)/x_1} \;=\; \begin{pmatrix} (x_2+r)/x_1 \\ -1 \end{pmatrix} .$$

Therefore, with formula (44), the gradient of integral (28) equals

$$\nabla_x \phi_q(x) \;=\; \int\limits_{q \le (x_2+r)/x_1 \le q+1} \frac{\partial}{\partial r}\Big(\varrho(r)\, H(x,r)\Big)\,\mathrm{d}r \;=\; \int\limits_{q \le (x_2+r)/x_1 \le q+1} a^q(x)\,\varrho(r)\,\mathrm{d}r \,,$$

where

$$a^q(x) \;=\; H(x,r)\,\frac{\partial}{\partial r}\mathrm{ln}\varrho(r) \;+\; \frac{\partial}{\partial r} H(x,r) \;=\; \begin{pmatrix} (x_2+r)/x_1 \\ -1 \end{pmatrix} \frac{\partial}{\partial r}\mathrm{ln}\varrho(r) + \begin{pmatrix} x_1^{-1} \\ 0 \end{pmatrix} .$$

Similar to (30), formula (39) in Appendix A implies that an unbiased estimate of the gradient $\nabla_x G(x)$ equals

$$K(\mathbf{q}_\ell(x), S)\, a^{\mathbf{q}_\ell}(x)\,, \tag{31}$$

and

$$\nabla_x G(x) \;=\; \mathbb{E}_\tau[K(\mathbf{q}_\ell(x), S)\, a^{\mathbf{q}_\ell}(x)]\,. \tag{32}$$

A similar estimate can be obtained with the Push Out approach [14]. Thus, we have two unbiased estimates, (29) and (31), for the gradient $\nabla_x G(x)$. For popular distributions such as the exponential or normal distribution, the derivative of the logarithm of the density function, $\frac{\partial}{\partial r}\ln\varrho(r)$, is a constant or a simple function; therefore $a^q(x)$ and the estimate (31) can be easily evaluated. Estimate (29) involves calculation of the density and cumulative distribution functions which may take more time than calculating the estimate (31). Nevertheless, estimate (29) could be preferable because the variance of this estimate is lower than the variance of the estimate (31).

5 Example calculations

This section calculates optimal parameters for an example system. We suppose that the cost and maintenance function per unit of product consists of three terms:

$$c\,(x, S) \;=\; c_1(x_1, x_2) + c_2(S, x_1) + c_3(x_1)\,.$$

The first term is an investment cost for the work station:

$$c_1(x_1, x_2) \;=\; \frac{C_1 x_1}{x_2}\,,$$

the second term is the buffer cost:

$$c_2(S, x_1) \;=\; C_2 S x_1\,,$$

and the third term is the maintenance cost

$$c_3(x_1) \;=\; C_3 x_1\,.$$

Constants C_1, C_2, and C_3 equal:

$$C_1 \;=\; 1\,, \quad C_2 \;=\; 0.1\,, \quad C_3 \;=\; 0.65\,.$$

The gain for the processed batch equals, $\beta = 2$, and the cost of a lost one equals, $\alpha = 3$. The parameter κ in the constraint (14) equals $\kappa = 0.1$ and parameter Q in formula (12) equals $Q = 30$. The mean time to failure of the work station equals $T = 20$. The maximum feasible buffer size S is 8, i.e., $S \in \{S_1, \dots, S_8\} = \{1, \dots, 8\}$.

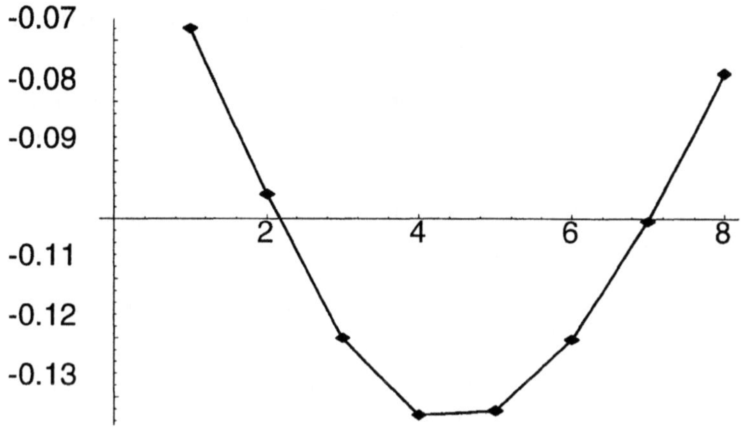

Figure 3: *Optimal values* $C(x_i^*, S_i)$ *for each buffer size* $S_i \in \{1, \dots, 8\}$.

The repair intervals are normally distributed with parameters $m = 2$, $\sigma = 1$. The normal distribution is truncated at the point 0 to avoid negative values for the repair intervals. Fixing the buffer size S_i reduces problem (13) to a nonlinear minimization problem with linear constraints w.r.t. x. For each buffer size $S_i \in \{1, \dots, 8\}$, we solved the problem

$$C(x, S_i) \quad \rightarrow \quad \min_{x \in X} \tag{33}$$

w.r.t. x and found an optimum vector x_i^*. We used MATHEMATICA code of the variable metric algorithm [19] running on PC 486. Optimal values for each $S_i \in \{1, \dots, 8\}$ are plotted in Fig. 3. This figure shows that the optimal buffer size equals $S^* = 4$. The performance function at the optimal point equals $C(x^*, C^*) = -0.133027$ and the optimal vector x^* equals $x^* = (0.47561, 0.37561)$.

References

[1] J.A. Buzacott (1982): "Optimal" operating rules for automated manufacturing systems. *IEEE - Transactions on Automatic Control* **27**, pp. 80-86.

[2] Ermoliev, Yu. (1988): *Stochastic Quasi-Gradient Methods.* In: "Numerical Techniques for Stochastic Optimization" Eds. Yu. Ermoliev and R.J-B Wets, Springer-Verlag,393–401.

[3] Yu. Ermoliev, S. Uryasev, and J. Wessels: (1992): *On Optimization of Dynamical Material Flow Systems Using Simulation.* International Institute for Applied Systems Analysis, Laxenburg, Austria, Report WP-92-76, 28 p.

[4] P. Glasserman (1991): *Gradient Estimation Via Perturbation Analysis.* Kluwer Academic Publishers, Boston-Dordrecht-London.

[5] W.B. Gong, Y.C. Ho (1987) Smoothed (conditional) perturbation analysis of discrete event dynamic systems. *IEEE - Transactions on Automatic Control* **32**, pp. 858-866.

[6] A.M. Gupal, L.T. Bazhenov. (1972): *A Stochastic Analog of the Methods of Conjugate Gradients.* Kibernetika, 1, 124–126, (in Russian).

[7] Ho Y.C. and X.R. Cao (1991): *Perturbation Analysis of Discrete Event Dynamic Systems.* Kluwer, Boston.

[8] M.B.M. de Koster (1989) *Capacity oriented analysis and design of production systems.* Springer-Verlag, Berlin (LNEMS 323).

[9] H.J. Kushner, Hai-Huang. (1981): *Asymptotic Properties of Stochastic Approximation with Constant Coefficients.* SIAM Journal on Control and Optimization, 19, 87–105.

[10] H.J. Kushner, G.G. Jin (1997): *Stochastic Approximation Algorithms and Applications.* Appl. Math. 35, Springer.

[11] Yin G. and Q. Zhang.(Eds) (1996): *Mathematics of Stochastic Manufacturing Systems.* Lectures in Applied Mathematics, Vol. 33.

[12] D. Mitra (1988) *Stochastic fluid models.* in: P.-J. Courtois, G. Latouche (eds.) Performance '87. Elsevier, Amsterdam. pp. 39-51.

[13] Pflug G. Ch. (1996): *Optimization of Stochastic Models, The Interface Between Simulation and Optimization.* Kluwer Academic Publishers, Boston-Dordrecht-London.

[14] Rubinstein, R. and A. Shapiro (1993): *Discrete Event Systems: Sensitivity Analysis and Stochastic Optimization via the Score Function Method.* Wiley, Chichester.

[15] Saridis, G.M. (1970): Learning applied to successive approximation algorithms. *IEEE Trans. Syst. Sci. Cybern.*, 1970, SSC-6, Apr., pp. 97–103.

[16] Bashyam, S. and M.S. Fu (1994): Application of Perturbation Analysis to a Class of Periodic Review (s,S) Inventory Systems. *Naval Research Logistics.* Vol. 41, pp.47-80.

[17] Syski, W. (1988): A Method of Stochastic Subgradients with Complete Feedback Stepsize Rule for Convex Stochastic Approximation Problems. *J. of Optim. Theory and Applic.* Vol. **39**, No. **2**, pp. 487–505.

[18] Uryasev, S.P. (1988): *Adaptive Stochastic Quasi–Gradient Methods.* In "Numerical Techniques for Stochastic Optimization". Eds. Yu. Ermoliev and R. J-B Wets. Springer Series in Computational Mathematics **10**, (1988), 373–384.

[19] Uryasev, S. (1991): New Variable-Metric Algorithms for Nondifferential Optimization Problems. *J. of Optim. Theory and Applic.* Vol. 71, No. 2, 1991, 359-388.

[20] Uryasev, S. (1992): A Stochastic Quasi–Gradient Algorithm with Variable Metric, *Annals of Operations Research.* 39, 251-267.

[21] Uryasev, S. (1994): Derivatives of Probability Functions and Integrals over Sets Given by Inequalities. *J. Computational and Applied Mathematics.* Vol. 56, 197-223.

[22] Uryasev, S. (1995): *Derivatives of Probability Functions and Some Applications.* Annals of Operations Research, Vol. 56, 287-311.

[23] Uryasev, S. (1997): Analytic Perturbation Analysis for DEDS with Discontinuous Sample-path Functions. *Stochastic Models.* Vol. 13, No. 3.

[24] J. Wijngaard (1979) The effect of interstage buffer storage on the output of two unreliable production units in series with different production rates. *AIIE - Transactions* **11**, pp. 42-47.

[25] S. Yeralan, E.J. Muth (1987) A general model of a production line with intermediate buffer and station breakdown. *AIIE - Transactions* **19**, pp. 130-139.

6 Appendix A. Description of the analytic perturbation analysis

Let $(\mathbb{P}, \mathcal{F}, \Omega)$ be a probability space, and $G(x) = \mathbb{E}\, g(x, \omega)$ be an expectation of the function $g(x, \omega)$ depending upon the control variables $x \in \mathbb{R}^n$ and a random element $\omega \in \Omega$.

We suppose that:

1. the set Ω can be split into subsets $\mu^q(x) \in \mathcal{F}$, $q = 1, 2, \ldots$

$$\Omega = \bigcup_{q=1}^{\infty} \mu^q(x) , \tag{34}$$

   and the function $g(x, \omega)$ is differentiable w.r.t. ω (or w.r.t. to some components of ω, if ω is a vector) on each subset $\mu^q(x)$, $q = 1, 2, \ldots$;

2. for any $q \neq j$

$$\mathbb{P}\Big(\mu^q(x) \bigcap \mu^j(x)\Big) = 0 ;$$

3. each subset $\mu^q(x)$, $q = 1, 2, \ldots$ can be represented by the system of inequalities

$$\mu^q(x) = \{\omega \in \Omega : f^q(x, \omega) \leq 0\}$$

$$\stackrel{\text{def}}{=} \{\omega \in \Omega : f_l^q(x, \omega) \leq 0 ,\ 1 \leq l \leq k^q\} ,$$

where $f_l^q : \mathbb{R}^n \times \Omega \to \mathbb{R}$, $1 \leq l \leq k^q$.

$I_{\{f^q(x,\omega)\leq 0\}}$ denotes an indicator function, which corresponds to the set $\mu^q(x)$

$$I_{\{f^q(x,\omega)\leq 0\}} = \begin{cases} 1 , & \text{if} , \quad f^q(x, \omega) \leq 0 ; \\ 0 , & \text{otherwise} . \end{cases}$$

With these definitions, the performance function $G(x) = \mathbb{E}\, g(x, \omega)$ can be represented as the sum

$$G(x) = \mathbb{E}\, g(x, \omega) = \mathbb{E}\Big[\sum_{q=1}^{\infty} I_{\{f^q(x,\omega)\leq 0\}}\, g(x, \omega)\Big] = \sum_{q=1}^{\infty} \mathbb{E}\Big[I_{\{f^q(x,\omega)\leq 0\}}\, g(x, \omega)\Big] . \quad (35)$$

The function

$$\phi_q(x) \stackrel{\text{def}}{=} \mathbb{E}\Big[I_{\{f^q(x,\omega)\leq 0\}}\, g(x, \omega)\Big]$$

is an integral of the function $g(x, \omega)$ over the set $\mu^q(x)$. This function is differentiable under general conditions (see formulas (41), (43), and (44)). The gradient $\nabla_x \phi_q(x)$ can be represented as an integral of another function $a^q(x, \omega)$ over the same set $\mu^q(x)$ plus an additional function $\psi_q(x)$, which is a surface integral, or equivalently, it can be represented as a mathematical expectation of the product $a^q(x, \omega) I_{\{f^q(x,\omega)\leq 0\}}$, plus $\psi_q(x)$, i.e.,

$$\nabla_x \phi_q(x) = \mathbb{E}\Big[I_{\{f^q(x,\omega)\leq 0\}}\, a^q(x, \omega)\Big] + \psi^q(x) . \quad (36)$$

The function $\phi_q(x)$ can be differentiated directly, or it can be written as

$$\phi_q(x) = \mathbb{E}\Big[I_{\{f^q(x,\omega)\leq 0\}}\, g(x, \omega)\Big] = \mathbb{E}\Big[\mathbb{E}_q\Big[I_{\{f^q(x,\omega)\leq 0\}}\, g(x, \omega)\Big]\Big] ,$$

where $\mathbb{E}_q$ is a conditional expectation such that the function

$$\mathbb{E}_q\Big[I_{\{f^q(x,\omega)\leq 0\}}\, g(x, \omega)\Big] \quad (37)$$

is smooth w.r.t. x. Further, we can interchange the gradient and expectation signs

$$\nabla_x \phi_q(x) = \nabla_x \mathbb{E}\Big[\mathbb{E}_q\big[I_{\{f^q(x,\boldsymbol{\omega})\le 0\}}\, g(x,\boldsymbol{\omega})\big]\Big] = \mathbb{E}\Big[\nabla_x \mathbb{E}_q\big[I_{\{f^q(x,\boldsymbol{\omega})\le 0\}}\, g(x,\boldsymbol{\omega})\big]\Big],$$

and apply differentiation formulas (41), (43), and (44) to the function (37). Thus, (35) and (36) imply

$$\begin{aligned}
\nabla_x G(x) &= \nabla_x \sum_{q=1}^{\infty} \phi_q(x) = \sum_{q=1}^{\infty} \nabla_x \phi_q(x) \\
&= \sum_{q=1}^{\infty} \mathbb{E}\big[I_{\{f^q(x,\boldsymbol{\omega})\le 0\}}\, a^q(x,\boldsymbol{\omega})\big] + \sum_{q=1}^{\infty} \psi^q(x) \\
&= \mathbb{E}\, a^{\mathbf{q}}(x,\boldsymbol{\omega}) + \sum_{q=1}^{\infty} \psi^q(x), \qquad (38)
\end{aligned}$$

where,

$$\mathbf{q} = \min\{\nu : \boldsymbol{\omega} \in \mu^\nu(x)\}.$$

If $\sum_{q=1}^{\infty} \psi^q(x) = 0$, then $a^{\mathbf{q}}(x,\boldsymbol{\omega})$ is an unbiased estimate of the gradient, which can be evaluated together with the sample-path function $g(x,\boldsymbol{\omega})$ during the same simulation run. If $\sum_{q=1}^{\infty} \psi^q(x) \neq 0$, it is desirable to find an expression for this term so that it can be calculated during the same simulation run, together with $g(x,\boldsymbol{\omega})$ and $a^{\mathbf{q}}(x,\boldsymbol{\omega})$. In some cases, it is possible to convert $\psi^q(x)$ using artificial variables to the integral over the volume

$$\psi^q(x) = \mathbb{E}\big[I_{\{f^q(x,\boldsymbol{\omega})\le 0\}}\, b^q(x,\boldsymbol{\omega})\big].$$

Then, with (38),

$$\nabla_x G(x) = \mathbb{E}\, a^{\mathbf{q}}(x,\boldsymbol{\omega}) + \sum_{q=1}^{\infty} \mathbb{E}\big[I_{\{f^q(x,\boldsymbol{\omega})\le 0\}}\, b^q(x,\boldsymbol{\omega})\big] = \mathbb{E}[a^{\mathbf{q}}(x,\boldsymbol{\omega}) + b^{\mathbf{q}}(x,\boldsymbol{\omega})]. \qquad (39)$$

Thus, $d^{\mathbf{q}}(x,\boldsymbol{\omega}) \triangleq a^{\mathbf{q}}(x,\boldsymbol{\omega}) + b^{\mathbf{q}}(x,\boldsymbol{\omega})$ is an unbiased estimate of the gradient. The random vector $d^{\mathbf{q}}(x,\boldsymbol{\omega})$ can be obtained with one Monte-Carlo simulation run of the model, analogous to the random value $g(x,\boldsymbol{\omega})$.

The estimate $d^{\mathbf{q}}(x,\boldsymbol{\omega})$ can be used in Monte-Carlo type simulations. Standard variance reduction techniques (see, for example [13]), such as conditioning, coupling, and importance sampling can be used to reduce the variance of this estimate. Let measure $\rho(x,\omega)$ dominate the measure $p(x,\omega)$. Importance sampling technique, changing the measure in expectation

$$\mathbb{E}[d^{\mathbf{q}}(x,\boldsymbol{\omega})] \triangleq \int d^{q(\omega)}(x,\omega)\, p(x,\omega)\, \mathrm{d}\omega =$$

$$\int d^{q(\omega)}(x,\omega)\,\frac{p(x,\omega)}{\rho(x,\omega)}\rho(x,\omega)\,\mathrm{d}\omega = \mathbb{E}_\rho[d^{q}(x,\omega)\frac{p(x,\omega)}{\rho(x,\omega)}]\,,$$

may significantly improve the variance.

7 Appendix B. Analytical derivatives of the integrals over sets given by inequalities

Let the function

$$F(x) = \int\limits_{f(x,y)\leq 0} p(x,y)\,\mathrm{d}y \tag{40}$$

be defined on the Euclidean space $\mathbb{R}^n$, where $f : \mathbb{R}^n \times \mathbb{R}^m \to \mathbb{R}^k$ and $p : \mathbb{R}^n \times \mathbb{R}^m \to \mathbb{R}$ are some functions. The inequality $f(x,y) \leq 0$ should be treated as a system of inequalities

$$f_i(x,y) \leq 0\,, \quad i = 1,\dots,k\,.$$

Further, we presented a general formula [21, 22] for the differentiation of integral (40). A gradient of the integral is represented as a sum of integrals taken over a volume and over a surface.

Let us introduce the following shorthand notations

$$f(x,y) = \begin{pmatrix} f_1(x,y) \\ \vdots \\ f_k(x,y) \end{pmatrix}, \quad f_{1l}(x,y) = \begin{pmatrix} f_1(x,y) \\ \vdots \\ f_l(x,y) \end{pmatrix},$$

$$\nabla_y f(x,y) = \begin{pmatrix} \frac{\partial f_1(x,y)}{\partial y_1}\,, & \dots & , \frac{\partial f_k(x,y)}{\partial y_1} \\ & \vdots & \\ \frac{\partial f_1(x,y)}{\partial y_m}\,, & \dots & , \frac{\partial f_k(x,y)}{\partial y_m} \end{pmatrix}.$$

Divergence for the matrix H is defined as

$$\mathrm{div}_y H = \begin{pmatrix} \sum_{i=1}^{m} \frac{\partial h_{1i}}{\partial y_i} \\ \vdots \\ \sum_{i=1}^{m} \frac{\partial h_{ni}}{\partial y_i} \end{pmatrix}, \quad H = \begin{pmatrix} h_{11}\,, & \dots & , h_{1m} \\ & \vdots & \\ h_{n1}\,, & \dots & , h_{nm} \end{pmatrix}.$$

We define

$$\mu(x) = \{\, y \in \mathbb{R}^m \,:\, f(x,y) \le 0 \,\} \stackrel{\text{def}}{=} \{\, y \in \mathbb{R}^m \,:\, f_l(x,y) \le 0\,,\ 1 \le l \le k \,\}\,,$$

and $\partial\mu(x)$ to be the surface of the set $\mu(x)$. Let us denote by $\partial_i\mu(x)$ a part of the surface which corresponds to the function $f_i(x,y)$

$$\partial_i\mu(x) \;=\; \mu(x) \bigcap \{\, y \in \mathbb{R}^m \,:\, f_i(x,y) = 0 \,\}\,.$$

If we split the set $K \stackrel{\text{def}}{=} \{1,\dots,k\}$ into two subsets K_1 and K_2, without a loss of generality, we can consider

$$K_1 \;=\; \{1,\dots,l\} \;\text{ and }\; K_2 \;=\; \{l+1,\dots,k\}\,.$$

There is freedom in the choice of the sets K_1 and K_2 and representation of the gradient of function (40). First, we consider the case when the subsets K_1 and K_2 are not empty. In this case, the derivative of integral (40) is given by the formula

$$\nabla_x F(x) \;=\; \int\limits_{\mu(x)} \Big[\, \nabla_x p(x,y) \;+\; \mathrm{div}_y\big(p(x,y) H_l(x,y)\big) \Big]\, \mathrm{d}y \;-$$

$$-\sum_{i=l+1}^{k}\; \int\limits_{\partial_i\mu(x)} \frac{p(x,y)}{\|\nabla_y f_i(x,y)\|} \Big[\, \nabla_x f_i(x,y) \;+\; H_l(x,y)\, \nabla_y f_i(x,y) \Big]\, \mathrm{d}S\,, \tag{41}$$

where the matrix function $H_l : \mathbb{R}^n \times \mathbb{R}^m \to \mathbb{R}^{n\times m}$ satisfies the equation

$$H_l(x,y)\, \nabla_y f_{1l}(x,y) \;+\; \nabla_x f_{1l}(x,y) \;=\; 0\,. \tag{42}$$

The last equation can have a lot of solutions and we can choose an arbitrary solution, differentiable with respect to the variable y.

Further, let us present the derivative of function (40) for the case with the empty set K_1. Then, matrix function H_l is absent and

$$\nabla_x F(x) \;=\; \int\limits_{\mu(x)} \nabla_x p(x,y)\, \mathrm{d}y \;-\; \sum_{i=1}^{k}\; \int\limits_{\partial_i\mu(x)} \frac{p(x,y)}{\|\nabla_y f_i(x,y)\|} \nabla_x f_i(x,y)\, \mathrm{d}S\,. \tag{43}$$

Finally, let us consider a formula for the derivative of function (40) for the case with the empty set K_2. The integral over the surface is absent and the derivative is represented as an integral over the volume

$$\nabla_x F(x) \;=\; \int\limits_{\mu(x)} \Big[\, \nabla_x p(x,y) \;+\; \mathrm{div}_y\big(p(x,y) H(x,y)\big) \Big]\, \mathrm{d}y\,, \tag{44}$$

where a matrix function $H : \mathbb{R}^n \times \mathbb{R}^m \rightarrow \mathbb{R}^{n \times m}$ satisfies the equation

$$H(x,y)\,\nabla_y f(x,y) + \nabla_x f(x,y) = 0\,. \tag{45}$$

In many cases, there is a simple way to solve equations (42) and (45) using a change of variables. Suppose that there is a change of variables

$$y = \gamma(x,z)$$

which eliminates vector x from the function $f(x,y)$, i.e., function $f(x,\gamma(x,z))$ does not depend upon variable x. Denote by $\gamma^{-1}(x,y)$ the inverse function, defined by the equation

$$\gamma^{-1}\big(x,\gamma(x,z)\big) = z\,.$$

In this case, the matrix

$$H(x,y) = \nabla_x \gamma(x,z)|_{z=\gamma^{-1}(x,y)}\,. \tag{46}$$

is a solution of equation (45).

Stochastic Optimization in Asset & Liability Management: A Model for Non-Maturing Accounts[1]

Karl Frauendorfer
University of St. Gallen
Institute of Operations Research
St. Gallen, Switzerland

Michael Schürle
University of St. Gallen
Institute of Operations Research
St. Gallen, Switzerland

Abstract

A multistage stochastic optimization model for the management of non-maturing account positions like savings deposits and variable-rate mortgages is introduced which takes the risks induced by uncertain future interest rates and customer behavior into account. Stochastic factors are discretized using the barycentric approximation technique. This generates two scenario trees whose associated deterministic equivalent programs provide exact upper and lower bounds to the original problem. Practical experience from the application in a major Swiss bank is reported.

Keywords: Stochastic programming, approximation, asset & liability management.

1 Introduction

Stochastic programming has received increasing attention from financial institutions recently since the shortcomings of traditional approaches that are widely used in

[1]Research for this paper was supported by the *Swiss National Science Foundation*, Grant No. 21-39'575.93.

S.P. Uryasev (ed.), Probabilistic Constrained Optimization, 67–101.
© 2000 *Kluwer Academic Publishers.*

practice came to light. For instance, in portfolio optimization the mean-variance framework due to Markowitz [48] captures the volatility and correlations among financial instruments but may generate solutions that do not seem like a reasonable mix to achieve the indicated risk and return (cf. [22]) and are highly sensitive to the input, i.e., expectations and covariances (cf. [3, 8]). Moreover, it does not take into account the possibility of future rebalancing transactions of the portfolio or additional in- and outflows of cash that occur during the planning period. In many problems in the field of asset and liability management (ALM), the increased volatility in financial markets since the 70s and the introduction of derivatives revealed severe deficiencies of popular portfolio immunization strategies which match the duration and, possibly, the convexity of both assets and liabilities (cf. [67]). These approaches hedge only against relatively small shifts in interest rates and are not appropriate to deal with the complex cash flow structures of new financial instruments (cf. [53]).

1.1 Stochastic optimization for financial decision making

For obvious reasons, stochastic optimization models seem to be a natural approach in order to address the requirements of a large number of financial planning problems (e.g., see [15, 49, 50] for an overview of applications, [4, 12, 14, 52] for general ALM model formulations or [23, 24, 36, 37, 40, 64, 65, 66, 67] for fixed-income problems). On one hand, the models allow the reflection of uncertainty in future prices, yields and exchange rates, volatilities etc. by generating scenarios of their possible future outcomes. These scenarios quantify the impact of changes in the underlying risk factors on the return of investment strategies or the deviation from a certain target position like an index, a benchmark portfolio, a liability etc.

On the other hand, a stochastic program reflects not only the dynamics of uncertain data but also of decisions more appropriately since transactions may take place at discrete points in time until the end of some predefined planning horizon. For all scenarios under consideration, a decision must be taken at each stage based on realizations of random data and earlier actions. This allows the correction of an initial policy, e.g., if it does not achieve the investment goal for certain scenarios.

In general, a stochastic optimization model yields a large-scale program since it has to include a high number of scenarios to reflect the entire universe of possible future outcomes of risk factors and cash flows. The first models for financial planning appearing in the 80s (cf. [44, 45]) could not meet this requirement due to limitations of the computational resources available at this time.

However, the dramatic improvement in powerful hardware as well as the development of efficient algorithms, in particular if they exploit the special structure and high sparsity inherent to stochastic programs, now provide the basis to solve problems where the number of scenarios is between some thousands and one million, depending on the problem structure and the system architecture (cf. [25]). Moreover, new theoretical models from the financial literature and related empirical evidence have sharpened the understanding of the dynamics of risk factors such as interest and exchange rates, prices of financial instruments etc. Both of these developments allow the modeling of complex problems more realistically. In this way, the fields of Finance and Optimization come closer together.

1.2 Review of current approaches

Meanwhile, a large number of stochastic optimization models has been introduced for various applications in Finance. Among them are the Russell-Yasuda-Kasai model due to Cariño et al. [4, 5], which can be seen as the first successful commercial application of multistage stochastic programming, the models for ALM and fixed-income portfolio management of Zenios et al. [36, 66, 67] and Dupačová et al. [18], or the multistage portfolio optimization models due to Dantzig and Infanger [11] and Steinbach [61], to mention just a few.

In general, one starts from assumptions about the distribution of risk factors which is typically continuous. Since scenarios are used as input for stochastic programs to describe the uncertainty, a discrete subset of possible realizations of the random data must be determined that is "representative" for the entire universe of future outcomes. Loosely speaking, this means that the solution of such an approximated problem comes "as close as possible" to the exact (but unknown) optimum of a problem with the true distribution.

Simulation is a widely exploited approach for the selection of scenarios and can be easily combined with decomposition methods (cf. [10, 38]). Since the solution depends heavily on the choice of the scenario set, a statistical analysis is essential to assess the accuracy and stability of the problem (cf. [16, 17]). Although the amount of computational effort is independent of the dimension of random data, Monte Carlo methods often suffer from a low convergence rate of the order $1/\sqrt{s}$ as implied by the central limit theorem, where s is the sample size. Therefore, variance reduction techniques like importance sampling are helpful to reduce the error of estimates for the objective. For multistage problems, the *expected value of perfect information (EVPI)* (cf. [7, 13]) is a useful criterion for the selection of an enhanced set of representative scenarios. In any case, simulation based approaches provide only probabilistic error bounds.

Approximation schemes are based on partitioning the domain of random data into cells and using representative points within them (cf. [1, 19, 20, 26, 32, 43]). Exploiting certain properties of the stochastic program, mainly convexity of value functions, allows the determination of exact lower and upper bounds to the original problem. As a consequence, the error induced by the approximation can be quantified more precisely, and the accuracy may be improved by deliberately adding new scenarios. Careful control of this process is necessary since the number of scenarios grows exponentially with the dimension size and accuracy.

1.3 Contribution of this paper

In this paper, a stochastic optimization model for an application from the field of ALM is introduced which was developed in co-operation with a major Swiss bank for the management of so-called "non-maturing account" positions. This includes savings deposits as well as a special type of non-fixed mortgages which is common in Switzerland. Their characteristic feature is that there exists no contractual maturity on such products but bank customers are allowed to withdraw their investments or prepay their mortgages, respectively, at any point in time at no penalty. As a consequence, the volume of both positions fluctuates heavily as customers react to changes in the market environment, e.g., rising or falling yields, or the relative attractiveness of alternative investment opportunities.

Therefore, in the formulation of the stochastic program uncertainty affects not only the coefficients in the objective (future interest rates) but also on the right-hand-side of constraints (volume change). Moreover, both may be correlated to reflect a dependency between interest rates and volume. All these aspects can be addressed by the *barycentric approximation* technique introduced in the sequel to derive exact bounds for value functions corresponding to convex multistage stochastic programs. An approximation technique is preferred for this type of application since statistical analysis implies that the dynamics of interest rates can be described by at most three factors in order to explain more than 95 % of the variance. These factors control level, curvature, and steepness of the yield curve (cf. [46]). This enables keeping the problem size relatively moderate even in a multistage model.

The remainder of the paper is organized as follows: The next section introduces the formulation of the optimization model for non-maturing accounts. In section 3, the structural properties of convex multistage stochastic programs are outlined. Section 4 introduces the barycentric approximation scheme that is used to derive scenario trees associated with upper and lower bounds to the original problem. Numerical results for the approximation are also given. Section 5 reports practical experience from the application of the model compared to traditional approaches. The main results are summarized in section 6 together with an outlook to possible improvements of the model as well as future directions of research.

2 A model for uncertain maturities

2.1 Problem characteristics

As outlined above, non-maturing accounts can be characterized as follows: (1) There is no contractual maturity on these positions since bank customers are allowed to withdraw or repay their investments and credits at any point in time. (2) The customer rate is not indexed to certain interest rates or prices of traded instruments but adjustable to market conditions as a matter of policy. The most common examples include some forms of savings accounts or non-fixed mortgages that are widespread in Europe and the U.S. The management of such account positions is a particularly ambitious task since these assets and liabilities are not only sensitive to changes in interest rates but have also embedded call or put options that may be exercised by the customer. For example, a homeowner has the option to prepay the outstanding balance of his mortgage and call the security.

It can be observed that customer behavior depends strongly on the current market environment. In case of variable-rate mortgages, changes in the total volume are positively correlated with interest rates. When the latter are low, there is a sharp drop in demand since customers switch to fixed-rate mortgages in order to hedge themselves against a future rise *(prepayment risk)*. In case of savings deposits, the volume increases since their yields are relatively attractive when compared to alternative short-term securities, and even institutional investors like pension funds prefer these deposits instead of direct investments in the money market. This results in a negative correlation between interest rates and volume change. In such a situation, it is difficult for financial management to find a combination of fixed-income instruments that provides a sufficient margin and takes into account the risk that a significant portion of the deposits is withdrawn *(withdrawal risk)*.

During a period of high interest rates, homeowners' demand for non-fixed mortgages rises significantly while investors shift their assets from variable-rate savings accounts to bonds with long maturities. As a consequence, the mortgages must be refinanced on the money and capital market at increased funding costs. Moreover, there is a political cap on the mortgage rate in Switzerland, and numerous banks were not able to refinance their mortgages at a positive margin at the beginning of the 90s. These difficulties caused a broad discussion among practitioners from the financial industry about the management of non-maturing account positions and the risks induced by the embedded options.

Clearly, the use of *duration matching* does not apply here since one cannot find adequate duration measures due to the volume fluctuations, beside other shortcomings of this concept. This has motivated the *replicating portfolio* approach which is based on the idea of mimicing the behavior of the target position in order to capture its characteristics. The objective is to find a portfolio of fixed-income securities whose return replicates the customer rate of the relevant asset or liability position plus a margin. Transaction costs remain low since liquid money market instruments and swaps are used that are held until maturity to avoid a rebalancing. Maturing funds are always renewed at the same maturity. Prepayment and withdrawal risks are implicitly taken into account as the volume of the replicating portfolio has to coincide with the volume of the target position at all points in time. The weights are determined through minimizing the tracking error for a historic sample period and remain constant over time.

By means of this approach, uncertain cash flows are transformed into (apparently) certain ones, allowing the bank to manage them like normal maturing accounts. These replicating portfolios are implemented as passive investment and refinancing strategies. However, the question arises whether a dynamic policy with active reactions to changes in the market environment and customer behavior could increase the bank's profit. In particular, it remains to be clarified if the correlation between interest rates and volume can be exploited more appropriately to manage the inherent risks. Clearly, a stochastic optimization model is able to address most of these requirements.

2.2 Model formulation

For simplicity, only the problem of reinvesting savings accounts on the market is investigated here since the model for refinancing mortgages is equivalent and can be derived easily from it. The formulation of the optimization model is straightforward: $\mathcal{D} = \{1, 2, \ldots, D\}$ denotes a set of maturity dates for fixed-income securities held in the portfolio. Investment opportunities are given by a set of traded standard maturities $\mathcal{D}^S \subset \mathcal{D}$. Let $\varphi_t^{d,+}$ be the discounted accrued interest payments for an investment of \$1 in maturity $d \in \mathcal{D}^S$ at time t. The model has also the option to raise funds which are reinvested in addition to the total savings volume. The sum of interest payments for \$1 of such a short position is given by $\varphi_t^{d,-}$. Clearly, except for $t = 0$ these coefficients depend on future interest rates and, hence, are uncertain.

The underlying interest rate model used here to describe the evolution of interest rates resembles the idea of *key rates* analogously to the duration model of Ho [39]. It is assumed that the yield curve can be segmented into K_t different sections where rates move in the same direction. These segments are represented by K_t key rates of different maturities whose dynamics can be described by correlated Brownian motions with (possibly time-dependent) drift. This results in normally distributed interest rate changes. Rates for the remaining maturities are interpolated.

The coefficients $\varphi_t^{d,+}(\eta_t)$ and $\varphi_t^{d,-}(\eta_t)$ for all $d \in \mathcal{D}^S$ are functions of this K_t-dimensional Brownian motion $(\eta_t; t = 1, \ldots, T)$ in discrete time driving the evolution of the yield curve. Here, $K_t = K = 3$ as a three factor model is sufficient to reflect a great variety of term structure movements. The functional relationship between η_t and $\varphi_t^{d,+}, \varphi_t^{d,-}$ incorporates the sensitivity of interest rates subject to changes in the risk factors, transactions costs, a bid-ask spread as well as the discount mechanism. Only payments within the planning horizon are considered, i.e., those that are induced by an investment or borrowing in $t \leq T$ but occur after $T+1$ are neglected. A formal specification of the relation between risk factors and coefficients in the objective is omitted here since the notation is rather cumbersome. Note that the current values of η_0 can be derived from market observations.

At each point in time $t = 0, \ldots, T$, decisions on the amount $x_t^{d,+} \geq 0$ of long and $x_t^{d,-} \geq 0$ of short positions in maturity d have to be made subject to budget constraints

$$x_t^d - x_{t-1}^{d+1} - x_t^{d,+} + x_t^{d,-} = 0 \qquad t = 0, \dots, T; \forall d \in \mathcal{D}^S$$
$$x_t^d - x_{t-1}^{d+1} = 0 \qquad t = 0, \dots, T; \forall d \in \mathcal{D} \setminus \mathcal{D}^S.$$

The latter constraint ensures that the sum of all long and short positions $x_t^d \in \mathbb{R}$ maturing after d periods is equal to the corresponding value in the previous period for non-traded maturity dates while the former corrects it by the new long and short sales in t for traded maturities. Note that x_{-1}^d indicates the amount of maturity d in the initial portfolio from decisions in the past. At time t, the portfolio has to match the total savings volume $v_t \in \mathbb{R}$:

$$v_t = \sum_{d \in \mathcal{D}} x_t^d \qquad t = 0, \dots, T.$$

The total savings volume is given by its value in the previous period $t-1$, corrected by the stochastic volume change ξ_t in t:

$$v_t = v_{t-1} + \xi_t \qquad t = 1, \dots, T.$$

Again, the volume change is modeled by a Brownian motion with drift in discrete time $(\xi_t; t = 1, \dots, T)$ of dimension $L_t = L = 1$. It may be correlated with the components of the stochastic process η_t to reflect a relation between changes in interest rates and volume.

Raising short positions and reinvesting them in addition to the savings volume may be viewed as some speculative strategy. Therefore, short sales can be restricted to an amount equal to the sum of funds maturing at $t+1, \dots, t+m$, where $m > 0$ is defined by the decision maker ($m = 1$ prohibits any borrowings):

$$\sum_{d \in \mathcal{D}^S} x_t^{d,+} \le \xi_t + \sum_{d=1}^{m} x_{t-1}^d \qquad t = 1, \dots, T.$$

Depending on the current interest rate curve and the amount that has to be reinvested, a situation might occur where the optimal investment strategy cannot be implemented due to liquidity restrictions in the Swiss market, in particular if the model finds a policy that is not broadly diversified over different maturities. This is addressed by imposing upper limits $\ell_t^{d,+}$, $\ell_t^{d,-}$ for investments and borrowings:

$$0 \le x_t^{d,+} \le \ell_t^{d,+} \qquad t = 0, \dots, T; \forall d \in \mathcal{D}^S$$
$$0 \le x_t^{d,-} \le \ell_t^{d,-} \qquad t = 0, \dots, T; \forall d \in \mathcal{D}^S.$$

The restrictions above must hold for all observations of η_t and ξ_t, $t = 1, \dots, T$. Moreover, decisions $x_t^{d,+}, x_t^{d,-}$ have to be made independent of future outcomes of $\eta_{t+1}, \dots, \eta_T$ and $\xi_{t+1}, \dots, \xi_T$ since these are unknown at time t. Hence, investment policies must not anticipate any information that becomes known in the future. This is incorporated in the optimization model by additional *nonanticipativity* constraints.

Finally, the objective is to maximize the expected present value of the income from all investments $\varphi_t^{d,+} \cdot x_t^{d,+}$ minus the costs for borrowings $\varphi_t^{d,-} \cdot x_t^{d,-}$ for all $d \in \mathcal{D}^S$ over the planning horizon T. The expectation is taken with respect to the joint probability measure P of (η, ξ) associated with time $t = 1, \ldots, T$, i.e., $\eta = (\eta_1, \ldots, \eta_T)$, $\xi = (\xi_1, \ldots, \xi_T)$. In the standard form of a minimization problem, the multistage stochastic program reads as:

$$
\begin{array}{llll}
\min \displaystyle\int \sum_{t=0}^{T} \sum_{d \in \mathcal{D}^S} (\varphi_t^{d,-}(\eta_t) \cdot x_t^{d,-} - \varphi_t^{d,+}(\eta_t) \cdot x_t^{d,+}) dP(\eta, \xi) & & & \\
\text{s.t.} \quad x_t^d - x_{t-1}^{d+1} - x_t^{d,+} + x_t^{d,-} & = & 0 & t = 0, 1, \ldots, T; \forall d \in \mathcal{D}^S \\
x_t^d - x_{t-1}^{d+1} & = & 0 & t = 0, 1, \ldots, T; \forall d \notin \mathcal{D}^S \\
v_t - \displaystyle\sum_{d \in \mathcal{D}} x_t^d & = & 0 & t = 0, 1, \ldots, T \\
v_t - v_{t-1} & = & \xi_t & t = 1, 2, \ldots, T \\
\displaystyle\sum_{d \in \mathcal{D}^S} x_t^{d,+} - \sum_{d=1}^{m} x_{t-1}^d & \leq & \xi_t & t = 0, 1, \ldots, T \\
0 \leq x_t^{d,+} \leq \ell_t^{d,+} & & & t = 0, 1, \ldots, T; \forall d \in \mathcal{D}^S \\
0 \leq x_t^{d,-} \leq \ell_t^{d,-} & & & t = 0, 1, \ldots, T; \forall d \in \mathcal{D}^S \\
x_t^{d,+}, x_t^{d,-} \quad \text{nonanticipative} & & & t = 0, 1, \ldots, T; \forall d \in \mathcal{D}^S \\
v_t, x_t^d \in \mathbb{R} \quad \text{nonanticipative} & & & t = 0, 1, \ldots, T; \forall d \in \mathcal{D}.
\end{array}
\tag{1}
$$

To evaluate the objective function for a particular policy, it is necessary to calculate an integral with respect to the measure P. At each stage t, the costs and profits induced by a decision can be quantified by a value function which is formally introduced in the next section. The latter is not given analytically, not even explicitly, but implicitly by the solutions of a multistage stochastic optimization problem with respect to the remaining stages $t + 1, \ldots, T$. As a consequence, the integration in (1) cannot be performed analytically, and numerical methods are required.

A common approach is to approximate the continuous distribution of random data in the original stochastic program by a discrete one. More precisely, the stochastic evolution of interest rates and volume change is approximated by two *scenario trees.* This yields two other optimization problems where one deals with a sum in the objective function which can be easily calculated. To minimize the error induced by the discretization, it is exploited that the value functions at stage t associated with (1) are convex-concave saddle functions in (η_t, ξ_t). The saddle property allows the determination of exact upper and lower bounds to the original problem and is discussed in the next section for a general formulation of multistage stochastic programs.

3 Multistage stochastic programs

3.1 Formal description

Formally, the evolution of uncertain data over the planning horizon T in a multistage stochastic program can be described by a multi-dimensional stochastic process $(\omega_t, t = 1, \ldots, T)$ in discrete time on a common Borel space $(\Omega, \mathcal{B}^M)$ with compact $\Omega \subset \mathbb{R}^M$ (cf. [27, 28, 29]). Let P represent the (regular) joint probability measure of $\omega := (\omega_1, \ldots, \omega_T)$. The associated conditional measure with respect to ω_t is denoted $P_t(\cdot|\omega^{t-1})$ for $t = 1, \ldots, T$. For reasons of compactness, $\omega^t := (\omega_1, \ldots, \omega_t)$ represents the sequence of observations of $\omega_t \in \Omega_t \subset \mathbb{R}^{M_t}$ up to time t, where $\Omega_1 \times \ldots \times \Omega_T = \Omega$, $M_1 + \ldots + M_T = M$. Note that ω_0 denotes those data that are currently observed and, hence, deterministic.

At time $t = 0$, a decision $u_0 \in \mathbb{R}^{n_0}$ is made without knowing ω_t for the subsequent stages $t = 1, \ldots, T$. After ω_t was observed at time $t > 0$, the initial policy may be corrected by a new decision $u_t \in \mathbb{R}^{n_t}$ based on the known history of observations ω^t and decisions $u^t := (u_0, u_1, \ldots, u_t) \in \mathbb{R}^{n^t}$, $n^t = n_0 + \ldots + n_t$. In particular, u_t has to be independent of future outcomes $\omega_{t+1}, \ldots, \omega_T$. Therefore, the solution of the underlying stochastic optimization problem is a *recourse function* with the property

$$u(\omega) = \big(u_0, u_1(\omega^1), \ldots, u_T(\omega^T)\big) \in \mathbb{R}^n, \qquad n = n_0 + n_1 + \ldots + n_T,$$

known as *nonanticipativity*. The initial decision u_0 induces some (non-random) costs ρ_0. For the subsequent stages $t = 1, \ldots, T$, the costs $\rho_t(u^t, \omega^t)$ are determined by the sequence of earlier decisions u^t and realizations of ω^t. The feasible set is assumed to be convex, compact, and non-empty for any ω. Again, it depends on previous decisions and observations for $t > 0$ and can be characterized by the system of inequalities

$$\begin{aligned} &f_0(u_0) \le 0 \\ &f_t(u^t, \omega^t) \le 0 \qquad t = 1, \ldots, T. \end{aligned} \tag{2}$$

$f_0(\cdot)$ and $f_t(\cdot,\cdot)$ are vector-valued and $\rho_0(\cdot)$ and $\rho_t(\cdot,\cdot)$ are real-valued functions defined on the corresponding Euclidian spaces. Furthermore, $\rho_t(\cdot,\cdot)$ are supposed to be convex in u^t for any random outcome ω^t. The objective is to find a nonanticipative recourse function $u(\cdot)$ that minimizes the expected total costs over the planning horizon and satisfies the constraints (2):

$$\begin{aligned} \min \quad & \Big\{\rho_0(u_0) + \textstyle\int_\Omega \big[\sum_{t=1}^T \rho_t(u^t, \omega^t)\big] dP(\omega)\Big\} \\ \text{s.t.} \quad & f_0(u_0) \le 0 \\ & f_t(u^t, \omega^t) \le 0, \qquad t = 1, \ldots, T, \\ & u(\cdot) \text{ nonanticipative.} \end{aligned} \tag{3}$$

The meaning of the last (nonanticipativity) constraint is: Let ω^t and u^t satisfy $f_0(u_0) \le 0, f_1(u^1, \omega^1) \le 0, \ldots, f_t(u^t, \omega^t) \le 0$, then there always exists a sequence

$u_{t+1}, \ldots, u_T$ for any $(\omega_{t+1}, \ldots, \omega_T)$, so that $u = (u_0, u_1, \ldots, u_T)$ is feasible with respect to (2). As a consequence, there is always a feasible completion of the problem (3) for the remaining stages $t+1, \ldots, T$ independent of the realizations of ω_τ, $\tau = t+1, \ldots, T$, provided that the decisions u_τ in $\tau = 0, 1, \ldots, t$ are feasible. This can be seen as a counterpart to the case of relatively complete recourse in two-stage stochastic programming (cf. [62]).

3.2 Saddle property of value functions

In order to distinguish those uncertain data that affect the objective from those influencing the constraints, the random vectors for $t = 1, \ldots, T$ are decomposed according to $\omega_t = (\eta_t, \xi_t)$, where $\eta_t \in \Theta_t \subset \mathbb{R}^{K_t}$, $\xi_t \in \Xi_t \subset \mathbb{R}^{L_t}$, $\Omega_t = \Theta_t \times \Xi_t$, $M_t = K_t + L_t$. The functions defining the feasible set in the second line of (2) can then be written as $f_1(u^1, \xi^1) \leq 0, \ldots, f_T(u^T, \xi^T) \leq 0$, and the costs are now of the form $\rho_t(u^t, \eta^t)$. In case such a decomposition of ω_t into η_t and ξ_t is not obvious, one may augment the probability space (for details, see [27]). In order to write the problem without stating the constraints explicitly, the function

$$g_t(u^t, \eta^t, \xi^t) := \begin{cases} \rho_t(u^t, \eta^t) & \text{if } f_t(u^t, \xi^t) \leq 0 \\ +\infty & \text{otherwise,} \end{cases}$$

is introduced. For the following analysis, it is useful to consider the dynamic version of problem (3) stated in terms of *recourse* or *value functions*. These can be obtained if the multistage program is written as a series of nested two-stage programs, starting in the last stage T with

$$\phi_T(u^{T-1}, \eta^T, \xi^T) = \min_{u_T \geq 0} g_T(u^{T-1}, u_T, \eta^T, \xi^T) \tag{4}$$

and then backwards for $t = T-1, \ldots, 0$

$$\begin{aligned} \phi_t(u^{t-1}, \eta^t, \xi^t) = \min_{u_t \geq 0} \{ & g_t(u^{t-1}, u_t, \eta^t, \xi^t) \\ & + \int_{\Theta_{t+1} \times \Xi_{t+1}} \phi_{t+1}(u^{t-1}, u_t, \eta^t, \xi^t, \eta_{t+1}, \xi_{t+1}) dP(\eta_{t+1}, \xi_{t+1} | \eta^t, \xi^t) \}. \end{aligned} \tag{5}$$

Again, (η^0, ξ^0) are currently observed data, whereas u^{-1} represents decisions from the past. Here, $u^t = (u^{t-1}, u_t)$ is decomposed in two parts to emphasize that decisions u^{t-1} were already made in the preceding stages and only u_t must be determined in (5). The optimal decision for the current stage represents a trade-off between the imminent costs ρ_t in t and the expected future costs for the remaining periods induced by u_t. According to [28], based on arguments in [54], to ensure that the problems (4) and (5) can be solved the following assumptions are required for $t = 1, \ldots, T$:

(*i*) $\Theta_t \times \Xi_t$ is compact, convex, and covers the support of (η_t, ξ_t).

(*ii*) $\rho_t(u^t, \eta^t)$ is a continuous saddle function on $\mathbb{R}^{n^t} \times \Theta^t$ which is convex in u^t and concave in η^t.

Note that this is satisfied, e.g., if ρ_t is bilinear.

(*iii*) The feasible sets $\{u_0 | f_0(u_0) \leq 0\}$ and $\{(u^t, \xi^t) | \xi \in \Xi, f_t(u^t, \xi^t) \leq 0\}$ are compact, convex subsets of $\mathbb{R}^{n^t} \times \Xi^t$.

In particular, this covers the case that the constraints can be written in the form

$$f_t(u^{t-1}, u_t, \xi^t) = d_t(u_t) - e_t(u^{t-1}, \xi^t)$$

with d_t convex and e_t linear affine.

As outlined before, approximation schemes are based on the convexity or, if uncertainty affects the objective and the right-hand-sides of constraints, saddle property of value functions which allows the derivation of lower and upper bounds based on the inequalities due to Jensen [42] and Edmundson-Madansky [21, 47]. Barycentric approximation is a generalization of these concepts for bounding the expectation of saddle functions in the case of dependent random variables which was introduced in [27] in the context of two-stage stochastic programming. It remains to clarify under which conditions it can be extended to the multistage case.

According to (*i*), the problem (4) for the final stage T is a convex optimization problem with parameters (u^{T-1}, η^T, ξ^T). Assumption (*ii*) implies that the objective function g_T in (4) is a lower closed saddle function. Therefore, the corresponding value function $\phi_T(u^T, \eta^T, \xi^T)$ is also a saddle function, convex in (x^{T-1}, ξ^T) and concave in η^T. In order to derive bounds for the expectation of (5), the saddle property of the value function in T must be "inherited" to the remaining stages $T-1, \ldots, 1$. When calculating the expected recourse costs

$$E_t\phi_t(u^{t-2}, u_{t-1}, \eta^{t-1}, \xi^{t-1}) = \int_{\Theta_t \times \Xi_t} \phi_t(u^{t-2}, u_{t-1}, \eta^{t-1}, \xi^{t-1}, \eta_t, \xi_t) dP_t(\eta_t, \xi_t | \eta^{t-1}, \xi^{t-1}), \quad (6)$$

the probability measure P_t depends on (η^{t-1}, ξ^{t-1}). As a consequence, the saddle property of $\phi_t(x^{t-2}, x_{t-1}, \eta^{t-1}, \xi^{t-1}, \eta_t, \xi_t)$ is *not* inherited in general due to the integration with respect to $P_t(\eta_t, \xi_t | \eta^{t-1}, \xi^{t-1})$. However, if the distribution functions are of the form

$$Q_t\big((\eta_t, \xi_t) + H_t(\eta^{t-1}, \xi^{t-1})\big), \quad (7)$$

where Q_t is a regular distribution function over $(\Theta_t \times \Xi_t, \mathcal{B}^{K_t + L_t})$ and H_t is a linear mapping, the integral in (6) can be written as

$$\int_{\Theta_t \times \Xi_t} \phi_t\big(u^{t-2}, u_{t-1}, \eta^{t-1}, \xi^{t-1}, (\eta_t, \xi_t) - H_t(\eta^{t-1}, \xi^{t-1})\big) dQ_t(\eta_t, \xi_t)$$

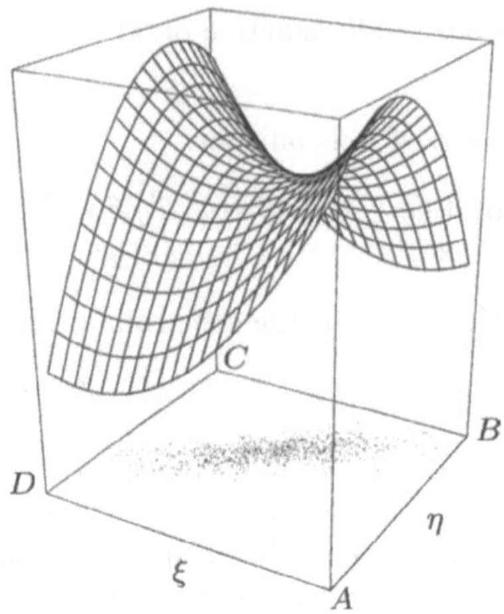

Figure 1: Saddle property of value functions

(note that the integration is now performed with respect to a measure independent of (η^{t-1}, ξ^{t-1})). Then, it can be easily verified that the expectation in (6) is a saddle function on its domain $\Theta^t \times \{(u^{t-1}, \xi^t) | \xi^t \in \Theta^t, f_\tau(u^\tau, \xi^\tau) \leq 0, \tau = 1, \ldots, t-1\}$ which is *convex* in $(u^{t-2}, u_{t-1}, \xi^{t-1})$ and *concave* in η^{t-1} (for details see [28]). Together with the convexity of $g_t(u^t, \eta^t)$ implied by (*ii*), this results in the saddle property of the objective function of problem (5),

$$g_t(u^{t-1}, u_t, \eta^t, \xi^t) + E_{t+1}\phi_{t+1}(u^{t-1}, u_t, \eta^t, \xi^t, \eta_{t+1}, \xi_{t+1}). \tag{8}$$

Hence, for stages $t = T-1, \ldots, 1$ the value functions $\phi_t(u^{t-1}, \eta^t, \xi^t)$ are lower closed saddle functions on their domain. An illustration is given in Figure 1 where $\phi_t(u^{t-1}, \eta^{t-1}, \eta_t, \xi^{t-1}, \xi_t)$ is shown for $(\eta_t, \xi_t) \in \Theta_t \times \Xi_t \subset \mathbb{R} \times \mathbb{R}$, i.e., one-dimensional distributions for the coefficients in objective and right-hand-sides. Note that the value function quantifies the imminent costs for the current stage t plus the expected future costs provided that the subsequent decisions are optimal. Therefore, it cannot be represented analytically but is given implicitly for each (u^{t-1}, η^t, ξ^t) as the solution of a multistage stochastic program with respect to the remaining stages $t+1, \ldots, T$.

3.3 Solvability of stochastic programs

To ensure that the multistage stochastic program can be solved, it is required that the corresponding value functions are continuous. A sufficient condition for this is that value functions are subdifferentiable which is given if the Slater condition holds: For the last stage T, there must be a point $\hat{u}_T$ depending on (u^{T-1}, ξ^T) with $f_T(u^{T-1}, \hat{u}_T, \xi^T) < 0$. Calculating the expectation in (6) preserves continuity since the

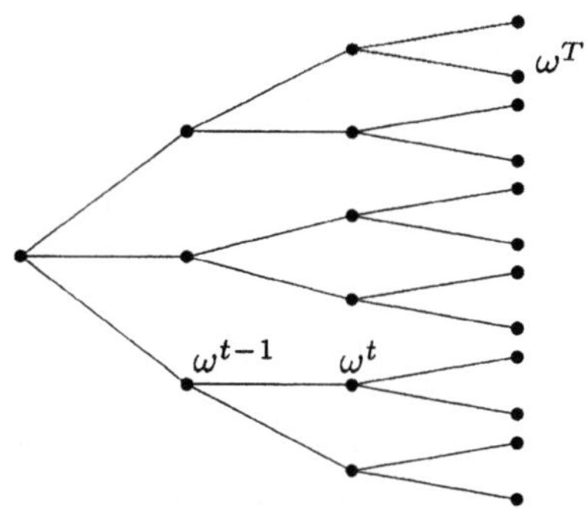

Figure 2: Illustration of a scenario tree

integration is done with respect to the compact region $\Theta_t \times \Xi_t$. To ensure continuity of the value functions for the preceding stages, again the existence of a Slater point is required which leads to the following additional assumption:

(*iv*) For any $\xi^t \in \Xi^t$ and any decision u^{t-1} that is feasible with respect to ξ^t, there exists some $\hat{u}_t$ depending on (u^{t-1}, ξ^t) with $f_t(u^{t-1}, \hat{u}_t, \xi^t) < 0$.

This is known as *strict nonanticipativity* which can be seen as a stronger version of the nonanticipativity property introduced in 3.1 with regard to the constraint multifunction. There, it was required that $\{u_t | f_t(u^{t-1}, u_t, \xi^{t-1}) \leq 0\} \neq \emptyset$, i.e., the feasible set in t depending on u^{t-1} and ξ^t is non-empty. Now, in addition it has to contain inner points: $\text{int}\{u_t | f_t(u^{t-1}, u_t, \xi^{t-1}) \leq 0\} \neq \emptyset$. Together with the results from above concerning the inheritance of the saddle property, one obtains:

(*v*) The expectation functionals $E_t\phi_t(u^{t-1}, u_t, \eta^{t-1}, \xi^{t-1})$ are continuous saddle functions – *convex* in $(u^{t-1}, u_t, \xi^{t-1})$ and *concave* in η^{t-1} with respect to their domain.

The numerical difficulty in solving a stochastic optimization problem of type (3) are the nested minimization and the multidimensional integration of the implicitly given value functions (6). As mentioned, approximation schemes partition the support of the original distribution in convex regions and use distinguished points within them. This is equivalent to a successive discretization of the conditional probability measure $P_t(\cdot|\omega^{t-1})$ for $t = 1, \ldots, T$, yielding *discrete* measures $Q_t(\cdot|\omega^{t-1})$ with support $\mathcal{A}_t(\omega^{t-1})$. As a result, one obtains a *scenario tree* $\mathcal{A}$ (see Figure 2) that can be formally defined as follows:

$$\mathcal{A} := \left\{(\eta, \xi) \in \Theta \times \Xi | (\eta_t, \xi_t) \in \mathcal{A}_t(\eta^{t-1}, \xi^{t-1}) \forall t > 0\right\}. \tag{9}$$

Each path in this tree represents a scenario for the evolution of η_t and ξ_t over the planning horizon T, and the associated probabilities are given by

$$q(\eta,\xi) := \prod_{t=1}^{T} q_t(\eta_t,\xi_t|\eta^{t-1},\xi^{t-1}). \tag{10}$$

Clearly, when the conditional probability measure is discrete in time and space, the stochastic two-stage program (5) has a characteristic block structure. As a consequence, the multistage problem (3) can be written as a mathematical program with dynamic block structure and high sparsity whose size depends on the number of scenarios within the tree. Powerful decomposition algorithms have been developed which exploit this special structure (see [2, 51, 55, 56, 57, 58, 59] for example).

4 Barycentric approximation

4.1 Discretization of distributions

In this section, it is shown how the original probability measure $P_t(\cdot|\eta^{t-1},\xi^{t-1})$ can be discretized using so-called *generalized barycenters.* These are calculated with respect to a cross-simplex (or briefly: $\times$-simplex), i.e., the Cartesian product of two simplices that cover the support of random data in the objective and the constraints, respectively. To this end, it is assumed that $\Theta_t(\eta^{t-1},\xi^{t-1}) \in \mathbb{R}^{K_t}$ and $\Xi_t(\eta^{t-1},\xi^{t-1}) \in \mathbb{R}^{L_t}$ are regular simplices covering the support of η_t and ξ_t (in the sequel, the dependency on previous observations may be omitted in the notation for simplicity). Their vertices are denoted

$$\begin{aligned} u_{\nu_t}(\eta^{t-1},\xi^{t-1}) &= \left(u_{1,\nu_t}(\eta^{t-1},\xi^{t-1}),\ldots,u_{K_t,\nu_t}(\eta^{t-1},\xi^{t-1})\right)' \in \Theta_t \\ v_{\mu_t}(\eta^{t-1},\xi^{t-1}) &= \left(v_{1,\mu_t}(\eta^{t-1},\xi^{t-1}),\ldots,v_{L_t,\mu_t}(\eta^{t-1},\xi^{t-1})\right)' \in \Xi_t. \end{aligned}$$

for $\nu_t = 0,\ldots,K_t$, $\mu_t = 0,\ldots,L_t$. The *barycentric weights*

$$\lambda_t(\eta_t|\eta^{t-1},\xi^{t-1}) = \left(\lambda_{t,0}(\eta_t|\eta^{t-1},\xi^{t-1}),\ldots,\lambda_{t,K_t}(\eta_t|\eta^{t-1},\xi^{t-1})\right)'$$

of η_t with respect to $\Theta_t(\eta^{t-1},\xi^{t-1})$ are those nonnegative barycentric coordinates that allow the representation of η_t as a linear combination of the vertices $u_{\nu_t}(\eta^{t-1},\xi^{t-1})$ and sum up to one:

$$\begin{aligned} \lambda_{t,0} + \lambda_{t,1} + \ldots + \lambda_{t,K_t} &= 1 \\ u_{t,0}\lambda_{t,0} + u_{t,1}\lambda_{t,1} + \ldots + u_{t,K_t}\lambda_{t,K_t} &= \eta_t. \end{aligned}$$

Analogously, the barycentric weights

$$\tau_t(\xi_t|\eta^{t-1},\xi^{t-1}) = \left(\tau_{t,0}(\xi_t|\eta^{t-1},\xi^{t-1}),\ldots,\tau_{t,L_t}(\xi_t|\eta^{t-1},\xi^{t-1})\right)'$$

$t = 2$. Analogously, in case of the lower scenario tree the vertices u_{ν_t} of the simplex covering the support of the uncertain data in the objective are combined with the barycenters ξ_{μ_t} for the stochastic right-hand-side coefficients.

4.3 Bounds for value functions

Recall the formulation of the original problem (3):

$$\min g_0(u_0) + \int_\Omega \Big[\sum_{t=1}^{T} g_t(u^t, \omega^t)\Big] dP(\omega). \tag{25}$$

Replacing the probability measure P by its discrete approximations Q^l and Q^u yields the multistage stochastic programs

$$\psi_0 = \min g_0(u_0) + \int_\Omega \Big[\sum_{t=1}^{T} g_t(u^t, \omega^t)\Big] dQ^l(\omega) \tag{26}$$

$$\Psi_0 = \min g_0(u_0) + \int_\Omega \Big[\sum_{t=1}^{T} g_t(u^t, \omega^t)\Big] dQ^u(\omega). \tag{27}$$

It is shown in [28] that for the corresponding value functions

$$\psi_t(u^{t-1}, \omega^t) = \min_{u_t} g_t(u^t, \omega^t) + \int_{\Omega_{t+1}} \psi_{t+1}(u^t, \omega^{t+1}) dQ^l_{t+1}(\omega_{t+1}|\omega^t)$$
$$\Psi_t(u^{t-1}, \omega^t) = \min_{u_t} g_t(u^t, \omega^t) + \int_{\Omega_{t+1}} \Psi_{t+1}(u^t, \omega^{t+1}) dQ^u_{t+1}(\omega_{t+1}|\omega^t)$$

for $t = 0, \dots, T$ with $\psi_{T+1}(\cdot) = \Psi_{T+1}(\cdot) := 0$, the following relation holds:

$$\psi_t(u^{t-1}, \omega^t) \le \phi_t(u^{t-1}, \omega^t) \le \Psi_t(u^{t-1}, \omega^t). \tag{28}$$

Therefore, (26) is a lower and (27) an upper approximation to the original problem (3). The situation described by the inequalities in (28) is illustrated in Figure 6. Again, for simplicity only the case $K = L = 1$ is considered and the time index omitted. For each stage t, the value function is supported by two bilinear functions. In particular, the supporting points for the minorant are the barycenters ξ_0 and ξ_1 while the majorant supports the value function in η_0 and η_1. The bilinearity of ψ_t and Ψ_t can be explained by the fact that the integrand $\lambda_{\mu_t}(\eta_t) \cdot \tau_{\nu_t}(\xi_t)$ in (15) and (17) is a bilinear function in (η_t, ξ_t).

Solving the approximated problems (26) and (27) yields policies $u^l := (u_0^l, \dots, u_T^l)$ and $u^u := (u_0^u, \dots, u_T^u)$, where u_t^l and u_t^u denote the decisions made after $\hat{\omega}_t^l \in \mathcal{A}_t^l$ and $\hat{\omega}_t^u \in \mathcal{A}_t^u$ are observed in $t = 1, \dots, T$. Clearly, lower and upper approximations of the value functions may result in different policies. This can lead to a situation

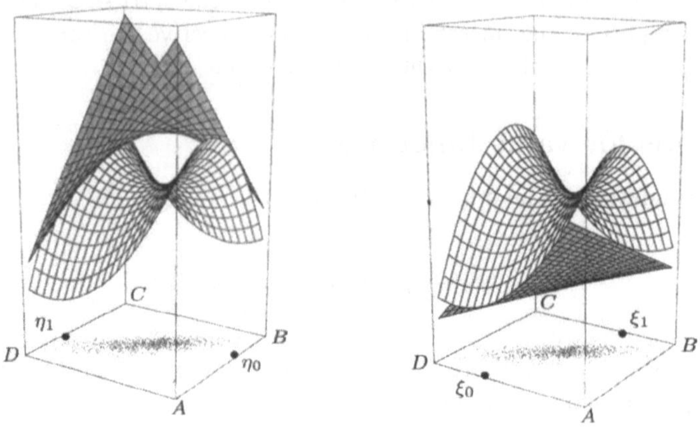

(a) Upper bound (b) Lower bound

Figure 6: Bilinear approximations of value functions

where the accuracy of the approximation has to be improved, in particular if the case $u_0^l \neq u_0^u$ occurs since for the decision maker only the decision at time $t = 0$ is of interest. The policies in $t > 0$ correspond to observations in the barycentric scenarios of the approximated problems and are not likely to be implemented.

For a *refinement* of the approximations (26) and/or (27), *epi-convergence* of the approximated value functions can be exploited, i.e., $\psi_t(\cdot)$ and $\Psi_t(\cdot)$ converge to $\phi_t(\cdot)$ if weak convergence of the discrete conditional probability measures $Q_t^l(\cdot|\hat{\omega}^{l,t-1})$ and $Q_t^u(\cdot|\hat{\omega}^{u,t-1})$ to $P_t(\cdot|\hat{\omega}^{l,t-1})$ and $P_t(\cdot|\hat{\omega}^{u,t-1})$, respectively, is ensured for $t = 1,\dots,T$ (see [28] for details). This requires the partition of the $\times$-simplices and the resulting sub-$\times$-simplices until they become arbitrarily small with respect to the diameters.

The refinement procedure can be outlined as follows: Starting from an initial scenario tree $\mathcal{A}$, the $\times$-simplex $\Omega_t(\omega^{t-1})$ covering the support of $\omega^{t-1} \in \mathcal{A}_{t-1}(\omega^{t-2})$ is split with respect to either the Θ_t- or the Ξ_t-component. Clearly, each of the resulting sub-$\times$-simplices may be divided again, and barycenters may be derived as described in the previous sections. For the $\ell_t(\omega^{t-1})$ sub-$\times$-simplices corresponding to a partition of the support $\mathrm{supp}(\omega_t|\omega^{t-1})$, the following conditions must be satisfied:

(1) $\bigcup_{i_t=1}^{\ell_t(\omega^{t-1})} \Omega_{t,i_t}(\omega^{t-1}) = \Omega_t(\omega^{t-1}) \supset \mathrm{supp}(\omega_t|\omega^{t-1})$,

(2) $\Omega_{t,i_t}(\omega^{t-1}) \cap \Omega_{t,j_t}(\omega^{t-1}) = \emptyset, \qquad i_t \neq j_t; i_t, j_t = 1,\dots,\ell_t(\omega^{t-1})$,

(3) $\Omega_{t,i_t}(\omega^{t-1})$ are regular $\times$-simplices for $i_t = 1,\dots,\ell(\omega^{t-1})$.

Each partition of a $\times$-simplex increases the number of scenarios $|\mathcal{A}|$ and, hence, the computational complexity of the associated deterministic equivalent. As a conse-

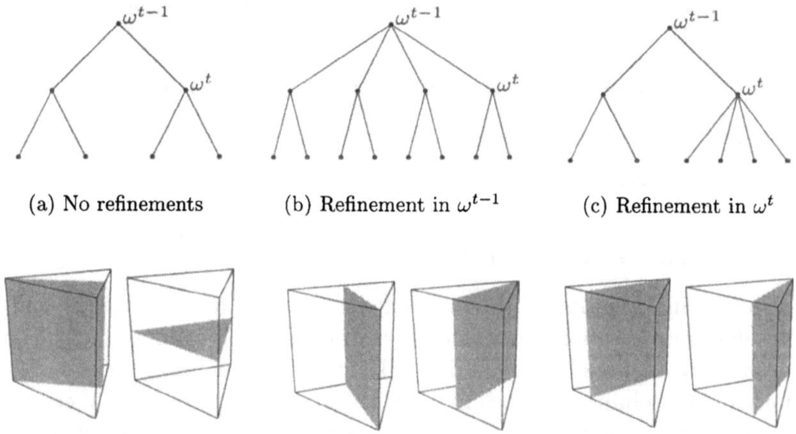

(a) No refinements (b) Refinement in ω^{t-1} (c) Refinement in ω^t

(d) Split of Θ or Ξ (e) Alternative edges (f) Alternative points

Figure 7: Possible refinements of scenario tree and $\times$-simplex

quence, Ω_t cannot be divided arbitrarily often in practice. The total number of nodes in the scenario trees at stage $t = 1, \ldots, T$ is given by

$$|\mathcal{A}^{l,t}| = \sum_{\hat{\omega}^{l,t-1} \in \mathcal{A}^{l,t-1}} \big(\ell_t(\hat{\omega}^{l,t-1}) - 1\big)(K_t + 1) \tag{29}$$

$$|\mathcal{A}^{u,t}| = \sum_{\hat{\omega}^{u,t-1} \in \mathcal{A}^{u,t-1}} \big(\ell_t(\hat{\omega}^{u,t-1}) - 1\big)(L_t + 1). \tag{30}$$

Therefore, the refinement process must be carefully monitored, particularly in the multistage case, in order to identify those nodes in the scenario trees where the largest approximation error

$$\epsilon_t(\omega^t) = \Psi_t(u^{t-1}, \omega^t) - \psi_t(u^{t-1}, \omega^t) \tag{31}$$

is observed. On the other hand, if $\epsilon_t(\cdot) = 0$ for a certain node, the approximation of the value function $\phi_t(\cdot)$ is exact and further refinements of the partition corresponding to this node will not improve the accuracy of the approximation. For an efficient implementation of refinement strategies, the following aspects must be considered:

(1) In which node should the scenario tree be refined (i.e., what is the amount of the approximation error to refine the existing partition; this has an immediate impact to the number of scenarios and, hence, the problem size, see Figure 7 (a)–(c)),

(2) does a division of Ω_t with respect to Θ_t or Ξ_t yield a higher accuracy (see Figure 7 (d)),

(3) which is the edge where the simplex is split (see Figure 7 (e)) and

(4) where does this edge have to be divided (see Figure 7 (f))?

For a detailed discussion and solution techniques, see [33].

4.4 Computational results

To complete the introduction of the barycentric approximation technique, some numerical results are presented. Two- and three-dimensional Brownian motions in discrete time are used to model the evolution of key rates η_t and one-dimensional Brownian motions for the volume change ξ_t in order to illustrate the influence of correlations and the dimension size on the accuracy of the approximation with respect to different refinement strategies. In particular, the distribution of (η_t, ξ_t) at time t induced by these Brownian motions are independent of the realizations in $t-1$. This case is covered by the general type of distribution functions (7) that is required to ensure the saddle property of value functions.

In the first case '2U', two uncorrelated processes for key rates of maturity 1 and 12 months are considered with $\sigma_1 = 0.179$ and $\sigma_{12} = 0.125$, rates for the remaining maturities are interpolated. Both risk factors are independent of the volume change ξ_t whose variance is given by $\sigma_V = 341'056$. The second case '2C' uses the same volatility for interest rate and volume changes, together with covariances of $\sigma_{1,12} = 0.117$ between both key rates as well as $\sigma_{1,V} = 44.247$ and $\sigma_{12,V} = 37.682$ between key rates and volume. Intuitively, such a model is able to reflect parallel and tilt movements of the yield curve.

Taking a third factor into account, e.g., the rate of an intermediate maturity, also allows the modeling of changes in the curvature of the term structure. In the last case '3C', the 3 month rate with variance $\sigma_3 = 0.141$ and covariances $\sigma_{1,3} = 0.151$, $\sigma_{3,12} = 0.121$, $\sigma_{3,V} = 43.2379$ is considered in addition to the 1 and 12 month key rates. These parameter estimates were taken from the description of a collection of test problems for multistage stochastic programs in [30] and are derived from money market rates for the Swiss Franc. All drifts of the Brownian motions are equal to zero. Note that in the context of a real savings application, the correlation between interest rates and volume has a negative sign.

In section 3.2, it was assumed that the $\times$-simplex $\Theta_t \times \Xi_t$ is compact and covers the support of (η_t, ξ_t). However, the normal distributions at each point in time associated with the Brownian motions under consideration have unbounded support. In this case, one must ensure that $P_t(\Theta_t \times \Xi_t | \eta^{t-1}, \xi^{t-1}) \geq 1 - \varepsilon$ for sufficiently small $\varepsilon > 0$ and substitute $P_t(\cdot | \eta^{t-1}, \xi^{t-1})$ by its normalized truncation (see [35]).

The procedure for the determination of a simplicial coverage is only conceptually outlined here. First, consider a K-dimensional standard normal distribution. A

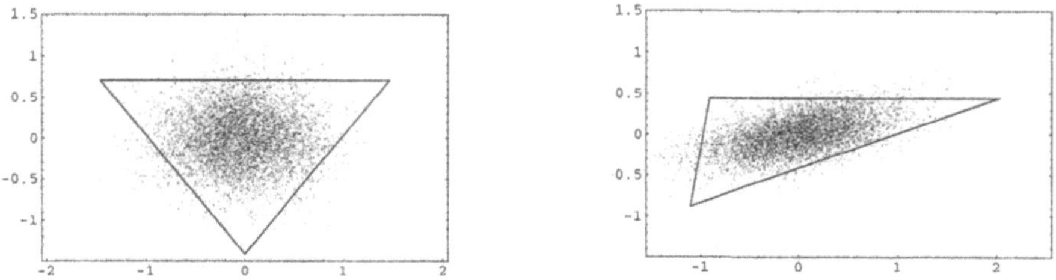

(a) Uncorrelated distribution '2U' (b) Correlated distribution '2C'

Figure 8: Simplicial coverage of two-dimensional distributions

sphere with radius δ around the origin contains a percentage of $2\Phi(\delta) - 1$ of the total mass distribution, where Φ denotes the c.d.f. This sphere can be covered by a simplex in $\mathbb{R}^K$ with $K+1$ vertices. In the one-dimensional case, the simplex reduces to an interval $[-\delta, \delta]$, and for $K = 2$ to a triangle with vertices $u_0 = (-\sqrt{3}\delta, \delta)'$, $u_1 = (\sqrt{3}\delta, \delta)'$ and $u_2 = (0, -2\delta)'$. It is well known that a standard normally distributed random variable $Z \in \mathbb{R}^K$ may be transformed into a $\mathcal{N}(\mu, \Sigma)$-distributed random variable $Y \in \mathbb{R}^K$ using the lower triangular matrix of the Cholesky decomposition L of the covariance matrix Σ, i.e., $\Sigma = L \cdot L'$. According to this rule, the vertices u_i^Y of the simplicial coverage for Y are given by

$$u_i^Y = \mu + L \cdot u_i^Z, \qquad i = 0, \ldots, K.$$

An example for the two-dimensional distributions '2U' and '2C' is shown in Figure 8 for $\delta = 3$, covering more than 99.7 % of the probability mass, i.e., $\varepsilon < 0.003$. Note how the correlation between the risk factors in the second case influences the shape of the simplex which, loosely speaking, results in a "stretched" triangle. For a more thorough description of $\times$-simplicial coverages, together with formulae for simplices in higher dimensions and a discussion of accuracy estimates, see [31].

Figure 9 summarizes numerical results for the savings problem introduced in (1) with $\mathcal{D}^S = \{1, 2, 3, 6, 12\}$ and $m = 1$. Since there are no short sales, to get positive values the objective function was multiplied by -1 for convenience which yields a maximization problem. In this case, the upper approximation is obtained from (26) while the lower is given by (27). The initial interest rates are 5.0 %, 5.1 %, 5.2 %, 5.5 % and 6.0 %. An amount of \$ 200'000 matures at all points in time. The graph shows the objective values of the upper and lower bounds depending on the number of refinements of the (sub-)simplices $\Theta_t(\cdot|\eta^{t-1}, \xi^{t-1})$ with respect to the risk factors for a planning horizon of $T = 4$. Increasing the number of stages yields similar results which are not presented here. For a more detailed discussion, including additional models for the dynamics of interest rates (e.g., the well-known Vasicek arbitrage-free

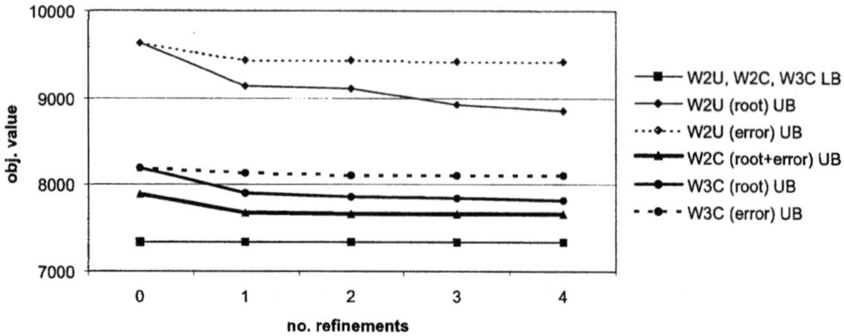

Figure 9: Objective values for different key rate distributions and refinement strategies

model [63] and a two-factor term structure model incorporating mean reversion), refer to the description of test problems in [30].

In case of all distributions under consideration, the lower bounds practically coincide. This can be explained by the fact that this approximation is formed by the vertices of the volume change ξ_t that remain the same in all cases and the barycenters of the risk factors η_t. The correlations used here are still too weak to cause more significant difference in their positions for the different distributions. Interestingly, the numerical accuracy of the uncorrelated distribution '2U' is significantly worse than those of its correlated counterpart '2C'. Incorporating a third factor does not deteriorate the accuracy much for the correlated distribution '3C' compared to the case of independent key rates. Moreover, partitioning the simplices in the root of the scenario trees always yields a much better improvement in the accuracy than splitting those Θ_t corresponding to the nodes with the highest approximation error except for '2C' which already exhibits a low gap between both approximations.

Another measure for the goodness of approximations in multistage stochastic programming is the local *expected value of perfect information (EVPI)* taken at the root node. It can be interpreted as the amount one would pay today to obtain perfect information about the uncertain future. In Figure 10, it is stated relative to the optimal objective value of the deterministic equivalent program corresponding to the upper approximation. Again, partitioning in nodes different from the root has only a slight impact on the refinement process, shown here for '2U' only. Both correlated distributions exhibit a significantly higher relative EVPI than the latter which leads to the following – rather surprising – conclusion: Although the larger EVPI implies a higher degree of stochasticity, it becomes easier to approximate the stochastic programs by the barycentric deterministic equivalent problems the higher the correlations are.

Taking the graphical illustrations in Figure 8 into account, an intuitive explanation can be seen in the simplicial coverage of original distributions. The distortion of the simplex that results from the correlations can yield "more extreme" scenarios due

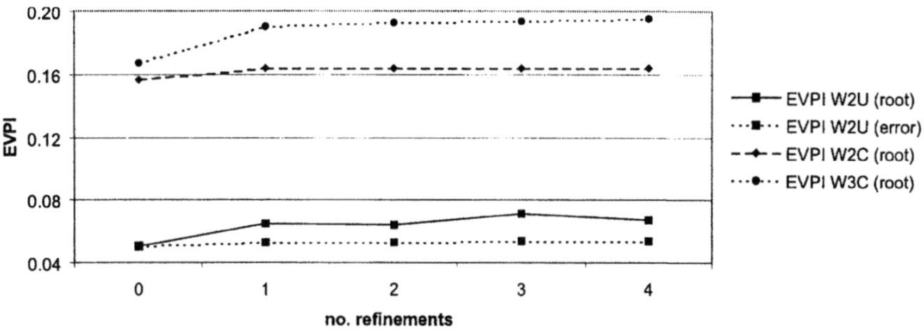

Figure 10: Number of refinements and relative EVPI

to extreme positions of some vertices. Within importance sampling procedures for multistage stochastic programs, a high EVPI as a measure for uncertainty is often used as a criterion to increase the number of scenarios. Here, the EVPI may grow with the number of partitions although the accuracy of the approximation is getting better. In particular, the approximation error can be zero but the EVPI might still have a positive value for some nodes. This questions its suitability as a measure for the control of the refinement process and underlines the advantage of quantifying the accuracy using exact error bounds.

4.5 Solution of deterministic equivalent problems

Barycentric scenario trees define multistage programs that minorize and majorize the given stochastic problem. It has been pointed out already that the corresponding deterministic equivalent programs are *large-scale problems* which limits the number of stages, i.e., the planning horizon T, the dimension M of the probability space Ω, and/or the accuracy that can be achieved by partitioning the simplicial coverage. In most cases, a trade-off between these quantities must be found when a practical situation is modeled to ensure that the optimization problem can be solved by the available computational resources.

The number of nodes at each stage of the scenario trees was derived in equations (29) and (30). It can easily be seen how the number of partitions and the dimension of the underlying distributions for η_t and ξ_t influence the problem size that grows exponentially with the planning horizon T *(curse of dimensionality)*. Let $\sum_{t=0}^{T} |\mathcal{A}^t|$ be the total number of nodes in the (lower or upper) scenario tree $\mathcal{A}$. Omitting liquidity limits of type (5a) and (5b), the size of the deterministic equivalent programs corresponding the savings problem (1) is given by $(D+3) \cdot \sum_{t=0}^{T} |\mathcal{A}^t|$ constraints, $(D + 2 \cdot |\mathcal{D}^S| + 1) \cdot \sum_{t=0}^{T} |\mathcal{A}^t|$ variables and $[3(|\mathcal{D}^S| + D + 1) + m] \cdot \sum_{t=0}^{T} |\mathcal{A}^t|$ nonzeros (i.e., '± 1'). Note that $K = 3$ and $L = 1$ and, hence, the deterministic equivalent corresponding to the upper tree is the larger of the two approximated problems.

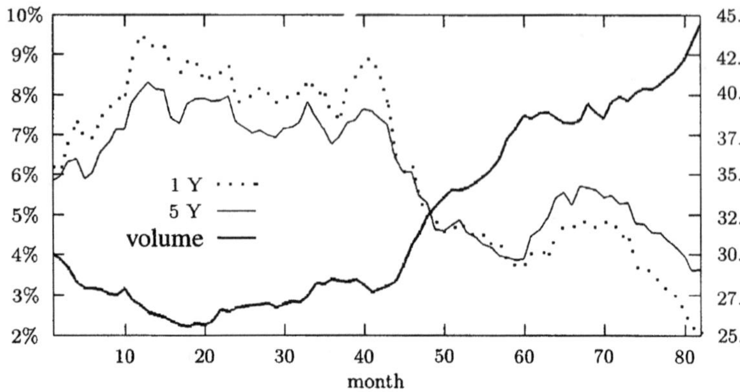

Figure 11: Interest rates (left) and savings volume (bio. CHF, right)

The formulae given above also illustrate the block structure and sparsity of the constraint matrix that can be exploited by powerful decomposition algorithms. A detailed discussion of numerical aspects for alternative solution techniques can be found in [34] where also results are presented comparing Cplex, regularized decomposition (cf. [56, 59]), nested decomposition (cf. [2]), and MSLiP-OSL (cf. [9]) for a set of standard problems similar to (1).

5 Reinvestment of savings accounts

The savings model (1) as well as an equivalent stochastic optimization model for refinancing variable-rate mortgages was developed and implemented in co-operation with a major Swiss bank. Before the bank decided to use them for the management of nonmaturing account positions, a case study was conducted to assess their performance for the savings application and compare it with the replicating portfolio approach that was employed up to this time. The study was based on monthly interest data (money market and swap rates) for the Swiss Franc from March 1989 to April 1996 and the corresponding volume of a savings position (see Figure 11). The latter was taken from the aggregated savings volume of all financial institutions in Switzerland and the Principality of Liechtenstein as stated in the monthly reports of the Swiss National Bank and scaled to an initial volume of 30 billion.

Investment opportunities consisted of money market and swap positions with maturities of 1, 2, 3, 4, 5, 7, and 10 years. In order to take liquidity restrictions into account that come close to conditions on the Swiss interbank market, upper bonds were set to 500 mio. for maturities up to 5 years and 200 mio. for 7 and 10 years. For the initial portfolio composition, it was assumed that a constant mix policy had been implemented before consisting of 50 % two-year and 50 % five-year instruments,

scen.	*key rate shifts*			*opt.*	*objective values*		...
	#1	*#2*	*#3*	*sol.*	*P1*	*P2*	
0	--	--	--	P1	**668'734.7**	668'672.2	...
1	-	-	-	P2	890'117.8	**890'209.2**	...
2	-	--	-	P2	902'151.4	**902'156.8**	...
3	-	-	0	P1	**975'132.0**	975'051.5	...
4	-	0	+	P3	**1'174'434.0**	1'174'193.0	...
5	0	0	0	P4	**1'152'769.0**	1'152'613.0	...
6	+	0	-	P4	**1'379'315.0**	1'378'796.0	...
7	+	+	0	P4	**1'421'833.0**	1'421'339.0	...
8	++	+	0	P4	**1'648'060.0**	1'647'201.0	...
9	+	+	+	P4	**1'421'833.0**	1'421'339.0	...
10	++	++	++	P4	**1'691'181.0**	1'690'350.0	...

Table 1: Stress scenarios and risk analysis of different investment policies

i.e., an amount of 875'000 matures in each of the first 24 months and 250'000 in each of the next 36 months.

It turned out that it is helpful to base the decisions not only on a single optimization where scenarios are determined with expected values of zero for the underlying multidimensional stochastic processes. Instead, a set of 11 stress scenarios characterized by various key rates drifts was defined which reflect shift and tilt movements of the yield curve. Performing the optimization for each of these expectations may result in different optimal investment policies and corresponding expected returns. This allows the assessment of the sensitivity of the solution with respect to different distribution assumptions and, in particular, to quantify the impact of "extremal events" that may be chosen by the decision maker.

An example for this set of stress scenarios is given in Table 1. Each '+' and '-' represents a decrease or increase in a key rate by one standard deviation. In general, the 11 optimization runs result in more than one solution. Then, the *first-stage decision* is fixed, and the optimization is repeated with respect to the remaining stages for all policies and all different key rate drifts. In other words, for the selected stress scenarios the consequences of suboptimal initial decisions and the costs for their correction in the subsequent periods are determined. This allows an analysis of the risk associated with the different solutions, taking into account non-anticipated shift and tilt movements of the term structure. In this way, one obtains a sort of profit and loss pattern which helps to identify dominant policies for the first-stage decision. For instance, in Table 1 policy P2 performs only slightly better than P1 for scenario 1 and 2 but is clearly inferior in all other cases. A pairwise comparison of the restricted solutions reveals the investment policy that is finally implemented.

The evolution of the margin between the return of the portfolio and the customer rate over the sample period is shown in Figure 12 for the dynamic policies determined by the stochastic optimization model with subsequent risk analysis and is compared to

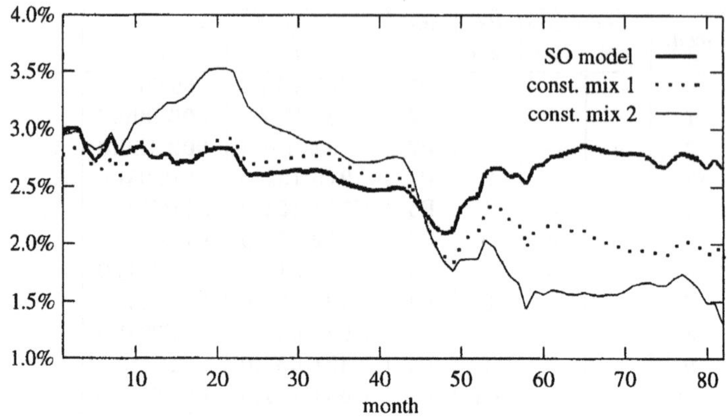

Figure 12: Evolution of margin for replicating portfolios and optimization model

two constant mix strategies. The latter were determined by the replicating portfolio approach. In case of CM2, portfolio weights were derived under the assumption of a constant volume, i.e., it tracks only the customer rate and not the total volume. Note that this results in a lower and more volatile margin. According to Table 2, the average margin could be improved by 25 basis points compared to the better constant mix policy. In particular, the standard deviation of the margin was reduced significantly although volatility is not considered explicitly in the model's objective. However, it is incorporated by the high number of scenarios in the tree of the stochastic program. Similar observations have been made for other stochastic optimization models (see [6] for example).

policy	*weights const. mix*			*margin [%]*	
	1 y	*2 yrs*	*5 yrs*	*mean*	*std. dev.*
SO model	–	–	–	2.659	0.188
const. mix 1	0.0	0.5	0.5	2.414	0.358
const. mix 2	0.35	0.35	0.3	2.399	0.696

Table 2: Comparison of dynamic policy with replicating portfolios

6 Conclusions and outlook

In this paper, a multistage stochastic optimization model was introduced for the management of non-maturing accounts like savings deposits. The solvability of the stochastic program suffers from the multidimensional integration and nested optimization of implicitly given value functions. In the convex case, which holds under certain restrictions for the underlying (conditional) probability distributions and the form of constraints, structural properties help to overcome these numerical difficulties. In particular, applying barycentric approximation yields distinguished scenario trees. The solution of the associated deterministic equivalent programs provides exact upper and lower bounds to the original problem. Moreover, the accuracy of the approximation may be quantified and improved by refinement techniques. An implementation of more efficient partitioning algorithms remains subject to future research.

The performance of the stochastic optimization model compared to traditional approaches encouraged a major Swiss bank to apply it for the management of their non-maturing account positions. Two important conclusions can be drawn from the case study presented above: First, the results indicate that a dynamic policy is superior to a static one since portfolios selected with the stochastic program clearly outperform constant mix strategies. Second, the stochastic optimization model hedges against the various sources of uncertainty inherent to non-maturing accounts like interest rate and prepayment/withdrawal risk more appropriately than the replicating portfolio approach. In particular, it is able to take the dependencies between interest rates and volume for dynamic portfolio strategies into account.

Stochastic programming is well suited for a broad class of problems within asset and liability management (ALM) that are characterized by cross and/or serial correlations between risk factors, e.g., cash management in insurance companies where premium payments exhibit a seasonal behavior. Other types of risk (credit, currency etc.) may also be considered if appropriate. Clearly, the Brownian motions exploited here to model the evolution of interest rates and volume may be viewed as a rather simple approach to model the uncertainty of the relevant risk factors. More sophisticated models have been proposed in the financial literature and are currently under investigation. Schürle [60] examines alternative term structure models for the generation of interest rate scenarios and a trend-stationary process for evolution of savings deposits. In the latter case, using the risk factors driving the yield curve as explanatory variables allows for a good description of the aggregated savings volume published by the Swiss National Bank (see also [41] and the references herein for a discussion of alternative processes for the evolution of demand deposits). It is expected that the integration of more sophisticated models for the generation of interest rate and volume scenarios in the stochastic program will yield an additional improvement in the performance. However, different types of distribution functions for the random data might not preserve the inheritance of the saddle property of value functions.

The savings account model introduced here may be seen as a first step towards a general ALM model for a bank's complete balance sheet to optimize investment and refinancing decisions with respect to interest rate and volume risk. Additional constraints may be imposed to limit the absolute risk exposure of certain positions in order to comply with regulatory restrictions concerning capital requirements. Multistage stochastic programming helps to overcome many difficulties in modeling dynamic decision making. It can be seen as a versatile tool for financial planning problems under uncertainty.

References

[1] J.R. Birge and R.J.-B. Wets. Designing approximation schemes for stochastic optimization problems, in particular for stochastic programs with recourse. *Mathematical Programming Study*, 27:54-102, 1986.

[2] J.R. Birge, C.J. Donohue, D.F. Holmes, and O.G. Svintsitski. A parallel implementation of the nested decomposition algorithm for multistage stochastic linear programs. *Mathematical Programming*, 75:327-352, 1995.

[3] M. Britton-Jones. The sampling error in estimates of mean-variance efficient portfolio weights. *Journal of Finance*, 54:655-671, 1999.

[4] D.R. Cariño, T. Kent, D.H. Myers, C. Stacy, M. Sylvanus, A.L. Turner, K. Watanabe, and W.T. Ziemba. The Russell-Yasuda Kasai model: An asset/liability model for a Japanese insurance company using multistage stochastic programming. *Interfaces*, 24:29-49, 1994.

[5] D.R. Cariño, D.H. Myers, and W.T. Ziemba. Conceps, technical issues, and uses of the Russell-Yasuda Kasai financial planning model. *Operations Research*, 46:450-462, 1998.

[6] D.R. Cariño and W.T. Ziemba. Formulation of the Russell-Yasuda Kasai financial planning model. *Operations Research*, 46:433-449, 1998.

[7] Z. Chen, G. Consigli, M.A.H. Dempster, and N. Hicks-Pedrón. Towards sequential sampling algorithms for dynamic portfolio management. In: C. Zopounidis (ed.), *Operational Tools in the Management of Financial Risks*, pp. 197-211. Kluwer, 1998.

[8] V.K. Chopra and W.T. Ziemba. The effect of errors in means, variances, and covariances on optimal portfolio choice. *Journal of Portfolio Management*, 20:6-11, 1993.

[9] G. Consigli and M.A.H. Dempster. Solving dynamic portfolio problems using stochastic programming. *Zeitschrift für Angewandte Mathematik und Mechanik (Supplement)*, 77:S 535-S 536, 1997.

[10] G.B. Dantzig and G. Infanger. Large-scale stochastic linear programs: Importance sampling and Benders decomposition. Technical Report SOL 91-4, Stanford University, 1991.

[11] G.B. Dantzig and G. Infanger. Multi-stage stochastic linear programs for portfolio optimization. *Annals of Operations Research*, 45:59-76, 1993.

[12] R. Dembo. Scenario optimization. *Annals of Operations Research*, 30:63-80, 1991.

[13] M.A.H. Dempster. The expected value of perfect information in the optimal evolution of stochastic problems. In: M. Arato, D. Vermes, and A.V. Balakrishnan (eds.), *Stochastic Differential Systems*, pp. 25-40. Springer, 1981.

[14] C.L. Dert. *Asset liability management for pension funds: A multistage chance constraint programming approach.* Phd thesis, Erasmus University, Rotterdam, 1995.

[15] J. Dupačová. Stochastic programming models in banking. Working paper, IIASA, Laxenburg, 1991.

[16] J. Dupačová. Postoptimality for multistage stochastic linear programs. *Annals of Operations Research*, 56:65-78, 1995.

[17] J. Dupačová. Scenario-based stochastic programs: Resistance with respect to sample. *Annals of Operations Research*, 64:21-38, 1996.

[18] J. Dupačová, M. Bertocchi, and V. Moriggia. Postoptimality for scenario based financial planning models with an application to bond portfolio management. In: W.T. Ziemba and J.M. Mulvey (eds.), *World Wide Asset and Liability Modeling*, pp. 263-285. Cambridge University Press, Cambridge, 1998.

[19] N.C.P. Edirisinghe. New second-order bounds on the expectation of saddle functions with applications to stochastic linear programming. *Operations Research*, 44:909-922, 1996.

[20] N.C.P. Edirisinghe and W.T. Ziemba. Bounds for two-stage stochastic programs with fixed recourse. *Mathematics of Operations Research*, 19:292-313, 1994.

[21] H.P. Edmundson. Bounds on the expectation of a convex function of a random variable. Technical Report 982, RAND Corporation, 1957.

[22] D. Eichhorn, F. Gupta, and E. Stubbs. Using constraints to improve the robustness of asset allocation. *Journal of Portfolio Management*, 24:41-48, 1998.

[23] S.-E. Fleten, K. Høyland, and S.W. Wallace. The performance of stochastic dynamic and fixed mix portfolio models. Working paper, Norwegian University of Science and Technology, Trondheim, 1998.

[24] B. Forrest, K. Frauendorfer, and M. Schürle. A stochastic optimization model for the investment of savings account deposits. In: P. Kischka et al. (eds.), *Operations Research Proceedings 1997*, pp. 382-387, Springer, 1998.

[25] E. Fragnière, J. Gondzio, and J.-P. Vial. A planning model with one million scenarios solved on an affordable parallel machine. Technical Report 1998.11, Logilab, University of Geneva, 1998.

[26] K. Frauendorfer. Solving SLP recourse problems with arbitrary multivariate distributions: The dependent case. *Mathematics of Operations Research*, 13:377-394, 1988.

[27] K. Frauendorfer. *Stochastic Two-Stage Programming.* Springer, 1992.

[28] K. Frauendorfer. Multistage stochastic programming: Error analysis for the convex case. *ZOR – Mathematical Methods of Operations Research*, 39:93-122, 1994.

[29] K. Frauendorfer. Barycentric scenario trees in convex multistage stochastic programming. *Mathematical Programming*, 75:277-293, 1996.

[30] K. Frauendorfer and G. Haarbrücker. Test problems in stochastic multistage programming. *Optimization*, to appear.

[31] K. Frauendorfer and F. Härtel. On the goodness of discretizing diffusion processes for stochastic programming. Working paper, Institute of Operations Research, University of St. Gallen, 1995.

[32] K. Frauendorfer and P. Kall. A solution method for SLP recourse problems with arbitrary multivariate distributions: The independent case. *Problems of Control and Information Theory*, 17:177-205, 1988.

[33] K. Frauendorfer and C. Marohn. Refinement issues in stochastic multistage linear programming. In: K. Marti and P. Kall (eds.), *Stochastic Programming Methods and Technical Applications (Proceedings of the 3rd GAMM/IFIP Workshop 1996)*, pp. 305-328, Springer, 1998.

[34] K. Frauendorfer, C. Marohn, and M. Schürle. SG-portfolio test problems for stochastic multistage linear programming (II). Working paper, Institute of Operations Research, University of St. Gallen, 1997.

[35] K. Frauendorfer and M. Schürle. Barycentric approximation of stochastic interest rate processes. In: W.T. Ziemba and J.M. Mulvey (eds.), *Worldwide Asset and Liability Modeling*, pp. 231-262. Cambridge University Press, Cambridge, 1998.

[36] B. Golub, M. Holmer, R. McKendall, L. Pohlman, and S.A. Zenios. A stochastic programming model for money management. *European Journal of Operational Research*, 85:282-296, 1995.

[37] T. Hakala. *A Stochastic Optimization Model for Multi-Currency Bond Portfolio Management.* Phd thesis, Helsinki School of Economics and Business Administration, 1996.

[38] J.L. Higle and S. Sen. *Stochastic Decomposition - A Statistical Method for Large Scale Stochastic Linear Programming.* Kluwer, 1996.

[39] T. Ho. Key rate durations: Measures of interest rate risks. *Journal of Fixed Income*, 2:29-44, 1992.

[40] M.R. Holmer. The asset-liability management strategy system at Fannie Mae. *Interfaces*, 24:3-21, 1994.

[41] R.A. Jarrow and D.R. van Deventer. The arbitrage-free valuation and hedging of demand deposits and credit card loans. *Journal of Banking and Finance*, 22:249-272, 1998.

[42] J.L. Jensen. Sur les fonctions convexes et les inégalités entre les valeurs moyennes. *Acta Mathematica*, 30:175-193, 1906.

[43] P. Kall, A. Ruszczyński, and K. Frauendorfer. Approximation techniques in stochastic programming. In: Y. Ermoliev and R.J.-B. Wets (eds.), *Numerical Techniques for Stochastic Optimization*, pp. 33-64. Springer, 1988.

[44] J.G. Kallberg, R.W. White, and W.T. Ziemba. Short term financial planning under uncertainty. *Management Science*, 28:670-682, 1982.

[45] M.I. Kusy and W.T. Ziemba. A bank asset and liability management model. *Operations Research*, 34:356-376, 1986.

[46] R. Litterman and J. Scheinkman. Common factors affecting bond returns. *Journal of Fixed Income*, 1:54-61, 1991.

[47] A. Madansky. Bounds on the expectation of a convex function of a multivariate random variable. *Annals of Mathematical Statistics*, 30:743-746, 1959.

[48] H.M. Markowitz. *Portfolio Selection: Efficient Diversification of Investment.* Wiley, 1959.

[49] J.M. Mulvey. Financial planning via multi-stage stochastic programs. In: J.R. Birge and K.G. Murty (eds.), *Mathematical Programming: State of the Art 1994*, pp. 151-171. University of Michigan, Ann Arbor, 1994.

[50] J.M. Mulvey. Multi-stage financial planning systems. In: R.L. D'Ecclesia and S.A. Zenios (eds.), *Operations Research Models in Quantitative Finance*, pp. 18-35. Physica, 1994.

[51] J.M. Mulvey and A. Ruszczyński. A new scenario decomposition method for large-scale stochastic optimization. *Operations Research*, 43:477-490, 1995.

[52] J.M. Mulvey and A.E. Thorlacius. The Towers Perrin global capital market scenario generation system. In: W.T. Ziemba and J.M. Mulvey (eds.), *Worldwide Asset and Liability Modeling*, pp. 286-312. Cambridge University Press, 1998.

[53] J.M. Mulvey and S.A. Zenios. Capturing the correlations of fixed-income instruments. *Management Science*, 40:1329-1342, 1994.

[54] R. T. Rockafellar and R.J.-B. Wets. Nonanticipativity and L^1-martingales in stochastic optimization problems. *Mathematical Programming Study*, 6:170-187, 1976.

[55] C.H. Rosa and A. Ruszczyński. On augmented Lagrangian decomposition methods for multistage stochastic programs. *Annals of Operations Research*, 64:289-309, 1996.

[56] A. Ruszczyński. On the regularized decomposition method for stochastic programming problems. In: K. Marti and P. Kall (eds.), *Stochastic Programming: Numerical Techniques and Engineering Applications*, pp. 93-108, Springer, 1993.

[57] A. Ruszczyński. Parallel decomposition of multistage stochastic programming problems. *Mathematical Programming*, 58:201-228, 1993.

[58] A. Ruszczyński. Decomposition methods in stochastic programming. *Mathematical Programming*, 79:333-353, 1997.

[59] A. Ruszczyński and A. Świętanowski. On the regularized decomposition method for two-stage stochastic linear problems. Working Paper WP-96-014, IIASA, Laxenburg, 1996.

[60] M. Schürle. *Zinsmodelle in der stochastischen Optimierung.* Haupt, 1998.

[61] M. Steinbach. Recursive direct algorithms for multistage stochastic programs in financial engineering. In: P. Kall and H.-J. Lüthi (eds.), *Operations Research Proceedings 1998*, pp. 241-250, Springer, 1999.

[62] P. Varaiya and R.J.-B. Wets. Stochastic dynamic optimization - approaches and computation. Working Paper WP-88-87, IIASA, Laxenburg, 1988.

[63] O. Vasicek. An equilibrium characterization of the term structure. *Journal of Financial Economics*, 5:177-188, 1977.

[64] C. Vassiadou-Zeniou and S.A. Zenios. Robust optimization models for managing callable bond portfolios. *European Journal of Operational Research*, 91:264-273, 1996.

[65] K.J. Worzel, C. Vassiadou-Zeniou, and S.A. Zenios. Integrated simulation and optimization models for tracking indices of fixed-income securities. *Operations Research*, 42:223-233, 1994.

[66] S.A. Zenios. A model for portfolio management with mortgage-backed securities. *Annals of Operations Research*, 43:337-356, 1993.

[67] S.A. Zenios. Asset/liability management under uncertainty for fixed-income securities. *Annals of Operations Research*, 59:77-97, 1995.

Optimization in the Space of Distribution Functions and Applications in the Bayes Analysis [1]

Alexandr N. Golodnikov (golodnik@d130.icyb.kiev.ua)
V.M. Glushkov Institute of Cybernetics
Prospect Glushkova,40, Kiev 252022, Ukraine

Pavel S. Knopov (knopov@d130.icyb.kiev.ua)
V.M. Glushkov Institute of Cybernetics
Prospect Glushkova,40, Kiev 252022, Ukraine

Panos M. Pardalos (pardalos@ufl.edu)
Center for Applied Optimization and ISE Department,
University of Florida, Gainesville, FL 3261-6595, USA

Stanislav P. Uryasev (uryasev@ise.ufl.edu)
ISE Department,
University of Florida, Gainesville, FL 3261-6595, USA

Abstract

This paper is stimulated by reliability estimation problems for safety systems of Nuclear Power Plants. A new approach for calculating robust Bayes estimators is considered. Upper and lower bounds for Bayes estimates, provided that a prior distribution satisfies available prior information, are constructed. The problems of calculating lower and upper bounds for Bayes estimates is reduced to optimization on a set of distributions satisfying available prior information. It was demonstrated that Bayes estimates of parameters are sensitive to the type of a prior distribution function. Analysis of the reliability data of a nuclear safety system was conducted using the developed methodology. The robust estimates were compared to Bayes estimates traditionally used in nuclear industry.

[1] Research was supported in part by a NATO grant (SST.CLG 975032)

S.P. Uryasev (ed.), Probabilistic Constrained Optimization, 102–131.
© 2000 *Kluwer Academic Publishers.*

1 Introduction

Analysis of nuclear safety data, as well as many other applications, rely on using the Bayes estimation techniques. For instance, Probabilistic Safety Assessments (PSAs) depend upon data to quantify the basic failure-related events and the initiating events frequencies. Plant-specific data are derived from plant records and are processed to estimate reliability parameters for use in PSA. Statistical techniques that are usually used for estimation of reliability parameters are based on sampling theory methods. However, sampling theory methods are inappropriate for those Nuclear Power Plant (NPP) components that have scarce data samples. Although there are plenty of historical operating data for other NPPs, these data cannot be incorporated in reliability parameter estimates within the sampling theory methods. In order to overcome this shortcoming of the sampling theory, the Bayesian approach can be used which requires less sample data to achieve the same quality of inferences than the methods based on sampling theory. The Bayesian methods permit the combining of plant-specific raw data with other relevant information available from reliability studies.

For a recent review of literature and collection of papers in the area of robust Bayesian estimates see Springer Verlag volume [5]. This volume includes state-of-the-art results, such as the paper on Linearization techniques in Bayesian Robustness [10]. On application of the optimization approaches in statistics see paper [11].

A major criticism of the Bayesian approach is related to the ability of the investigator to select a true distribution function. The justification of a prior distribution frequently is a practical difficulty in the application of the Bayesian approach. This shortcoming of the Bayesian approach is usually overcome by using the Empirical Bayesian decision procedures. The Empirical Bayesian decision procedure is an efficient tool for combining either existing sets of reliability data or reliability parameter estimates from various sources. One of the advantages of such methods is their asymptotic optimality. However, the practical rate of convergence of the Empirical Bayesian risk to the minimum Bayesian risk can be quite slow. Therefore, in cases when small datasets of reliability data or reliability parameter estimates are available, the accuracy of the approximation of the Bayesian estimator by the Empirical Bayesian estimators is never really known. In such cases, it is desirable to estimate how far the calculated Bayesian estimate is from the true Bayesian estimate.

Similar problems take place in calculating two-sided Bayes probability intervals, which are used in uncertainty evaluation. Since this requires knowledge of the true posterior distribution, which depends on the true prior distribution, there can be situations where the two-sided Bayes probability interval derived using the assumed prior distribution differs from that derived using the true prior distribution.

Since in practice the true prior distribution is never known, it is reasonable to consider that any distribution function that corresponds to the same available prior information has the same justification to be used as a prior distribution. Using the beta distribution as a prior in binomial sampling and the gamma distribution as a prior in Poisson sampling is often justified by their mathematical tractability and versatility. However, this reasoning is not so important now because of availability of high-speed computational resources.

Therefore, when prior information is insufficient to accurately specify a prior distribution, the most attractive Bayes estimate is a robust Bayes estimate derived under an assumption that the true prior distribution corresponds to the available prior information.

In this connection, the following problems have been addressed: (1) development of methods of calculation of upper and lower bounds for Bayes estimates provided that a prior distribution satisfies available prior information; (2) development of methods of calculation of robust Bayes estimates; and (3) development of methods of calculation of a robust two-sided Bayes probability interval provided that a prior distribution satisfies available prior information.

This paper presents mathematical statements of the problems mentioned above, numerical methods for their solving, example of failure data analysis, and comparison of obtained results with traditionally used Bayes estimates. This paper is based on the approach developed in [4, 3].

2 Sensitivity to a prior distribution in a binomial sampling

Let us consider the following slightly modified example, borrowed from [9]. The WASH-1400 "Reactor Safety Study" (1975) reported the following data on the number of pump failures observed in 1972 in eight pressurized water reactors (PWRs) in commercial operation in the United States.

Table 1: Data on the number of pumps failures.

j	n_j	s_j	x_j	$t_j(h)$	$\frac{x_j}{n_j}$	$\frac{s_j}{n_j}$
1	50	12	38	438	0.76	0.24
2	50	2	48	438	0.96	0.04
3	50	1	49	438	0.98	0.02
4	50	5	45	438	0.9	0.1
5	50	6	44	438	0.88	0.12
6	50	0	50	438	1	0
7	50	1	49	438	0.98	0.02
8	50	3	47	438	0.94	0.06
Total:	400	30	370	3504	7.4	0.6

Here s_j is the observed number of failures in n_j pumps, and t_j is the total test time in hours. The data in rows of the table 1 can be treated as results in series of $N\,(N = 8)$ repeated trials.

The situation where n is fixed in advance, and the number of failures is left to chance, is known as binomial sampling. The probability distribution of the number of failures is the binomial distribution given by

$$Pr\{s \text{ failures will occur in } n \text{ trials }\} = f(s|p) = \frac{n!}{(n-s)!s!}p^s(1-p)^{n-s}, \quad (1)$$

where $s = 0, 1, \ldots, n$, $0 < p < 1$; p is a constant probability of failure in each trial. In the sampling theory approach, the maximum likelihood estimator of unknown parameter p in jth series of repeated trials is

$$\hat{p}_j = \frac{s_j}{n_j}. \quad (2)$$

In the Bayesian approach, the parameter p is assumed to be a random value and $p_1, p_2, \ldots, p_N$ are its statistically independent realization for the sequence of trials. The most widely used prior distribution for p is the beta distribution $\mathcal{B}(s_0, n_0)$ with the following probability density function

$$g(r) = \begin{cases} \frac{\Gamma(n_0)}{\Gamma(s_0)\Gamma(n_0-s_0)} r^{s_0-1}(1-r)^{n_0-s_0-1}, & 0 \le r \le 1 \\ 0, \quad \text{otherwise,} \end{cases}$$

where $\Gamma(n) = \int_0^\infty x^{n-1}e^{-x}dx$ is the gamma function. Using the beta distribution as a prior is often justified by its mathematical tractability and versatility.

Let us consider a family of distributions defined on [0,1] with the same mean μ and variance σ^2. First, let us express the mean and variance of a prior distribution in terms of sample moments by using the method of moments [9]. The conditional mean and variance of $\hat{p}_j$, conditioned upon the unknown value of p_j in jth trial, are

$$E(\hat{p}_j|p_j) = p_j \quad (3)$$

$$Var(\hat{p}_j|p_j) = \frac{p_j(1-p_j)}{n_j}. \quad (4)$$

The unconditional expectation of $\hat{p}_j$ is

$$E(\hat{p}_j) = E_{p_j}[E_{\hat{p}_j|p_j}(\hat{p}_j|p_j)] = E_{p_j}[p_j] = \mu. \quad (5)$$

The relation between unconditional and conditional variance is given by

$$Var(\hat{p}_j) = E_{p_j}[Var_{\hat{p}_j|p_j}(\hat{p}_j|p_j)] + Var_{p_j}[E_{\hat{p}_j|p_j}(\hat{p}_j|p_j)].$$

By substituting (3) and (4) in the previous formula we have

$$\begin{aligned} Var(\hat{p}_j) &= E\left[\frac{p_j(1-p_j)}{n_j}\right] + Var(p_j) = \\ &= \frac{1}{n_j}E(p_j) - \frac{1}{n_j}E(p_j^2) + Var(p_j) = \\ &= \frac{1}{n_j}E(p_j) - \frac{1}{n_j}[Var(p_j) + E^2(p_j)] + Var(p_j). \end{aligned}$$

Therefore,

$$Var(\hat{p}_j) = \frac{1}{n_j}\mu - \frac{1}{n_j}[\sigma^2 + \mu^2] + \sigma^2. \tag{6}$$

The method of moments equates the sample moments to their expected values and solves for the μ and σ^2. The first weighted sample moment is

$$\bar{p}_w = \frac{\sum_{j=1}^N n_j \hat{p}_j}{\sum_{j=1}^N n_j} = \frac{1}{N^*}\sum_{j=1}^N s_j, \quad N^* = \sum_{j=1}^N n_j. \tag{7}$$

And the second weighted sample moment is

$$m_w^2 = \frac{1}{N^*}\sum_{j=1}^N n_j \hat{p}_j^2. \tag{8}$$

The expected value of the first weighted sample moment is

$$E\bar{p}_w = \frac{1}{N^*}\sum_{j=1}^N Es_j = \frac{\mu}{N^*}\sum_{j=1}^N n_j = \mu. \tag{9}$$

The expected value of the second weighted sample moment is

$$\begin{aligned} E(m_w^2) &= \frac{1}{N^*}\sum_{j=1}^N n_j E(\hat{p}_j^2) = \\ &= \frac{1}{N^*}\sum_{j=1}^N n_j [Var(\hat{p}_j) + E^2(\hat{p}_j)] \end{aligned} \tag{10}$$

By substituting (5) and (6) in (10) we have

$$E(m_w^2) = \frac{1}{N^*}\sum_{j=1}^N (\mu - [\sigma^2 + \mu^2] + n_j[\sigma^2 + \mu^2]) = \frac{N}{N^*}(\mu - [\sigma^2 + \mu^2]) + \sigma^2 + \mu^2. \tag{11}$$

By equating (9) and (11) to $\bar{p}_w$ and m_w^2, respectively and solving for μ and σ^2, we obtain

$$\mu = \bar{p}_w \tag{12}$$

and

$$\sigma^2 = \frac{N^*(m_w^2 - \bar{p}_w^2) - N(\bar{p}_w - \bar{p}_w^2]}{N^* - N}. \tag{13}$$

The first weighted sample moment of the prior distribution for data from the Table 1 is $\bar{p}_w = 0.075$ and the second weighted sample moment is $m_w^2 = 0.011$. Therefore, estimates of mean and variance of the prior distribution of the parameter p obtained by using (12), (13), and data from Table 1 equal

$$\mu = 0.075, \quad \sigma^2 = 0.004069. \tag{14}$$

2.1 Beta distribution is used as a prior.

Taking into consideration that the mean and variance of the beta distribution $\mathcal{B}(s_0, n_0)$ are

$$\mu = \frac{s_0}{n_0}$$

and

$$\sigma^2 = \frac{s_0(n_0 - s_0)}{n_0^2(n_0 + 1)},$$

and solving for s_0 and n_0, we obtain

$$n_0 = \frac{\mu(1-\mu)}{\sigma^2} - 1 = 16.05016\,,$$

$$s_0 = \mu n_0 = 1.203762\,.$$

According to [9], the Bayesian point estimator for p under a squared-error loss function and beta prior distribution is

$$\tilde{p}_{\mathcal{B}} = E(p|s) = \frac{s + s_0}{n + n_0}, \tag{15}$$

2.2 Step-function with 3 steps is used as a prior.

Suppose that the true prior is one of the following step functions that have steps at three points:

$$G_0(p) = \begin{cases} 0, & \text{if } 0 \le p < 0.07, \\ 0.396773, & \text{if } 0.07 \le p < 0.071, \\ 0.995267, & \text{if } 0.071 \le p < 1, \\ 1, & \text{if } p = 1, \end{cases} \tag{16}$$

$$G_1(p) = \begin{cases} 0, & \text{if } 0 \le p < 0.001, \\ 0.120358, & \text{if } 0.001 \le p < 0.081, \\ 0.996052, & \text{if } 0.081 \le p < 1, \\ 1, & \text{if } p = 1, \end{cases} \tag{17}$$

$$G_2(p) = \begin{cases} 0, & \text{if } 0 \le p < 0.001, \\ 0.426296, & \text{if } 0.001 \le p < 0.129, \\ 0.434103, & \text{if } 0.129 \le p < 0.13, \\ 1, & \text{if } 0.13 \le p \le 1, \end{cases} \tag{18}$$

$$G_3(p) = \begin{cases} 0, & \text{if } 0 \le p < 0.003, \\ 0.439741, & \text{if } 0.003 \le p < 0.131, \\ 0.713175, & \text{if } 0.131 \le p < 0.132, \\ 1, & \text{if } 0.132 \le p \le 1, \end{cases} \tag{19}$$

It can be verified that each of these distribution functions has the same mean $\mu = 0.075$ and variance $\sigma^2 = 0.004069$.

For each of these distributions, Bayes estimates were calculated for number of failures $s = 0, 1, 2, 3$. Results obtained are compared with the corresponding Bayes estimates derived under an assumption that beta distribution is a true prior (see Table 2.) In the Table 2 each column corresponds to Bayes estimates calculated under an assumption that the true prior is a distribution function exhibited in the first row.

Table 2: Bayes estimates calculated for different prior distributions

s	*BETA*	$G_0(p)$	$G_1(p)$	$G_2(p)$	$G_3(p)$
0	0.018225	0.070588	0.00906024	0.001173	0.003165
1	0.033365	0.070592	0.07363717	0.022494	0.010802
2	0.048505	0.070596	0.08090801	0.125804	0.101286
3	0.063645	0.070599	0.08099895	0.129957	0.130724

Data in the Table 2 demonstrate that Bayes estimates are sensitive to the type of a prior distribution function that has the same first two moments.

3 Problem statement

As a rule, in PSA applications, sources of prior information are plant-specific data that are derived from plant records; observational data from previous comparable experiments; and expert estimates. Important sources of prior information are reports of US Nuclear Regulatory Commission (NUREGs) and generic data books where parameters of a prior distribution such as the mean, the 5th, 50th and 95th percentile values are presented. In practical applications, the range of possible values of an estimated parameter can be restricted by a finite interval $[a, b]$, and prior data from both objective and subjective sources can be expressed in the form of equalities and inequalities of the following types

$$\int_a^b dH(\omega) = 1, \tag{20}$$

$$\int_a^b f_i(\omega) dH(\omega) \le d_i, \tag{21}$$

$$i = 1, 2, \ldots, m,$$

or

$$\int_a^b dH(\omega) = 1, \tag{22}$$

$$\int_a^b \omega^i dH(\omega) = \mu_i, \tag{23}$$

$$i = 1, \ldots, m,$$

where functions $f_i(\omega)$, $i = 1, 2, \ldots, m$, are not necessarily continuous. Let K denote one of the classes of distribution functions defined by (20)-(21) or (22)-(23). By means of expressions presented above, we can describe sufficiently general prior information about prior distributions. For example, they can be used to define a class of distribution functions with fixed first m moments, restrictions on q quantiles or their support. Obviously, in the later two cases, functions $f_i(\omega)$, $i = 1, 2, \ldots, m$, are not continuous.

Given sample data x, the Bayes estimate δ according to Bayesian procedures can be found by minimizing a posterior risk

$$\min_{\delta} \int_a^b L(\omega, \delta) dG(\omega|x) = \min_{\delta} \frac{\int_a^b L(\omega, \delta) f(x|\omega) dH(\omega)}{,} \int_a^b f(x|\omega) dH(\omega) \qquad (24)$$

where $f(x|\omega)$ is the conditional probability density function of x given ω, which describes the sampling model; $G(\omega|x)$ is the posterior distribution of ω given x, which is calculated by using Bayes' theorem; and $L(\omega, \delta)$ is a loss function for given estimate δ.

For the squared-error loss function $L(\omega, \delta)$, the Bayes estimate is the posterior mean of ω given x

$$\delta = \frac{\int_a^b \omega f(x|\omega) dH(\omega)}{\int_a^b f(x|\omega) dH(\omega)} \, . \qquad (25)$$

As it was demonstrated in the previous section, for given sample data x, there are ranges of values of the first two moments where Bayes estimates are sensitive to the type of the prior distribution function that has the same first two moments. Since any distribution function that corresponds to the same available prior information has the same justification to be used as a prior distribution, it is important to find a range of Bayes estimates that can be derived for all distributions from class K. Its upper, c^*, and lower, c_*, bounds can be used to determine how far the calculated Bayes estimate is from the true Bayes estimate. The range of possible values of Bayes estimates, $[c_*, c^*]$, can be treated as a measure of sensitivity of Bayes estimates with respect to selection of a prior distribution under available prior information. Since points of the segment $[c_*, c^*]$ are derived under the same prior information, any of them can be used as a Bayes estimate in PSA applications. According to the conservative approach, it is reasonable to use the largest value, c^*, of the range of possible Bayes estimates in PSA applications. This value is a conservative Bayes estimate because any other Bayes estimates derived under another distribution satisfying the same prior information can only reduce the estimate of a failure probability and thereby improve PSA estimates.

In order to find such bounds for given sample data x, it is necessary to solve the following optimization problems

$$\frac{\int_a^b \omega f(x|\omega) dH(\omega)}{\int_a^b f(x|\omega) dH(\omega)} \rightarrow min, \qquad (26)$$

subject to (20) - (21), or (22) - (23), and

$$\frac{\int_a^b \omega f(x|\omega)dH(\omega)}{\int_a^b f(x|\omega)dH(\omega)} \to max, \tag{27}$$

subject to (20) - (21), or (22) - (23).

The reasons presented above change the point of view on estimating the Bayes probability interval, which is used in uncertainty evaluation. According to the Bayesian approach, once the posterior distribution of ω given x has been obtained, a symmetric $100(1-\gamma)\%$ Bayes probability interval estimate of ω is obtained by solving the two equations

$$\frac{\int_a^{\omega_*} f(x|\omega)dH(\omega)}{\int_a^b f(x|\omega)dH(\omega)} = \frac{\gamma}{2}, \tag{28}$$

and

$$\frac{\int_{\omega^*}^b f(x|\omega)dH(\omega)}{\int_a^b f(x|\omega)dH(\omega)} = \frac{\gamma}{2}, \tag{29}$$

for the lower limit ω_* and the upper limit ω^*, so that $Pr(\omega_* \leq \omega \leq \omega^*|x) = 1-\gamma$.

If there is no accurate information about the true prior distribution, determining the Bayes probability interval by using equations (28) - (28) for any prior distribution from the class K will no longer guarantee that the true unknown parameter belongs to it with probability $1-\gamma$. In this case, the new procedure should be applied to determine a 'generalized' Bayes probability interval, which would guarantee that the true unknown parameter belongs to it with probability at least $1-\gamma$.

For this reason, if the only prior information available is that the prior distribution belongs to the class K, a symmetric $100(1-\gamma)\%$ Bayes probability interval estimate of ω can be obtained by solving the following two problems.

$$\omega_* \to \sup \tag{30}$$

subject to

$$\frac{\int_a^{\omega_*} f(x|\omega)dH(\omega)}{\int_a^b f(x|\omega)dH(\omega)} \leq \frac{\gamma}{2}, \tag{31}$$

for all $H(\omega) \in K$, and

$$\omega^* \to \inf \tag{32}$$

subject to

$$\frac{\int_{\omega^*}^b f(x|\omega)dH(\omega)}{\int_a^b f(x|\omega)dH(\omega)} \leq \frac{\gamma}{2}, \tag{33}$$

for all $H(\omega) \in K$.

The problems (30)-(31) and (32)-(33) can be expressed in the following equivalent forms

$$\omega_* \to \sup \tag{34}$$

subject to

$$\sup_{H(\omega)\in K} \frac{\int_a^{\omega_*} f(x|\omega)dH(\omega)}{\int_a^b f(x|\omega)dH(\omega)} \leq \frac{\gamma}{2}, \tag{35}$$

and

$$\omega^* \to \inf \tag{36}$$

subject to

$$\inf_{H(\omega)\in K} \frac{\int_a^{\omega^*} f(x|\omega)dH(\omega)}{\int_a^b f(x|\omega)dH(\omega)} \geq 1 - \frac{\gamma}{2}. \tag{37}$$

Another definition of the sensitivity of the Bayes estimator can be given in terms of the Bayes risk. The measure of sensitivity of Bayes estimator $\delta_0(x)$ derived under an assumed prior distribution $H_0(\omega) \in K$ to the type of the prior distributions $H(\omega) \in K$ that correspond to the same available prior information can be expressed as follows:

$$\chi(\delta_0(x)) = \max_{H(\omega)\in K} \frac{r_H(\delta_0(x)) - r_{H_0}(\delta_0(x))}{r_{H_0}(\delta_0(x))}, \tag{38}$$

where

$r_{H_0}(\delta_0(x))$ is the Bayes risk that corresponds to the Bayes estimator $\delta_0(x)$ provided that distribution $H_0(\omega)$ is the true prior;

$r_H(\delta_0(x))$ is the Bayes risk that corresponds to the Bayes estimator $\delta_0(x)$ provided that distribution $H(\omega) \in K$ is the true prior.

In order to calculate the value of $\chi(\delta_0(x))$, it is necessary to solve the following problem

$$\max_{H(\omega)\in K} r_H(\delta_0(x)) = \max_{H(\omega)\in K} \int_a^b \int_X L(\omega, \delta_0(x)) f(x|\omega) dx dH(\omega),$$

where

$L(\omega, \delta_0(x))$ is a loss function for given decision function (estimator) $\delta_0(x)$;

$f(x|\omega)$ is the conditional probability density function (sampling model). Let

$$f_0(\omega) = \int_X L(\omega, \delta_0(x)) f(x|\omega) dx \tag{39}$$

Then, the problem of calculating the sensitivity of a given Bayes estimator $\delta_0(x)$ is reduced to the following problem of optimization of a linear functional over prior distributions from the class K of distribution functions described by available prior information.

$$\int_a^b f_0(\omega) dH(\omega) \to \max, \tag{40}$$

subject to (20) - (21), or (22) - (23).

It is also desirable to estimate how far the calculated Bayes risk derived under assumed prior distribution is from the true Bayes risk. The problems of calculating lower and upper bounds for Bayes risk are stated as the problem of minimization and maximization of functional (40) over distributions that satisfy (20) - (21), or (22) - (23).

This section considers stochastic programming problems where it is required to optimize linear or linear-fractional functionals over one-dimensional distribution functions that belong to some class, defined by restrictions (20) - (21), or (22) - (23). The next sections presents numerical methods for solving such problems for a general multi-dimensional case.

4 Optimization linear functional under constraints of inequality type

Consider the following optimization problem:

$$\varphi(H) = \int_X f_0(x)dH(x) \to \inf, \tag{41}$$

subject to

$$\int_X dH(x) = 1, \tag{42}$$

$$\psi_i(H) = \int_X f_i(x)dH(x) \leq a_i, \tag{43}$$

$$i = 1, 2, \ldots, m,$$

where X is a compact set from R^n.

Denote by $K_m(X)$ the set of distribution functions that satisfy (42)-(43). Let us consider two cases:

1. Functions $f_\nu(x)$ are lower semi-continuous, $\nu = 0, 1, \ldots, m$.
2. Functions $f_\nu(x)$ are upper semi-continuous, $\nu = 0, 1, \ldots, m$.

Case 1. Let functions $f_\nu(x)$ be lower semi-continuous, $\nu = 0, 1, \ldots, m$ and a_0 is a number such that for some distribution function $H(x)$, which satisfies (42)-(43), the following inequality holds

$$\varphi(H) = \int_X f_0(x)dH(x) \leq a_0. \tag{44}$$

Then the problem (41)-(43) can be written in the following form:

$$\varphi(H) = \int_X f_0(x)dH(x) \to \inf, \tag{45}$$

subject to

$$\int_X dH(x) = 1, \tag{46}$$

$$\int_X f_\nu(x)dH(x) \leq a_\nu, \tag{47}$$

$$\nu = 0, 1, \ldots, m.$$

Following [2], [7], and [8] we show that the functional (41) achieves its minimal value and the solution of the problem (41)-(43) is a step-function with at most $m+1$ steps.

Let

$$z_\nu = f_\nu(x), \qquad x \in X, \qquad \nu = 0, 1, \ldots, m,$$

Define

$$Z = \{z : z = (f_0(x), f_1(x), \ldots, f_m(x)), x \in X, f_\nu(x) \le a_\nu, \nu = 0, 1, \ldots, m\}. \quad (48)$$

The following theorem holds.

Theorem 1.

Assume that functions $f_\nu(x), \nu = 0, 1, \ldots, m$, are lower semi-continuous. Then the set Z is closed.

Proof. Let $z^k \to z^*$, where $z^k = (f_0(x^k), f_1(x^k), \ldots, f_m(x^k))$ and $z^k \in Z$. Since the set X is compact then from the sequence $\{x^k\}$ we can select a convergent subsequence. Without any loss of generality, assume that $x^k \to x^* \in X$. Since functions $f_\nu(x), \nu = 0, 1, \ldots, m$, are lower semi-continuous on the set X, then $\forall \varepsilon > 0 \ \exists N$ such that $\forall k \ge N$ the following inequalities hold

$$f_\nu(x^*) - f_\nu(x^k) < \varepsilon, \qquad \nu = 0, 1, \ldots, m.$$

From these inequalities and (48), we obtain

$$f_\nu(x^*) = f_\nu(x^*) - f_\nu(x^k) + f_\nu(x^k) \le a_\nu + \varepsilon,$$

$$\nu = 0, 1, \ldots, m.$$

Since ε is arbitrary positive value, these inequalities imply that $z^* \in Z$ and therefore the set Z is closed, which proves the theorem. ◇

Consider the convex hull of Z:

$$coZ = \{z : z = \sum_{k=1}^{r} p_k z^k, z^k \in Z, \sum_{k=1}^{r} p_k = 1, p_k \ge 0, k = 1, 2, \ldots, r\},$$

where r is an arbitrary positive integer value.

Since Z is a closed set, then coZ is a closed set as well. Denote by G a set of vectors $Q = (\varphi(H), \psi_1(H), \ldots, \psi_m(H))$ whose components satisfy (47) for any $H(x)$ such that $\int_X dH(x) = 1$. Since step-functions with a finite number of steps belong to the set of distribution functions, then $coZ \subset G$.

Any distribution function $H(x)$ can be approximated by step-functions with a sufficiently large number N of steps and $\int_X f_\nu(x) dH(x)$ can be approximated with sums $\sum_{k=1}^{N} f_\nu(x^k) p_k$, $\nu = 0, 1, \ldots, m$, which belong to coZ. Since coZ is a closed set then limits of these sums belong to coZ. Therefore $G = coZ$.

Now, the problem (41)-(43) can be written in the following form:

$$z_0 \to \min \tag{49}$$

subject to

$$z = (z_0, z_1, \ldots, z_m) \in coZ. \tag{50}$$

In (49) we substitute 'inf' by 'min' because the set coZ is closed and therefore $\min z_0$ is achieved on the set coZ. A solution of the problems (49)-(50) belongs to the boundary of coZ from $m+1$-dimensional space. According to the Caratheodory theorem, each point on the boundary of coZ can be represented as a convex combination of at most $m+1$ points from Z. Therefore the functional (41) achieves its minimal value and the solution of the problem (41)-(43) is a step-function with at most $m+1$ steps.

Assume that functions $f_\nu(x)$, $\nu = 0, 1, 2, \ldots, m$ have finite number of surfaces of discontinuity $A_1, A_2, \ldots, A_r$. Suppose that functions $f_\nu(x)$, $\nu = 0, 1, \ldots, m$ are continuous on $A_1, A_2, \ldots, A_r$, i.e. if A is the surface of discontinuity of function $f_\nu(x)$, then $\forall x' \in A$ and $\forall \varepsilon > 0$ $\exists \delta > 0$ such that $\forall x \in U_\delta(x') \bigcap A$ the following inequality holds

$$| f_\nu(x') - f_\nu(x) | < \varepsilon,$$

where $U_\delta(x') = \{x : \|x - x'\| < \delta\}$.

If $X \subset R^1$, a surface of discontinuity is a point of discontinuity.

Algorithm.

The idea of the algorithm is to reduce the initial infinite programming problem to a sequence of finite-dimensional problems. At each iteration we shall search for a solution of the problem (41)-(43) in a set of step-functions of finite support. The number of points in support sets increase from iteration to iteration.

On the s-th iteration construct partitions $R_s^i = (x_{1,i}^s, x_{2,i}^s, \ldots, x_{l_s,i}^s)$ of sets A_i, $i = 0, 1, \ldots, r$, such that $A_i \subset \bigcup_{j=1}^{l_s} U_{\lambda_s}(x_{j,i}^s)$, $R_s^i \subset R_{s+1}^i$, $i = 0, 1, \ldots, r$; $\lambda_s \to 0$ as $s \to \infty$, where $A_0 = X \setminus \bigcup_{i=1}^r A_i)$; $U_{\lambda_s}(x_{j,i}^s) = \{x : \|x - x_{j,i}^s\| < \lambda_s\}$.

Let $R_s = \bigcup_{i=0}^r R_s^i$, $R_s = (x_1^s, x_2^s, \ldots, x_{n_s}^s)$, where $n_s = (r+1) \cdot l_s$, $x_{(j+1)\cdot l_s + i}^s = x_{i,j}^s$, $i = 1, 2, \ldots, l_s$; $j = 0, 1, \ldots, r$.

Then, we solve the following linear programming problem:

$$\sum_{j=1}^{n_s} f_0(x_j) p_j \to \min \tag{51}$$

subject to

$$\sum_{j=1}^{n_s} p_j = 1, \tag{52}$$

$$\sum_{j=1}^{n_s} f_i(x_j) p_j \le a_i \qquad i = 1, 2, \ldots, m, \tag{53}$$

$$p_j \geq 0, \qquad x_j \in R_s, \qquad j = 1, 2, \ldots, n_s.$$

Let $p^s = (p_1^s, p_2^s, \ldots, p_{n_s}^s)$ be a solution of the problem (51)-(53). Since the number of basic variables in the problem (51)-(53) equals to $m+1$ then at most $m+1$ components of the vector p^s can be nonzero. Without any loss of generality, we can assume that the first $m+1$ components $p_1^s, p_2^s, \ldots, p_{m+1}^s$ are nonzero. Therefore, the solution of the problem (51)-(53), we can represent as

$$H_s(x) = (x_1^s, \ldots, x_{m+1}^s, p_1^s, \ldots, p_{m-1}^s), \tag{54}$$

where basic variable p_i^s corresponds to point x_i^s and appropriate basic column

$$(1, f_1(x_i^s), f_2(x_i^s), \ldots, f_m(x_i^s))'.$$

On the other hand, (54) can be considered as a representation of the distribution function, where the step-function $H_s(x)$ has step p_i^s at the point x_i^s, $i = 1, 2, \ldots, m$.

The solution $H_s(x)$ found at the sth iteration will be used as initial basic variables in solving the problem (51)-(53) at the $s+1$th iteration.

The following theorem holds.

Theorem 2.

Assume that

1. Functions $f_\nu(x)$ are lower semi-continuous and bounded on the set X, $\nu = 0, 1, 2, \ldots, m$.
2. X is a compact set from R^n.
3. $\exists \tilde{H}(x)$ such that the following restrictions hold

$$\int_X d\tilde{H}(x) = 1, \tag{55}$$

$$\psi_i(\tilde{H}) = \int_X f_i(x) d\tilde{H}(x) < a_i, \tag{56}$$
$$i = 1, 2, \ldots, m,$$

Then, limit of any convergent subsequence $\{H_{s_k}\}$ of the sequence $\{H_s\}$ belongs to the set X^*, where

$$X^* = \{H^*(x) : \varphi(H^*) = \min_{H(x) \in K_m(X)} \varphi(H)\}$$

and

$$\lim_{s \to \infty} \varphi(H_s) = \varphi(H^*).$$

Proof. As it was shown earlier, the solution of the problem (41)-(43) is a step-function $H^*(x)$ with at most $m+1$ steps. Suppose that $\varphi(H_s)$ does not converge to $\varphi(H^*)$. Then there exists an $\varepsilon_0 > 0$ and sequence of indexes $\{s_k\}$ such that the following inequality holds

$$\varphi(H_{s_k}) - \varphi(H^*) \geq \varepsilon_0 > 0 \tag{57}$$

Denote

$$H^*(x) = (x_1^*, \ldots, x_{m+1}^*, p_1^*, \ldots, p_{m+1}^*)$$

and

$$H_{s_k}(x) = (x_1^{s_k}, \ldots, x_{m+1}^{s_k}, p_1^{s_k}, \ldots, p_{m+1}^{s_k}),$$

where p_i^* and $p_i^{s_k}$ are the steps of the distribution functions $H^*(x)$ and $H_{s_k}(x)$ at the corresponding points x_i^* and $x_i^{s_k}$.

Since the point $Q = (\varphi(\tilde{H}), \psi_1(\tilde{H}), \ldots, \psi_m(\tilde{H}))$ belongs to $coZ \subset R^m$ (not necessarily a boundary point of coZ), according to the Caratheodory theorem, it can be presented as a convex combination of at most $m+2$ points from Z. Therefore, there exists a step-function with at most $m+2$ steps that satisfies (55)-(56). Without any loss of generality we can assume that $\tilde{H}(x)$ is a step-function with at most $m+2$ steps. Denote

$$\tilde{H}(x) = (\tilde{x}_1, \ldots, \tilde{x}_{m+2}, \tilde{p}_1, \ldots, \tilde{p}_{m+2}).$$

For a sufficiently small positive value δ $(0 < \delta < 1)$, we construct the following distribution function

$$\bar{H}(x) = (1-\delta)H^*(x) + \delta\tilde{H}(x).$$

Denote

$$\bar{H}(x) = (\bar{x}_1, \ldots, \bar{x}_{2m+3}, \bar{p}_1, \ldots, \bar{p}_{2m+3}),$$

where

$$\bar{x}_i = x_i^*, \bar{p}_i = (1-\delta)p_i^*, i = 1, \ldots, m+1,$$

$$\bar{x}_{m+1+i} = \tilde{x}_i, \bar{p}_{m+1+i} = \delta\tilde{p}_i, i = 1, \ldots, m+2.$$

Since $H^*(x) \in K_m(X)$ and $\tilde{H}(x)$ satisfy (56), the distribution function $\bar{H}(x)$ satisfies (56) and for a sufficiently small $\delta > 0$, the following inequality holds

$$\begin{aligned} \varphi(\bar{H}) - \varphi(H^*) &= (1-\delta)\sum_{i=1}^{m+1} f_0(x_i^*)p_i^* + \delta\sum_{i=1}^{m+2} f_0(\tilde{x}_i)\tilde{p}_i - \sum_{i=1}^{m+1} f_0(x_i^*)p_i^* = \\ &= \delta\Big[\sum_{i=1}^{m+2} f_0(\tilde{x}_i)\tilde{p}_i - \sum_{i=1}^{m+1} f_0(x_i^*)p_i^*\Big] < \frac{\varepsilon_0}{8}. \end{aligned} \tag{58}$$

Let us construct the following distribution function

$$H'(x) = (x_1', \ldots, x_{2m+3}', \bar{p}_1, \ldots, \bar{p}_{2m+3}),$$

where x_i' is an arbitrary point from $U_\varepsilon(\bar{x}_i) \bigcap A_j$, if $\bar{x}_i \in A_j, j = 0, 1, \ldots, r$; $i = 1, \ldots, 2m+3$.

Since functions $f_i(x)$ are continuous on sets A_j, $i = 0, 1, \ldots, m$; $j = 0, 1, \ldots, r$ and $\bar{H}(x)$ satisfy (56), for sufficiently small $\varepsilon > 0$ the distribution function $H'(x)$ satisfies (56).

From the definition of sets A_j, $j = 0, 1, \dots, r$, and the construction of the distribution function $H'(x)$, for a sufficiently small $\varepsilon > 0$, it follows that

$$\begin{aligned}\varphi(H') - \varphi(\bar{H}) &= \sum_{i=1}^{2m+3} f_0(x_i')\bar{p}_i - \sum_{i=1}^{2m+3} f_0(\bar{x}_i)\bar{p}_i = \\ &= \sum_{i=1}^{2m+3} \bar{p}_i[f_0(x_i') - f_0(\bar{x}_i)] < \frac{\varepsilon_0}{8}. \end{aligned} \tag{59}$$

Since $\lambda_s \to 0$ as $s \to \infty$, then for a sufficiently large s, the inequality holds $\lambda_s < \varepsilon$. Now let x_i' be a point $x_i^s \in R_j^s$ (if $\bar{x}_i \in A_j$), which is nearest to $\bar{x}_i$. Then, the distribution function $H'(x)$ satisfies (53). Since the distribution function $H_s(x)$ is a solution of the problem (51)-(53), the following inequality holds

$$\varphi(H_s) - \varphi(H') \leq 0. \tag{60}$$

From (57)-(60) for all sufficiently large s_k, we have the following inequalities

$$\begin{aligned}0 < \varepsilon_0 &\leq \varphi(H_{s_k}) - \varphi(H^*) = \varphi(H_{s_k}) - \varphi(H') + \varphi(H') - \\ &- \varphi(\bar{H}) + \varphi(\bar{H}) - \varphi(H^*) < \frac{\varepsilon_0}{8} + \frac{\varepsilon_0}{8} = \frac{\varepsilon_0}{4}. \end{aligned} \tag{61}$$

From the contradiction obtained it follows that

$$\lim_{s \to \infty} \varphi(H_s) = \varphi(H^*). \tag{62}$$

The last limit equality (62) implies that the limit of any convergent subsequence $\{H_{s_k}\}$ belongs to the set X^*. Indeed, if the following had been the case

$$\lim_{k \to \infty} H_{s_k}(x) = H'(x) \notin X^*,$$

then the inequality $\varphi(H') > \varphi(H^*)$ would have held. Since $f_0(x)$ is lower semi-continuous, we have

$$\lim_{k \to \infty} \varphi(H_{s_k}) \geq \varphi(H') > \varphi(H^*).$$

The later contradicts to (62) and this completes the proof. ◇

Case 2. Let functions $f_\nu(x)$ be upper semi-continuous, $\nu = 0, 1, \dots, m$. In this case, the set Z defined by (48) is not closed. Therefore, infinimum (41) under constraints (42)-(43) may not be reached. However, step-functions with at most $m + 1$ steps can approximate the infinimum (41) as accurately as required. The following theorem holds.

Theorem 3.

Assume that

1. Functions $f_\nu(x)$ are upper semi-continuous and bounded on the set X, $\nu = 0, 1, 2, \dots, m$.

2. X is a compact set from R^n.
3. $\exists \tilde{H}(x)$ such that the restrictions (55)-(56) hold.
Then

$$\lim_{s\to\infty} \varphi(H_s) = \inf_{H(x)\in K_m(X)} \varphi(H).$$

Proof. Suppose that $\varphi(H_s)$ does not converge to $\inf_{H(x)\in K_m(X)} \varphi(H)$. Then, there exists a $\varepsilon_0 > 0$ and a sequence of indexes $\{s_k\}$ such that the following inequality holds

$$\varphi(H_{s_k}) - \inf_{H(x)\in K_m(X)} \varphi(H) \geq 2\varepsilon_0 > 0\,. \tag{63}$$

Besides, there exists $H_{\varepsilon_0}(x) \in K_m(X)$, such that the following inequality holds

$$\varphi(H_{\varepsilon_0}) - \inf_{H(x)\in K_m(X)} \varphi(H) \leq \frac{\varepsilon_0}{2}. \tag{64}$$

(63) and (64) implies that

$$\varphi(H_{s_k}) - \varphi(H_{\varepsilon_0}) \geq \frac{3\varepsilon_0}{2} > 0. \tag{65}$$

For a sufficiently small positive value δ $(0 < \delta < 1)$, we construct the following distribution function

$$\tilde{H}_{\varepsilon_0}(x) = (1-\delta)H_{\varepsilon_0}(x) + \delta\tilde{H}(x).$$

Since $H_{\varepsilon_0}(x) \in K_m(X)$, and $\tilde{H}(x)$ satisfies (56), the distribution function $\tilde{H}_{\varepsilon_0}(x)$ satisfies (56) and for a sufficiently small $\delta > 0$, the following inequality holds

$$\begin{aligned}\varphi(\tilde{H}_{\varepsilon_0}) - \varphi(H_{\varepsilon_0}) &= (1-\delta)\int_X f_0(x)dH_{\varepsilon_0}(x) + \delta\int_X f_0(x)d\tilde{H}(x) - \\ &- \int_X f_0(x)dH_{\varepsilon_0}(x) = \delta\Big[\int_X f_0(x)d\tilde{H}(x) - \\ &- \int_X f_0(x)dH_{\varepsilon_0}(x)\Big] \leq \frac{\varepsilon_0}{8}.\end{aligned} \tag{66}$$

Integrals $\int_X f_\nu(x)d\tilde{H}_{\varepsilon_0}(x)$ can be approximated by sums

$$\sum_{k=1}^{N} f_\nu(x_k)p_k, \quad \sum_{k=1}^{N} p_k = 1, \quad p_k \geq 0$$

as accurate as required.

Therefore, there exists a step-function $\bar{H}(x)$ that satisfies (56) and

$$\varphi(\bar{H}) - \varphi(\tilde{H}_{\varepsilon_0}) \leq \frac{\varepsilon_0}{8}. \tag{67}$$

Since the distribution function $\bar{H}(x)$ is a step-function, the point $Q = (\varphi(\bar{H}), \psi_1(\bar{H}), \ldots, \psi_m(\bar{H})) \in coZ$ and can be represented as a convex combination of at most $m+2$ points from Z. Therefore, we can suggest that $\bar{H}(x)$ is a step-function with at most $m+2$ steps.

Denote

$$\bar{H}(x) = (\bar{x}_1, \ldots, \bar{x}_{m+2}, \bar{p}_1, \ldots, \bar{p}_{m+2})$$

and

$$H'(x) = (x'_1, \ldots, x'_{m+2}, \bar{p}_1, \ldots, \bar{p}_{m+2}),$$

where x'_i is an arbitrary point from the set $U_\varepsilon(\bar{x}_i) \bigcap X, i = 1, \ldots, m+2$. Since the distribution function $\bar{H}(x)$ satisfies (56) and all functions $f_i(x)$, $i = 1, \ldots, m$, are upper semi-continuous at points $\bar{x}_i$, for a sufficiently small $\varepsilon > 0$, the following inequality holds

$$\begin{aligned}\psi_i(H') &= \sum_{j=1}^{m+2} f_i(x'_j)\bar{p}_j = \sum_{j=1}^{m+2} f_i(x'_j)\bar{p}_j - \sum_{j=1}^{m+2} f_i(\bar{x}_j)\bar{p}_j + \psi_i(\bar{H}) = \\ &= \sum_{j=1}^{m+2} \bar{p}_j[f_i(x'_j) - f_i(\bar{x}_j)] + \psi_i(\bar{H}) < a_i, \quad i = 1, \ldots, m. \end{aligned} \tag{68}$$

Therefore, $H'(x) \in K_m(X)$, and since the function $f_0(x)$ is upper semi-continuous at the points $\bar{x}_i$, $i = 1, \ldots, m+2$, the following inequality holds

$$\begin{aligned}\varphi(H') - \varphi(\bar{H}) &= \sum_{j=1}^{m+2} f_0(x'_j)\bar{p}_j - \sum_{j=1}^{m+2} f_0(\bar{x}_j)\bar{p}_j = \\ &= \sum_{j=1}^{m+2} \bar{p}_i[f_0(x'_j) - f_0(\bar{x}_j)] < \frac{\varepsilon_0}{8}. \end{aligned} \tag{69}$$

Since $\lambda_s \to 0$ as $s \to \infty$, for a sufficiently large s the inequality holds $\lambda_s < \varepsilon$. Now let x'_i be a point $x^s_i \in R^s$, which is nearest to $\bar{x}_i$. Then, the distribution function $H'(x)$ satisfies (53).

Since the distribution function $H_s(x)$ is a solution of the problem (51)-(53), then the following inequality holds

$$\varphi(H_s) - \varphi(H') \leq 0. \tag{70}$$

From (65)-(70) for all sufficiently large s_k, we have the following inequalities

$$\begin{aligned} 0 < \frac{3\varepsilon_0}{2} &\leq \varphi(H_{s_k}) - \varphi(H_{\varepsilon_0}) = \varphi(H_{s_k}) - \varphi(H') + \varphi(H') - \varphi(\bar{H}) + \\ &+ \varphi(\bar{H}) - \varphi(\tilde{H}_{\varepsilon_0}) + \varphi(\tilde{H}_{\varepsilon_0}) - \varphi(H_{\varepsilon_0}) < \\ &< \frac{\varepsilon_0}{8} + \frac{\varepsilon_0}{8} + \frac{\varepsilon_0}{8} = \frac{3\varepsilon_0}{8}. \end{aligned} \tag{71}$$

The contradiction obtained implies that

$$\lim_{s \to \infty} \varphi(H_s) = \inf_{H(x) \in K_m(X)} \varphi(H).$$

The theorem is proved. ◇

5 Optimization of linear functional under fixed moments

Theorems in the previous section provides the algorithm only when the set $K_m(X)$ contains an inner point $\tilde{H}(x)$. However, it is often required to optimize (41) under the condition that m first moments of a distribution function $H(x)$ are fixed. This is the case when the third condition of Theorems 2 and 3 are violated. Since this class of problems is very important, let us investigate in detail the numerical technique for solving such problems.

Let us consider the following problem: it is required to minimize

$$\varphi(H) = \int_a^b f_0(x)dH(x) \tag{72}$$

subject to

$$\int_a^b dH(x) = 1,$$

$$\int_a^b x^i dH(x) = \mu_i, \tag{73}$$

$$i = 1, \ldots, m, \quad -\infty < a < b < \infty.$$

Denote by $K_m(a,b)$ the set of distribution functions that satisfy (73). We consider only the case when $f_0(x)$ is lower semi-continuous. As in the previous section, it can be shown that a solution of the problem (72)-(72) is achieved at a step-function with at most $m+1$ steps.

Algorithm.

On the sth iteration, construct partition $R_s = (x_1, x_2, \ldots, x_{n_s})$ of the segment $[a,b]$ such that intervals of length r_s with centers in points $x_i \in [a,b]. i = 1, \ldots, n_s$ would cover the segment $[a,b]$; $a = x_1 \le x_2 \le \ldots \le x_{n_s} = b$; $R_s \subset R_{s+1}$ and $r_s \to 0$ as $s \to \infty$.

Then, solve the following linear programming problem:

$$\sum_{j=1}^{n_s} f_0(x_j)p_j \to \min \tag{74}$$

subject to

$$\sum_{j=1}^{n_s} p_j = 1,$$

$$\sum_{j=1}^{n_s} x_j^i p_j = \mu_i \quad i = 1, 2, \ldots, m, \tag{75}$$

$$p_j \ge 0, \qquad x_j \in R_s, \qquad j = 1, 2, \ldots, n_s.$$

Let $p^s = (p_1^s, p_2^s, \ldots, p_{n_s}^s)$ be a solution of the problem (74)-(75). Since the number of basic variables in the problem (74)-(75) equals $m+1$, at most $m+1$ components of the vector p^s can be nonzero. Without any loss of generality, we can assume that the first $m+1$ components $p_1^s, p_2^s, \ldots, p_{m+1}^s$ are nonzero. Therefore, we can present the solution of the problem (74)-(75) as follows

$$H_s(x) = (x_1^s, \ldots, x_{m+1}^s, p_1^s, \ldots, p_{m+1}^s), \tag{76}$$

where the basic variable p_i^s corresponds to point x_i^s and appropriate basic column

$$(1, x_i^s, (x_i^s)^2, \ldots, (x_i^s)^m)'.$$

On the other hand, (76) can be considered as a presentation of the distribution function, where the step-function $H_s(x)$ has the step p_i^s at the point x_i^s, $i = 1, 2, \ldots, m$.

The solution $H_s(x)$ found at the s-th iteration will be used as initial basic variables in solving the problem (74)- (75) at the $s+1$-th iteration.

Before formulating the algorithm discussed above, let us consider some properties of the moment space (see [7], [8], [6]).

Let D denote the collection of all distribution functions whose support is the segment $[a, b]$, and D_A denote the collection of all step-functions whose support is the segment $[a, b]$. Let $I(t - t_0)$ denote the following one-step distribution function

$$I(t - t_0) = \begin{cases} 1, \text{ if } & t \geq t_0 \\ 0, \text{ if } & t < t_0. \end{cases}$$

One-step distribution functions are extreme points of D_A. Any step-function can be presented as a convex combination of one-step distribution functions.

Definition 1. Moment space D^m is said to be the set of points $x = (x_1, \ldots, x_m) \in E_m$ whose coordinates are moments $\mu_1(H), \mu_2(H), \ldots, \mu_m(H)$ for at least one distribution function from D.

A point in D^m, which corresponds to the one-step distribution function $I(t - t_1)$ is denoted by $x(t_1) = (t_1, t_1^2, \ldots, t_1^m)$.

Let C^m denote a curve $x(t_1)$, $a \leq t_1 \leq b$. The following well known theorems hold.

Theorem 4.

D^m is a convex set.

Theorem 5.

A set of extreme points of D^m for $m \geq 2$ coincides with C^m.

Theorem 7.

Presentation of a point $x \in D^m$ in the form of a convex combination of extreme points is unique if and only if x belongs to boundary of D^m.

From Theorem 7 the following theorem follows.

Theorem 8.

The extreme point of D^m and only it corresponds to the unique distribution function from D.

It is obvious that for any $k \geq 1$, any inner point $x \in D^m$ can be presented as a convex combination of r $(r > k)$ points from C^m. Hence, for any k and any system of moments $(\mu_1, \ldots, \mu_m)$, which is an inner point of D^m, for some $r > k$ there exists step-function $H(x)$ with r steps and moments $\mu_1, \ldots, \mu_m$.

Theorem 8 implies that if $(\mu_1, \ldots, \mu_m)$ is a boundary point of D^m, then, the problem (72)-(73) is reduced to the searching of a unique distribution function that satisfies (73). We assume that $(\mu_1, \ldots, \mu_m)$ is an inner point of D^m. The following theorem holds.

Theorem 9.

Assume that

1. Function $f_0(x)$ is lower semi-continuous and bounded on the segment [a,b] with a finite number of points of discontinuity.

2. Point $(\mu_1, \ldots, \mu_m)$ is an inner point of the moment space D^m.

Then,

$$\lim_{s \to \infty} \varphi(H_s) = \varphi(H^*).$$

and limit of any convergent subsequence $\{H_{s_k}\}$ of the sequence $\{H_s\}$ belongs to the set X^*, where

$$X^* = \{H^*(x) : \varphi(H^*) = \min_{H(x) \in K_m(a,b)} \varphi(H)\}$$

Proof. Suppose that $\varphi(H_s)$ does not converge to $\varphi(H^*)$. Then, there exists a $\varepsilon_0 > 0$ and a sequence of indexes $\{s_k\}$ such that the following inequality holds

$$\varphi(H_{s_k}) - \varphi(H^*) \geq \varepsilon_0 > 0. \tag{77}$$

The distribution function $H^*(x) \in X^*$ has k steps $(k \leq m+1)$. As it was pointed out above, there exists the distribution function $\tilde{H}(x)$, which has moments $(\mu_1, \ldots, \mu_m)$ and r steps $(r + k \geq m + 1)$.

Without a loss of generality, we can assume that points in which distribution functions $H^*(x)$ and $\tilde{H}(x)$ have steps are different. Consider the following distribution function

$$\bar{H}(x) = (1 - \alpha) H^*(x) + \alpha \tilde{H}(x),$$

where α is a sufficiently small positive value $(0 < \alpha < 1)$.

Denote

$$H^*(x) = (x_1^*, \ldots, x_k^*, p_1^*, \ldots, p_k^*),$$

$$\tilde{H}(x) = (\tilde{x}_1, \ldots, \tilde{x}_r, \tilde{p}_1, \ldots, \tilde{p}_r),$$

$$\bar{H}(x) = (\bar{x}_1, \ldots, \bar{x}_{k+r}, \bar{p}_1, \ldots, \bar{p}_{k+r}),$$

where

$$\bar{x}_i = x_i^*, \quad \bar{p}_i = (1 - \alpha) p_i^*, \; p_i^* > 0, \; i = 1, \ldots, k,$$

$$\bar{x}_{k+i} = \tilde{x}_i, \quad \bar{p}_{k+i} = \alpha\tilde{p}_i, \, \tilde{p}_i > 0, \, i = 1, \ldots, r.$$

Since the function $f_0(x)$ is bounded on $[a, b]$, for a sufficiently small $\alpha > 0$, the following inequality holds

$$\begin{aligned} \varphi(\bar{H}) - \varphi(H^*) &= (1-\alpha)\sum_{i=1}^{k} f_0(x_i^*)p_i^* + \alpha\sum_{i=1}^{r} f_0(\tilde{x}_i)\tilde{p}_i - \sum_{i=1}^{k} f_0(x_i^*)p_i^* = \\ &= \alpha\Big[\sum_{i=1}^{r} f_0(\tilde{x}_i)\tilde{p}_i - \sum_{i=1}^{k} f_0(x_i^*)p_i^*\Big] < \frac{\varepsilon_0}{8}\,. \end{aligned} \tag{78}$$

Since distribution functions $H^*(x)$ and $\tilde{H}(x)$ have the same m first moments, the distribution function $\bar{H}(x)$ satisfies (73)

$$\sum_{j=1}^{m+1} \bar{p}_j = 1 - \sum_{j=m+2}^{k+r} \bar{p}_j,$$

$$\sum_{j=1}^{m+1} \bar{x}_j^i \bar{p}_j = \mu_i - \sum_{j=m+2}^{k+r} \bar{x}_j^i \bar{p}_j, \tag{79}$$

$$i = 1, \ldots, m, \quad \bar{p}_j > 0, \quad j = 1, \ldots, k+r.$$

Since $det B \neq 0$, where

$$B = \begin{pmatrix} 1 & 1 & \ldots & 1 \\ \bar{x}_1 & \bar{x}_2 & \ldots & \bar{x}_{m+1} \\ \ldots & \ldots & \ldots & \ldots \\ \bar{x}_1^m & \bar{x}_2^m & \ldots & \bar{x}_{m+1}^m, \end{pmatrix}$$

from equations (79), we can express $p_1, p_2, \ldots, p_{m+1}$ as functions of $(x_1, \ldots, x_{k+r}, p_{m+2}, \ldots, p_{k+r})$, which are continuous in the point $(\bar{x}_1, \ldots, \bar{x}_{k+r}, \bar{p}_{m+2}, \ldots, \bar{p}_{k+r})$.

Since

$$\bar{p}_i = g_i(\bar{x}_1, \ldots, \bar{x}_{k+r}, \bar{p}_{m+2}, \ldots, \bar{p}_{k+r}) > 0,$$

$$i = 1, \ldots, m+1,$$

then for sufficiently small $\delta > 0$ and any $x_i' \in U_\delta(\bar{x}_i) \bigcap [a, b]$

$$p_i' = g_i(x_1', \ldots, x_{k+r}', \bar{p}_{m+2}, \ldots, \bar{p}_{k+r}) > 0,$$

$$i = 1, \ldots, m+1,$$

and the distribution function equals

$$H'(x) = (x_1', \ldots, x_{k+r}', p_1', \ldots, p_{k+r}'),$$

where

$$p'_i = g_i(x'_1, \dots, x'_{k+r}, \bar{p}_{m+2}, \dots, \bar{p}_{k+r}),$$

$$i = 1, \dots, m+1,$$

$$p'_i = \bar{p}_i,\ i = m+2, \dots, k+r,$$

satisfies (73).

Let us select points x'_i as follows:

if $f_0(\bar{x}_i) = f_0(\bar{x}_i + 0)$, then x'_i - any point from $(\bar{x}_i, \bar{x}_i + \delta) \bigcap [a, b]$;

if $f_0(\bar{x}_i) = f_0(\bar{x}_i - 0)$, then x'_i - any point from $(\bar{x}_i - \delta, \bar{x}_i) \bigcap [a, b]$.

Then, the first condition of the theorem for sufficiently small $\delta > 0$ implies

$$|f_0(x'_i) - f_0(\bar{x}_i)| < \frac{\varepsilon_0}{16C(k+r)}, \tag{80}$$

$$i = 1, \dots, k+r.$$

From the continuity of functions $g_i(x_1, \dots, x_{k+r}, \bar{p}_{m+2}, \dots, \bar{p}_{k+r})$ in the point $(\bar{x}_1, \dots, \bar{x}_{k+r}, \bar{p}_{m+2}, \dots, \bar{p}_{k+r})$ and from (80), for a sufficiently small $\delta > 0$, we have

$$\begin{aligned}
\varphi(H') &- \varphi(\bar{H}) = \sum_{i=1}^{m+1} f_0(x'_i) g_i(x'_1, \dots, x'_{k+r}, \bar{p}_{m+2}, \dots, \bar{p}_{k+r}) + \\
&+ \sum_{i=m+2}^{k+r} f_0(x'_i)\bar{p}_i - \sum_{i=1}^{m+1} f_0(\bar{x}_i) g_i(\bar{x}_1, \dots, \bar{x}_{k+r}, \bar{p}_{m+2}, \dots, \bar{p}_{k+r}) - \\
&- \sum_{i=m+2}^{k+r} f_0(\bar{x}_i)\bar{p}_i = \sum_{i=1}^{m+1} [f_0(x'_i) - f_0(\bar{x}_i)] g_i(x'_1, \dots, x'_{k+r}, \bar{p}_{m+2}, \dots, \bar{p}_{k+r}) + \\
&+ \sum_{i=m+2}^{k+r} [f_0(x'_i) - f_0(\bar{x}_i)]\bar{p}_i + \sum_{i=1}^{m+1} f_0(\bar{x}_i)[g_i(x'_1, \dots, x'_{k+r}, \bar{p}_{m+2}, \dots, \bar{p}_{k+r}) - \\
&- g_i(\bar{x}_1, \dots, \bar{x}_{k+r}, \bar{p}_{m+2}, \dots, \bar{p}_{k+r})] < \\
&< C(k+r)\frac{\varepsilon_0}{16C(k+r)} + \frac{\varepsilon_0}{16} = \frac{\varepsilon_0}{8}.
\end{aligned} \tag{81}$$

Since $r_s \to 0$ as $s \to \infty$, for a sufficiently large s, the inequality $r_s < \delta$ holds. Select point x'_i by using the following rule.

If the function $f_0(x)$ is left semi-continuous in the point $\bar{x}_i$, then select point $x^s_i \in R_s$ nearest from the left to $\bar{x}_i$

If the function $f_0(x)$ is right semi-continuous in the point $\bar{x}_i$, then select point $x^s_i \in R_s$ nearest from the right to $\bar{x}_i$.

Thus, the distribution function $H'(x)$ satisfies (75). Since the distribution function $H_s(x)$ is a solution of the problem (74)-(75), the following inequality holds

$$\varphi(H_s) - \varphi(H') \le 0. \tag{82}$$

From (77)-(82) for all sufficiently large s_k, we have

$$0 < \varepsilon_0 \le \varphi(H_{s_k}) - \varphi(H^*) = \varphi(H_{s_k}) - \varphi(H') + \varphi(H') - \\ - \varphi(\bar{H}) + \varphi(\bar{H}) - \varphi(H^*) < \frac{\varepsilon_0}{8} + \frac{\varepsilon_0}{8} = \frac{\varepsilon_0}{4}. \tag{83}$$

From the obtained contradiction it follows that

$$\lim_{s\to\infty} \varphi(H_s) = \varphi(H^*). \tag{84}$$

(84) implies that the limit of any convergent subsequence $\{H_{s_k}\}$ belongs to the set X^* and this completes the proof. ◇

6 Optimization of linear-fractional functional under constraints of inequality type

Now, let us consider the following problem

$$\varphi(H) = \frac{\int_X g_1(x)dH(x)}{\int_X g_2(x)dH(x)} \to \inf_{H(X)\in K_m(X)} \tag{85}$$

where $K_m(X)$ is the set of distribution functions which satisfy (42)-(43). Let us denote $J_1(H) = \int_X g_1(x)dH(x)$ and $J_2(H) = \int_X g_2(x)dH(x)$. We assume that $J_2(H) > \gamma > 0$ for all distribution functions $H(x) \in K_m(X)$, where γ is a fixed positive number. The following theorem holds.

Theorem 10. The problem (85) is equivalent to a problem of searching for a number t_* such that the following equality holds

$$\inf_{H\in K_m(X)} [J_1(H) - t_* J_2(H)] = 0. \tag{86}$$

Proof. Let (86) holds. Then, for any distribution function $H(x) \in K_m(X)$, we have

$$J_1(H) - t_* J_2(H) \ge 0.$$

Therefore, for $\forall\, H(x) \in K_m(X)$ we have

$$t_* \le \frac{J_1(H)}{J_2(H)}. \tag{87}$$

Besides, for $\forall\, \varepsilon > 0$ there exists distribution function $H_\varepsilon(x) \in K_m(X)$ such that

$$J_1(H_\varepsilon) - t_* J_2(H_\varepsilon) \le \varepsilon\gamma.$$

Therefore,

$$t_* \ge \frac{J_1(H_\varepsilon)}{J_2(H_\varepsilon)} - \varepsilon\frac{\gamma}{J_2(H_\varepsilon)} \ge \frac{J_1(H_\varepsilon)}{J_2(H_\varepsilon)} - \varepsilon. \tag{88}$$

From (87)-(88), we have

$$t_* = \inf_{H(x)\in K_m(X)} \frac{J_1(H)}{J_2(H)}.$$

The same reasoning can be used to prove the theorem in the reverse direction.

From the results of section 1, it follows that if functions $g_1(x), f_1(x), \dots, f_m(x)$ are lower semi-continuous and the function $g_2(x)$ is upper semi-continuous, then, the minimum of the functional (86) is achieved on a step-function with at most $m+1$ steps. Since the problem (86) is equivalent to the problem (85), the minimum of the functional (85) is achieved at a step-function with at most $m+1$ steps.

Suppose that functions $g_1(x), g_2(x), f_i(x), i = 1, \dots, m$, have a finite number of surfaces of discontinuity $A_1, A_2, \dots, A_q$. Suppose that these functions are continuous on $A_1, A_2, \dots, A_q$.

Algorithm.

On the sth iteration, construct partitions $R_s^i = (x_{1,i}^s, x_{2,i}^s, \dots, x_{l_s,i}^s)$ of sets $A_i, i = 0, 1, \dots, q$, such that $A_i \subset \bigcup_{j=1}^{l_s} U_{r_s}(x_{j,i}^s)$, $R_s^i \subset R_{s+1}^i$, $i = 0, 1, \dots, q$; $r_s \to 0$ as $s \to \infty$, where $A_0 = X \setminus \bigcup_{i=1}^q A_i)$; $U_{r_s}(x_{j,i}^s) = \{x : \|x - x_{j,i}^s\| < r_s\}$.

Let $R_s = \bigcup_{i=0}^q R_s^i$, $R_s = (x_1^s, x_2^s, \dots, x_{n_s}^s)$, where $n_s = (q+1) \cdot l_s$, $x_{(j+1)\cdot l_s + i}^s = x_{i,j}^s$, $i = 1, 2, \dots, l_s$; $j = 0, 1, \dots, q$.

Then, let us solve the following linear programming problem

$$\sum_{j=1}^{n_s} [g_1(x_j) - t_{s-1} g_2(x_j)] p_j \to \min \tag{89}$$

subject to

$$\sum_{j=1}^{n_s} p_j = 1, \tag{90}$$

$$\sum_{j=1}^{n_s} f_i(x_j) p_j \le a_i \qquad i = 1, 2, \dots, m, \tag{91}$$

$$p_j \ge 0, \qquad x_j \in R_s, \qquad j = 1, 2, \dots, n_s.$$

Let a distribution function $H_s(x)$ be a solution of this problem. Then, let us calculate

$$t_s = \frac{J_1(H_s)}{J_2(H_s)} \tag{92}$$

and go to the $s+1$th iteration.

As an initial approximation, take the value

$$t_0 = \frac{J_1(H_0)}{J_2(H_0)},$$

where H_0 is an arbitrary distribution function from $K_m(X)$ with at most $m+1$ steps.

The following theorem holds.

Theorem 11.

Assume that

1. Functions $g_1(x), f_i(x)$, $i = 1, \dots, m$, are lower semi-continuous and $g_2(x)$ is upper semi-continuous and bounded on the set X.

2. X is a compact set from R^n.

3. $\exists \tilde{H}(x)$ such that the following restrictions hold

$$\int_X d\tilde{H}(x) = 1,$$

$$\psi_i(\tilde{H}) = \int_X f_i(x) d\tilde{H}(x) < a_i, \tag{93}$$
$$i = 1, 2, \dots, m,$$

Then, limit of any convergent subsequence $\{H_{s_k}\}$ of the sequence $\{H_s\}$ belongs to the set X^*, where

$$X^* = \{H^*(x) : \varphi(H^*) = \min_{H(x) \in K_m(X)} \varphi(H)\}$$

and

$$\lim_{s \to \infty} t_s = \min_{H(x) \in K_m(X)} \frac{\int_X g_1(x) dH(x)}{\int_X g_2(x) dH(x)}$$

Proof.

The sequence $t_1, \dots . t_s, \dots$ is lower bounded since for all s the following inequality holds

$$t_s \geq \inf_{H(x) \in K_m(X)} \frac{J_1(H}{J_2(H)}.$$

Besides, the sequence is non-increasing. Indeed, from (89) and (92), we have

$$J_1(H_s) - t_{s-1} J_2(H_s) \leq J_1(H_{s-1}) - t_{s-1} J_2(H_{s-1}) = 0.$$

Since, according to our assumption, $J_2(H_s) \geq \gamma > 0$,

$$t_{s-1} \geq \frac{J_1(H_s)}{J_2(H_s)} = t_s.$$

Therefore, there exists the limit

$$\lim_{s \to \infty} t_s = t_*.$$

If we show that

$$\min_{H(x) \in K_m(X)} [J_1(H) - t_* J_2(H)] = J_1(H^*) - t_* J_2(H^*) = 0\,,$$

then

$$t_* = \min_{H(x) \in K_m(X)} \frac{J_1(H)}{J_2(H)} = \frac{J_1(H^*)}{J_2(H^*)}.$$

Suppose the opposite is true, then

$$J_1(H^*) - t_* J_2(H^*) < 0.$$

Consequently, there exists a value $\varepsilon_0 > 0$ such that

$$J_1(H^*) - t_* J_2(H^*) \leq -\varepsilon_0. \tag{94}$$

According to our assumptions and the structure of the algorithm, we can show in a way similar to that of section 1 that for sufficiently large s there exist distribution functions $\bar{H}_s(x) \in K_m(R_s)$, such that

$$J_1(H^*) - t_* J_2(H^*) > J_1(\bar{H}_s) - t_* J_2(\bar{H}_s) - \frac{\varepsilon_0}{8}, \tag{95}$$

where $K_m(R_s) \subset K_m(X)$ and $K_m(R_s)$ - the set of all distribution functions whose support belong to R_s.

Since $t_s \to t_*$ as $s \to \infty$, for sufficiently large s, we have

$$J_1(\bar{H}_s) - t_* J_2(\bar{H}_s) > J_1(\bar{H}_s) - t_{s-1} J_2(\bar{H}_s) - \frac{\varepsilon_0}{8}. \tag{96}$$

Since $H_s(x)$ is a solution of the problem

$$\min_{H(x)\in K_m(R_s)} [J_1(H) - t_{s-1} J_2(H)],$$

for sufficiently large numbers s, the following inequality holds

$$\begin{aligned} J_1(\bar{H}_s) &- t_{s-1} J_2(\bar{H}_s) \geq J_1(H_s) - t_{s-1} J_2(H_s) \geq \\ &\geq J_1(H_s) - t_s J_2(H_s) - \frac{\varepsilon_0}{8} = -\frac{\varepsilon_0}{8} \end{aligned} \tag{97}$$

Therefore, we have

$$\begin{aligned} 0 > -\varepsilon_0 &\geq J_1(H^*) - t_* J_2(H^*) > J_1(\bar{H}_s) - t_* J_2(\bar{H}_s) - \frac{\varepsilon_0}{8} > \\ &> J_1(\bar{H}_s) - t_{s-1} J_2(\bar{H}_s) - \frac{2\varepsilon_0}{8} > J_1(H_s) - t_{s-1} J_2(H_s) - \frac{2\varepsilon_0}{8} > -\frac{3\varepsilon_0}{8} \end{aligned} \tag{98}$$

From the obtained contradiction, we have

$$\min_{H(x)\in K_m(X)} [J_1(H) - t_* J_2(H)] = 0.$$

Therefore,

$$t_* = \min_{H(x)\in K_m(X)} \frac{J_1(H)}{J_2(H)}.$$

Thus,

$$\lim_{s\to\infty} t_s = \lim_{s\to\infty} \varphi(H_s) = \min_{H\in K_m(X)} \varphi(H). \tag{99}$$

(99) implies that the limit of any convergent subsequence $\{H_{s_k}\}$ of the sequence $\{H_s\}$ belongs to the set X^*. The theorem is proved. ◇

Remarks.

1. If at least one of the functions $f_i(x)\ i = 1, \ldots, m$, or the function $\frac{g_1(x)}{g_2(x)}$ is upper semi-continuous, then the algorithm converges to the infinimum of the functional $\varphi(H)$.

2. The algorithm can be slightly modified to optimize the functional (85) under m fixed first moments of distribution functions $H(x)$. For convergence of appropriate algorithm, it is necessary that the point $(\mu_1, \ldots, \mu_m)$ be an inner point of the moment space D^m. If the function $\frac{g_1(x)}{g_2(x)}$ is lower semi-continuous, then the algorithm converges to the minimum of the functional (85), and if it is upper semi-continuous, then the algorithm converges to infinimum of the (85).

Proof of these statements can be carried out in the same manner as in the previous sections.

7 Results of numerical calculations

Further, for a particular application, we calculate upper and lower bounds for Bayes estimates corresponding to the same available prior information. The case study was done for binomial sampling. We suppose that available prior information allows us to estimate the values of the first and the second moments of the prior distribution.

In order to calculate the upper and lower bounds for the Bayes estimates, we should solve the problems (26), (22) - (23) and (27), (22) - (23), where $a = 0$, $b = 1$, $i = 2$ and μ_1 is the first moment and μ_2 is the second moment of the prior distribution. Given μ_1 and σ^2, the second moment is calculated as follows

$$\mu_2 = mu_1^2 + \sigma^2.$$

Calculations were performed for different values of the mean μ_1 and variance σ^2 of the prior distribution, the number of trials n, and the number of failures s. Upper and lower bounds for Bayes estimates calculated for fixed values of these parameters were compared with the Bayes estimate (15) derived under the beta prior distribution, which fits selected values of parameters.

Range of values of the parameters was chosen in such a manner that they were close to that encountered in PSA of Nuclear Power Plants.

Table 3 gives results of numerical calculations of upper and lower bounds for Bayes estimates derived under any prior distribution having the same mean $\mu = 0.01$ and variance $\sigma^2 = 1.0E - 4$. The number of trials was set $n = 220$ and the number of failures s is variated from 0 to 10. Obtained results are compared with the Bayes estimate (beta estimate) derived under the beta prior distribution, which has the same mean and variance.

Table 3: Upper and lower bounds for Bayes estimates and Bayes estimate derived under beta prior distribution (beta estimate) for $\mu = 0.01$, $\sigma^2 = 1.0E-4$ and $n = 220$

s	*lower bound*	*beta estimate*	*upper bound*
0	4.23E-04	3.08E-03	9.90E-03
1	3.46E-03	6.23E-03	1.28E-02
2	5.85E-03	9.37E-03	1.73E-02
3	7.20E-03	1.25E-02	2.09E-02
4	7.97E-03	1.57E-02	2.33E-02
5	8.42E-03	1.88E-02	2.73E-02
6	8.79E-03	2.19E-02	3.29E-02
7	8.97E-03	2.51E-02	4.08E-02
8	9.06E-03	2.82E-02	5.09E-02
9	9.30E-03	3.14E-02	6.24E-02
10	9.47E-03	3.45E-02	7.56E-02

Table 3 demonstrates that for small numbers of failures ($s = 0, 1$), the range between upper and lower bounds is large and the discrepancy between upper bounds and Bayes estimates derived under beta prior distribution is the largest.

The dependence of upper and lower bounds as well as the beta estimate of the variance is presented in Table 4.

Table 4: Dependence of upper and lower bounds as well as the beta estimate of the variance for $\mu = 0.01, n = 100$ and $s = 1$

Variance	1.00E-06	5.00E-06	1.00E-05	5.00E-05	1.00E-04
Lower bound	9.9715E-03	9.8565E-03	9.7104E-03	8.5987E-03	7.3750E-03
Beta estimate	1.0000E-02	1.0000E-02	1.0000E-02	1.0000E-02	1.0000E-02
Upper bound	1.0076E-02	1.0393E-02	1.0805E-02	1.3428E-02	1.4983E-02

Table 4 demonstrates that the range between upper and lower bounds as well as the discrepancy between upper bounds and Bayes estimates derived under the beta prior distribution increases with the increase of variance σ^2.

Results of calculations presented in the tables show that there are ranges of parameters of practical significance where Bayes estimates are sensitive to the type of a prior distribution function that has the same first two moments. In these cases, using the beta distribution as a prior may lead to a significant underestimation of the probability of failure in binomial sampling. For this reason, the conservative Robust Bayes estimate should be used.

References

[1] Yu.Ermoliev, Method for Stochastic Programming in Randomized Strategies, Kibernetica,1 (1970), pp.3-9. (In Russian).

[2] Yu.Ermoliev,Methods of Stochastic Programming, Nauka, Moscow, 1976. (In Russian)

[3] A.N. Golodnikov and L.S. Stoikova, The Determination of the Optimal Period of Preventive Replacement on the Basis of Information on Mathematical Expectation and Time Dispersion of the Trouble-Free Operation of a System. (Russian) Kibernetika (Kiev) 1978, No.3, 110-118, English translation: Cybernetics 14 (1978), No.3, 431-440 (1979).

[4] A.N. Golodnikov, Minimax Approach to Bayes Estimation. Operation Research (Models, Systems, Solutions) No 7, pp. 36-41, 1, Akad. Nauk. SSSR, Vycisl. Centr. Moscow (1979) (In Russian).

[5] D.R. Insua and F. Ruggeri, Eds., Robust Bayesian Analysis. Lecture Notes in Statistics. Springer Verlag (2000).

[6] S.Karlin and W.J.Studden, Tchebycheff Systems: With Applications in Analysis and Statistics. Wiley Interscience. New York, 1966.

[7] J.H.B.Kemperman, The General Moment Problem, a Geometric Approach. Ann. Math. Statist., 39 (1968), pp.93-122.

[8] M.Krein and A.Nudelman, The Markov Moment Problem and Extremal Problems, Trans. Math. Monographs, American Mathematical Society, Providence, RI,1977

[9] Harry F. Martz and Ray A. Waller. Bayesian reliability Analysis. Krieger Publishing Company, Malabar, Florida, 1991.

[10] M. Lavine, at al., Linearization techniques in Bayesian Robustness. In.: Robust Bayesian Analysis. D.R. Insua and F. Ruggeri, Eds., Lecture Notes in Statistics. Springer Verlag (2000).

[11] R. J-B. Wets, Statistical estimation from an optimization viewpoint. Annals of Operations Research 85 (1999)79-101.

Sensitivity Analysis of Worst-Case Distribution for Probability Optimization Problems[1]

Yu.S.Kan
Space Research Institute,
Russian Academy of Scienses

A. I. Kibzun (kibzun@k804.mainet.msk.su)
Space Research Institute,
Russian Academy of Scienses

Abstract

The paper summarizes the known results on the uniformity principle. The principle establishes that under certain assumptions the worst-case distribution in probability optimization models is uniform. The uniformity principle has an important significance for engineering practice. The paper also presents new results concerning the sensitivity of the uniformity principle with respect to slight violation of basic assumptions for the validity of the principle.

1 Introduction

The stochastic programming models include random parameters whose distribution can be incompletely known in practice. For this reason there is a problem of chosing a distribution which is least favourable for optimization of a criterion. In this situation the researcher needs to take into account the worst case when solving the stochastic programming problem. Similar problems are studied in theory of statistical decisions [13]. Traditionally the performance index for problems studied in that theory is expectation of a loss function. But in applications another statistical characteristic namely the probability that some constraints are satisfied is often used. For example,

[1]Supported by Russian Science Foundation, grant number 99-01-01033

S.P. Uryasev (ed.), Probabilistic Constrained Optimization, 132–147.
© 2000 *Kluwer Academic Publishers.*

probability of a successful landing of the aircraft, probability of a successful guidance of the controllable object and probability of obtaining a desired profit are such criteria for optimization models [7].

Sure, the probability can be written as expectation by using the set indicator function however the discontinuity of that function leads to the necessaty of using special research methods.

In practice the researcher often knows only that the random parameters belong to a bounded feasible domain. In this case one usually believes that the uniform distribution over that domain is "worst" for his problem. The satisfactory mathematical justification of such a choice was absent up to the present time. Moreover, in theory of statistical decisions there are known examples where the worst-case distribution for calculation of the expected loss function is concentrated in boundary points of the feasible domain [13].

Nevertheless different classes of problems with probabilistic objectives were described in [2, 6] where the worst-case distribution was established to be uniform, i.e. *a uniformity principle* was to be valid. This term was first introduced in [2]. In those papers the uniformity principle was proved under different assumptions about the symmetry and monotonocity of both the unknown probability density functions (p.d.f.) and the loss function. The original proof of the uniformity principle presented in [2] assumes implicitly that the unknown p.d.f. is a proper function. In [5] the main results of [2] were generalized for an improper case. All the known results on the subject are summarized below in section 3. Note that the finite-dimensional *maximization* problem for the probability function under similar assumptions about both an integration domain (given by the loss function) and the p.d.f. properties was investigated in [1, 4, 10] where some explicit solutions and probability inequalities were found. We deal below with another problem, namely with an infinite-dimensional *minimization* problem on a non-parametric functional class of feasible densities. The solution to that minimization problem is called *the worst-case distribution*. In application to results obtained in [1, 4, 10] the worst-case distribution can be considered as a component of a solution to a minimax problem corresponding to the statistically uncertain situation. In that minimax problem the maximization is performed with respect to some controllable parameters, and the minimization does with respect to unknown distribution.

Since the conditions for the validity of the uniformity principle are exotic for practice, its sensitivity analysis with respect to change of those conditions seems to be important. Some results in this direction were obtained in [5]. Section 4 of the paper presents new results when insignificant violation of those conditions leads to an insignificant change of the guaranteed value of the probabilistic criterion.

2 Problem statement

Let Π be a family of densities $p(x): \mathbb{R}^n \to \mathbb{R}^1$ and $A \subset \mathbb{R}^n$ be a bounded measurable set such that $p(x) = 0$ for every $x \in \mathbb{R}^n \backslash A$ and every $p(\cdot) \in \Pi$. In the sequel we call A the *support* for brevity. Note that $p(x)$ may be equal to 0 for some $x \in A$. Let S be a subset of A and $\mathcal{P}$ be a probability measure which is generated by $p(x)$. Consider the probability objective function $P_S(p(\cdot)) \stackrel{\text{def}}{=} \mathcal{P}(S)$. Suppose that the p.d.f. $p(x)$ is unknown, i.e. $p(\cdot) \in \Pi$. If we intend to maximize this probability with respect to some parameter $v \in V$ then we need to deal with a worst-case p.d.f. $p^*(\cdot) \in \Pi$ such that

$$P_S(p^*(\cdot)) \leq P_S(p(\cdot)) \quad \text{for all } p(\cdot) \in \Pi.$$

If such a p.d.f. $p^*(x)$ was found, we would know the most unfavourable case and could solve the following optimization problem

$$P_S(p^*(\cdot), v) \longrightarrow \min_{v \in V}.$$

In [2, 6] some conditions are obtained under which the uniformity principle is valid, i.e. the worst-case p.d.f. $p^*(x)$ coincides with the p.d.f. of the uniform distribution over A :

$$p^*(x) = u(x) \stackrel{\text{def}}{=} 1/\mu_n(A),$$

where μ_n denotes the Lebesgue measure in $\mathbb{R}^n$.

We shall study the sensitivity of the uniformity principle with respect to violation of conditions from [2, 6], i.e. evaluate

$$\delta \stackrel{\text{def}}{=} P_G(u(\cdot)) - \inf_{p(\cdot) \in \widetilde{\Pi}} P_G(p(\cdot)),$$

where the set G differs slightly from S satisfying conditions from [2, 6], i.e. $\mu_n(G \triangle S) < \varepsilon_2$, $G \triangle S \stackrel{\text{def}}{=} (G \backslash S) \cup (S \backslash G)$, where ε_2 is a small positive number. The family $\widetilde{\Pi}$ of densities $p(\cdot)$ also differs insignificantly from the family Π which satisfies conditions from [2, 6]. In particular, the support A can differ slightly from the cube

$$I_n \stackrel{\text{def}}{=} \{x \in \mathbb{R}^n : \ |x_i| \leq 1/2, \ i = \overline{1, n}\},$$

where $\mu_n(A \triangle I_n) = \varepsilon_3$.

3 The uniformity principle

In this section all the known results concerning the uniformity principle are collected. This principle means that the worst-case distribution in the maximization problem for the probability function is uniform on the support A. The analysis and comparison of the mentioned results are performed. This section is preliminary in order to analyse the sensitivity of the uniformity principle in the next section with respect to violation of the conditions of its validity.

Everywhere in this section we assume that the support A is the cube

$$A = I_n \stackrel{\text{def}}{=} \{x \in \mathbb{R}^n : \ |x_i| \leq 1/2, \ i = \overline{1,n}\}. \tag{1}$$

We first formulate an auxiliary result from [5] which generalizes an assertion from [2].

Lemma 1. *Let the following conditions hold:*

(i) *S is a convex subset of the support I_n;*

(ii) *every p.d.f. $p(\cdot) \in \Pi$ has the structure of the product*

$$p(x) = p_1(x_1) \cdots p_n(x_n), \tag{2}$$

where each univariate p.d.f. $p_i(x_i)$ is even and quasi-concave on $I_1 \stackrel{\text{def}}{=} [-1/2, 1/2]$.

Then

$$\inf_{p(\cdot)\in\Pi} P_S(p(\cdot)) = \inf_{p(\cdot)\in U} P_S(p(\cdot)), \tag{3}$$

where the family $U \subset \Pi$ differs from Π so that each p.d.f. $p_i(x_i)$ in expression (2) is the p.d.f. of a uniform distribution over a symmetric interval belonging to I_1.

Remark 1. We recall that a function $p(x)$ is *quasi-concave* on a convex set S if

$$p(\lambda x_1 + (1-\lambda)x_2) \geq \min\{p(x_1), p(x_2)\}$$

for all $\lambda \in [0,1]$, $x_1 \in S$ and $x_2 \in S$. The function $p(x)$ is *even* if $p(x) = p(-x)$ for all $x \in \mathbb{R}^1$.

Remark 2. Relation (3) can be symbolically denoted by $\Pi \Rightarrow U$, since the essence of lemma 1 is replacement of the initial class Π by the more simple family U. The original proof of lemma 1 is presented in [2] and based on the following implications:

$$\Pi \Rightarrow \Pi_a \Rightarrow U,$$

where $\Pi_a \subset \Pi$ differs from Π by the fact that each p.d.f. $p_i(x_i)$ in (2) is piece-wise constant. We should emphasize that justification of the implication $\Pi \Rightarrow \Pi_a$ is based on piece-wise constant approximations of even quasi-concave functions $p_i(x_i)$. Authors of [2] asserted that such functions are integrable in the Riemann sense and therefore their piece-wise constant approximations are admissible. A confusion is possible here because the following univariate p.d.f.

$$f(x) = \begin{cases} \frac{1}{2\sqrt{2|x|}}, & \text{if } x \in I_1; \\ 0, & \text{if } x \notin I_1 \end{cases} \tag{4}$$

is even and quasi-concave. However $p(x)$ is unbounded and therefore non-integrable in the Riemann sense [8]. The deal is that $f(0) = +\infty$, i.e. $f(x)$ is an improper function. Such functions are the usual subject under consideration in Convex Analysis [12]. The question about their piece-wise constant approximation can be solved but requires the special study. In [2] lemma 1 is formulated in another way. Instead of conditions of evenness and quasi-concavity there is written in the original assertion that each p.d.f. $p_i(x_i)$ has the form

$$p_i(x_i) = q_i(|x_i|), \tag{5}$$

where $q_i(\cdot)$ is a non-increasing function. Any notion of Convex Analysis is not used here and for this reason the improper case is excluded by default out the consideration. Formally speaking, function (4) is not feasible for the original formulation [2] of lemma 1. Nevertheless, lemma 1 is valid if the improper case is taken into account. The proof of lemma 1 in the formulation of the present paper with assumption that functions $p_i(x_i)$ can be improper of type (4) without the intermediate implication $\Pi \Rightarrow \Pi_a$ is presented in [5].

Remark 3. Let $N = \{1, 2, \ldots, n\}$ be the set of subscripts and $M = \{i_1, \ldots, i_k\} \subset N$. Consider the *section* $S_M(c_1, \ldots, c_k)$ of the set S with respect to variables $x_{i_1}, \ldots, x_{i_k}$. This section is the projection of the set

$$\{x : \ x \in S, \ x_i = c_i, \ i \in M\}$$

onto the space $\mathbb{R}^{n-k}$ of variables x_j, $j \in N \backslash M$. In [5] the author found that lemma 1 is also valid if the convexity condition for the set S is replaced by the assumption that

$$S = \{x \in \mathbb{R}^n : \ g(x) \le 0\},$$

where $g(x)$ is a continuous function such that for each $k = \overline{0, n-1}$, each subset $M \subset N$ containing exactly k elements and every parameters $c_1, \ldots, c_k$ it follows that

$$\mu_{n-k} B_M(c_1, \ldots, c_k) = 0$$

with

$$B = \{x \in \mathbb{R}^n : \ g(x) = 0\}.$$

Now we formulate an assertion from [5] which generalizes a theorem from the paper [2] where the notion "uniformity principle" was first introduced.

Theorem 1. *Let the following conditions hold:*

(i) *S is a convex subset of the support I_n;*

(ii) *$S = -S$, i.e. S is symmetric with respect to the origin;*

(iii) *every function of the family Π has structure (2), where each function $p_i(x_i)$ is even and quasi-concave.*

Then

$$\inf_{p(\cdot)\in\Pi} P_S(p(\cdot)) = P_S(u(\cdot)), \tag{6}$$

where $u(x)$ is the p.d.f. of the uniform distribution over I_n.

Remark 4. The original proof of theorem 1 is based on application of lemma 1 and the Brunn-Minkowski inequality [9] and the one is enough complicated. The author of [6] shown that for $n \leq 2$ theorem 1 follows easily from the following lemma.

Lemma 2. *Let the following conditions hold:*

(i) *the function $p(x): [0,1/2] \to [-\infty,+\infty]$ belongs to the class Π of the non-negative non-increasing functions defined on $[0,1/2]$ such that*

$$\int_0^{1/2} p(x)\,\mathrm{d}x = c = \mathrm{const}; \tag{7}$$

(ii) *the function $f(x): [0,1/2] \to \mathbb{R}^1$ is non-negative and non-increasing on $[0,1/2]$.*

Then for every $s \in [0,1/2]$ the function $u(x) \equiv c$ on $[0,1/2]$ is a solution to the problem

$$\int_0^s u(x)f(x)\,\mathrm{d}x = \min_{p(\cdot)\in\Pi} \int_0^s p(x)f(x)\,\mathrm{d}x. \tag{8}$$

Remark 5. Any even quasi-concave function is non-decreasing on $[-1/2,0]$ and non-increasing on $[0,1/2]$. Hence, for $n=1$ we have

$$P_S(p_1(\cdot)) = 2\int_0^s p_1(x_1)\,\mathrm{d}x_1,$$

where $s = \sup\{x_1 : x_1 \in S\}$. Therefore theorem 1 easily follows from lemma 2 for $n=1$ with $f(x) \equiv 1$.

Let now $n=2$. In this case we have

$$P_S(p(\cdot)) = \int\!\!\int_S p_1(x_1)p_2(x_2)\,\mathrm{d}x_1\mathrm{d}x_2 = \int_{-s}^{s} p_1(x_1)\,\mathrm{d}x_1 \int_{S_M(x_1)} p_2(x_2)\,\mathrm{d}x_2, \tag{9}$$

where $M = \{1\}$, $s = \sup\{x_1 : x \in S\}$. By the virtue of the symmetry of S we have $S_M(x_1) = -S_M(-x_1)$. Hence, from the evenness of $p_2(x_2)$ it follows that the function

$$f(x_1) \stackrel{\mathrm{def}}{=} \int_{S_M(x_1)} p_2(x_2)\,\mathrm{d}x_2 \tag{10}$$

is also even. This function is a probability measure of the section $S_M(x_1)$. Therefore $f(x_1)$ is bounded. In [6] a geometrically visual proof of the fact that the function

$f(x_1)$ is not decreasing with respect to $x_1 \geq 0$ was presented. Thus theorem 1 is an easy consequence of lemma 2 also for the case $n = 2$.

A simple proof of theorem 1 is presented in [5] for an arbitrary n. Let us illustrate its essence for $n = 2$. According to lemma 1, in (6) we can deal with the class U instead of Π ($\Pi \Rightarrow U$). Since $U \subset \Pi$, it follows that the above conclusion on the evenness and boundedness of the function $f(x_1)$ is valid. Furthermore, without loss of generality we can assume that the set S is closed. Hence

$$S_M(x_1) = \{x_2 : \gamma_S(x_1, x_2) \leq 1\},$$

where $\gamma_S(x_1, x_2) = \min\{\lambda \geq 0 : x \in \lambda S\}$ is the Minkowski function. Recall that the set S is convex and centrally symmetric. In this case $\gamma_S(x_1, x_2)$ is convex jointly in x_1 and x_2 [8]. We can write

$$f(x_1) = \int_{\gamma_S(x_1,x_2)\leq 1} p_2(x_2)\, dx_2,$$

i.e. $f(x_1)$ is a probability function. The p.d.f. $p_2(\cdot) \in U$ corresponds to the uniform distribution over a convex set and according to [3] generates a quasi-concave probability measure $\mathcal{P}$, i.e. such that

$$\mathcal{P}(\lambda S + (1 - \lambda)D) \geq \min\{\mathcal{P}(S), \mathcal{P}(D)\}$$

for all convex sets $S, D \subset \mathbb{R}^n$ and every $\lambda \in [0, 1]$. For this reason the function $f(x_1)$ is quasi-concave [11] (see also [7]). Thus $f(x_1)$ is even and quasi-concave. Hence $f(x_1)$ is non-increasing in $x_1 \geq 0$ and application of lemma 2 completes the proof.

Remark 6. Theorem 1 requires hard restrictions to the geometry of the set S and to the structure of the p.d.f. $p(x)$. The author of [6] shown that the uniformity principle is valid also for other classes of sets S and densities $p(x)$. Let us introduce some notions.

Definition 1 [7]. A set $S \subset \mathbb{R}^n$ is called *convex with respect to co-ordinate* x_j if the one-dimensional section $S_M(x_1, \ldots, x_{j-1}, x_{j+1}, \ldots, x_n)$ is convex for all $x \in S$, where $M = N\backslash\{j\}$.

Remark 7. A similar notion (the *unirectangular set*) was introduced in [2].

Definition 2 [7]. A function $g(x)$ defined on a set S which is convex with respect to co-ordinate x_j is called *quasi-concave in* x_j *on* S if for every $x \in S$ the univariate function

$$f(y) = g(x_1, \ldots, x_{j-1}, y, x_{j+1}, \ldots, x_n)$$

is quasi-concave on $S_M(x_1, \ldots, x_{j-1}, x_{j+1}, \ldots, x_n)$, where $M = N\backslash\{j\}$.

Theorem 2 [6]. *Let the following conditions hold:*

(i) *the set $S \subset I_n$ is convex in each co-ordinate and symmetric with respect to each co-ordinate axis;*

(ii) *every p.d.f. $p(\cdot) \in \Pi$ is quasi-concave in each coordinate x_i on the support I_n and for every $x \in I_n$ the function*

$$f(y) = p(x_1, \dots, x_{j-1}, y, x_{j+1}, \dots, x_n)$$

attains a maximum at point $y = 0$.

Then the p.d.f. $u(x)$ of the uniform distribution over I_n solves problem (6).

Remark 8. We can observe that the conditions of theorem 2 don't require the p.d.f. $p(\cdot) \in \Pi$ to be the product of independent densities and $p(\cdot)$ to be even in each variable. Moreover, the set S can be non-convex, e.g. the set

$$S = \{(x_1, x_2) : |x_1| \leq 1/2, |x_2| \leq 1/4\} \cup \{(x_1, x_2) : |x_1| \leq 1/4, |x_2| \leq 1/2\}$$

is feasible but not convex.

Remark 9. We should also emphasize that theorem 1 is not a consequence of theorem 2, since the set S in theorem 1 can be asymmetric with respect to the co-ordinate axes.

Theorem 3 [6]. *Let the following conditions hold:*

(i) *the set $S \subset I_n$ is convex with respect to each co-ordinate and*

$$0 \in S_M(x_1, \dots, x_{j-1}, x_{j+1}, \dots, x_n)$$

for each $j \in N$ and every $x \in S$, where $M = N \backslash \{j\}$;

(ii) *every p.d.f. $p(\cdot) \in \Pi$ is even and quasi-concave in each co-ordinate on the support I_n.*

Then the p.d.f. $u(x)$ of the uniform distribution over I_n solves problem (6).

Remark 10. Theorem 3 differs from theorem 2 by absence of the symmetry reguirement for S. However the evenness of $p(x)$ with respect to each co-ordinate is additionally assumed as well as in theorem 1.

Remark 11. Unlike theorem 1, theorem 3 does not require $p(x)$ to have the product structure (2) of densities, as well as theorem 2. For example, the function

$$p(x_1, x_2) = \begin{cases} c \cdot e^{-|x_1 x_2|}, & (x_1, x_2) \in S; \\ 0, & (x_1, x_2) \notin S \end{cases}$$

for appropriate value of the normalizing constant c satisfies condition **(ii)** of both theorem 2 and theorem 3 but $p(x_1, x_2)$ cannot be expressed as product of independent densities.

Remark 12. The set S in theorem 3 can be convex and symmetric with respect to the origin as is necessary for theorem 1, e.g.

$$S = \{(x_1, x_2) : \ |x_1| \le 1/2, \ |x_2| \le 1/4, \ |x_1 + 2x_2| \le 1/2\}.$$

Remark 13. Meanwhile, in theorem 1 the requirements to the product structure (2) can hardly be weakened. For example, let the set S have the form

$$S = \{(x_1, x_2) : \ |x_1| \le 1/2, \ |x_2| \le 1/2, \ |x_1 + x_2| \le 1/4\}$$

satisfying condition **(i)** of theorem 1 but not satisfying condition **(i)** of theorem 2. Then for the p.d.f. $u(x)$ of the uniform distribution over I_n we obtain

$$P_S(u(\cdot)) = \mu_2(S) = \frac{7}{16}.$$

Consider now the p.d.f. $\tilde{u}(x)$ of the uniform distribution over the set

$$D = \{(x_1, x_2) : \ |x_1| \le 1/2, \ |x_2| \le 1/2, \ |x_1 - x_2| \le 1/4\}.$$

In this case

$$P_S(\tilde{u}(\cdot)) = \frac{\mu_2(S \cap D)}{\mu_2(D)} = \frac{2}{7},$$

where $S \cap D$ is the square $\{(x_1, x_2) : \ |x_1 - x_2| \le 1/4, \ |x_1 + x_2| \le 1/4\}$. Thus the p.d.f. $\tilde{u}(x)$ is "worse" than $u(x)$ with respet to the probabilistic criterion under consideration.

Remark 14. In conclusion of this section we note that theorems 1–3 can be generalized for the case where the conditions of these theorems are satisfied for new variables obtained by a non-degenerated linear transformation (see [6]).

4 Sensitivity analysis

Let us study the sensitivity of the uniformity principle with respect to weak violations of the conditions of theorems 1–3.

Lemma 3. *Let the following conditions hold:*

(i) *the set S and the family Π of densities $p(\cdot)$ satisfy the conditions of any above theorem;*

(ii) *the class* $\widetilde{\Pi}$ *of functions* $\widetilde{p}(\cdot)$ *is defined in the following way:*

$$\widetilde{p}(x) = (1-\varepsilon_1)p(x) + \varepsilon_1 q(x), \tag{11}$$

where $p(\cdot) \in \Pi$, $0 < \varepsilon_1 < 1$, $q(\cdot) \in \Delta\Pi$ *and* $\Delta\Pi$ *consists of arbitrary densities* $q(\cdot)$ *such that* $q(x) = 0$ *for all* $x \in \mathbb{R}^n \backslash I_n$.

Then

$$\delta_1 \stackrel{\text{def}}{=} P_S(u(\cdot)) - \inf_{\widetilde{p}(\cdot)\in\widetilde{\Pi}} P_S(\widetilde{p}(\cdot)) \le \varepsilon_1 \mu_n(S) \le \varepsilon_1,$$

where $u(x)$ *is the p.d.f. of the uniform distribution over* I_n.

Proof. It is evident that every function $\widetilde{p}(\cdot) \in \widetilde{\Pi}$ is a p.d.f., since $\widetilde{p}(\cdot)$ is non-negative and satisfies the normalization condition. We have

$$\inf_{\widetilde{p}(\cdot)\in\widetilde{\Pi}} P_S(\widetilde{p}(\cdot)) = (1-\varepsilon_1) \inf_{p(\cdot)\in\Pi} P_S(p(\cdot)) + \varepsilon_1 \inf_{q(\cdot)\in\Delta\Pi} P_S(q(\cdot)). \tag{12}$$

Let, for example, the conditions of theorem 1 hold for S and Π. Then a minimum in the first summand of the right-hand side of (12) is attained at the uniform p.d.f. $u(\cdot)$. It follows that

$$\inf_{p(\cdot)\in\Pi} P_S(p(\cdot)) = P_S(u(\cdot)) = \mu_n(S).$$

Since $P_S(q(\cdot)) \ge 0$ for all $q(\cdot) \in \Delta\Pi$, we conclude that

$$\delta_1 \le P_S(u(\cdot)) - (1-\varepsilon_1)P_S(u(\cdot)) = \varepsilon_1 \mu_n(S).$$

Finally, taking into account the inequality $\mu_n(S) \le 1$ which holds by the virtue of the inclusion $S \subset I_n$ we obtain the assertion of lemma 3.

Remark 15. Let now the class $\Delta\Pi$ consists of bounded densities $q(\cdot)$ such that

$$q(x) \le c \le \frac{1}{\mu_n(I_n \backslash S)} \quad \text{for all } x \in I_n,$$

where $\mu_n(I_n \backslash S) > 0$. Denote by $q^*(\cdot)$ a solution to the optimization problem

$$P_S(q^*(\cdot)) = \min_{q(\cdot)\in\Delta\Pi} P_S(q(\cdot)).$$

Then from the normalization condition for $q^*(x)$ on I_n it follows that $q^*(x) \equiv c$ for all $x \in I_n \backslash S$. Hence

$$P_S(q^*(\cdot)) = 1 - c\mu_n(I_n \backslash S).$$

In this case we have

$$\delta_1 = P_S(u(\cdot)) - (1-\varepsilon_1)P_S(u(\cdot)) - \varepsilon_1 P_S(q^*(\cdot)) = \varepsilon_1(\mu_n(S) + c\mu_n(I_n \backslash S) - 1).$$

In particular, if $c = 1$ then $\delta_1 = 0$. If $c = 1/\mu_n(I_n \backslash S)$ then $\delta_1 = \varepsilon_1 \mu_n(S)$.

Remark 16. Another model for the class $\widetilde{\Pi}$ is considered in [5], where S is assumed to satisfy the conditions of theorem 1. The author of [5] suggested to disturb each univariate p.d.f. $p_i(x_i)$ in (2), i.e.

$$\widetilde{p}_i(x_i) = (1-\varepsilon)p_i(x_i) + \varepsilon q_i(x_i),$$

where $q_i(x_i)$ is an arbitrary p.d.f. which is equal to 0 outside of I_1. The corresponding disturbed class of densities

$$\widetilde{p}(x) = \widetilde{p}_1(x_1) \cdots \widetilde{p}_n(x_n)$$

is denoted by $\widetilde{\Pi}$. In this case

$$\delta_1 = P_S(u(\cdot)) - \inf_{\widetilde{p}(\cdot) \in \widetilde{\Pi}} P_S(\widetilde{p}(\cdot)) \le \varepsilon n \cdot \mu_n(S) + o(\varepsilon), \tag{13}$$

where $\frac{o(\varepsilon)}{\varepsilon} \to 0$, as $\varepsilon \to 0$, and

$$\int_{-1/2}^{1/2} |p_i^*(x_i) - 1| \, dx_i \le 2\varepsilon,$$

where $p^*(x) = p_1^*(x_1) \cdots p_n^*(x_n)$ is a generalized p.d.f. at which a minimum in the left-hand side of (13) is attained.

Lemma 4. *Let the following conditions hold:*

(i) *the set S and the family Π of densities satisfy the conditions of any above theorem;*

(ii) *a set G is such that $S \subset G \subset I_n$ and $\mu_n(G \backslash S) < \varepsilon_2$.*

Then

$$\delta_2 \stackrel{\text{def}}{=} P_G(u(\cdot)) - \inf_{p(\cdot) \in \Pi} P_G(p(\cdot)) \le \mu_n(G \backslash S) < \varepsilon_2.$$

Proof. Since $S \subset G$, it follows that

$$P_G(p(\cdot)) \ge P_S(p(\cdot))$$

for every $p(\cdot) \in \Pi$. Hence

$$\inf_{p(\cdot) \in \Pi} P_G(p(\cdot)) \ge \inf_{p(\cdot) \in \Pi} P_S(p(\cdot)). \tag{14}$$

By the virtue of **(i)** the right-hand side of (14) is equal to $P_S(u(\cdot))$. Therefore from (14) it follows that

$$\delta_2 = P_G(u(\cdot)) - \inf_{p(\cdot) \in \Pi} P_G(p(\cdot)) \le P_G(u(\cdot)) - P_S(u(\cdot)) = P_{G \backslash S}(u(\cdot)) = \mu_n(G \backslash S) < \varepsilon_2.$$

The lemma is proved.

Remark 17. As shown in [5], if in addition to the conditions of lemma 4 a set H satisfying all the requirements to S is known and $S \subset G \subset H$ then the bound δ_2 can be evaluated by

$$\delta_2 \leq \mu_n(H \backslash S).$$

Theorem 4. *Let the following conditions hold:*

(i) *the set S and the family Π of densities satisfy the conditions of theorem 1;*

(ii) *the class $\widetilde{\Pi}$ satisfies condition* **(ii)** *of lemma 3;*

(iii) *a set G is such that $S \subset G \subset I_n$ and $\mu_n(G \backslash S) < \varepsilon_2$.*

Then

$$\delta_3 \stackrel{\text{def}}{=} P_G(u(\cdot)) - \inf_{\widetilde{p}(\cdot) \in \widetilde{\Pi}} P_G(\widetilde{p}(\cdot)) \leq \varepsilon_1 + \varepsilon_2,$$

where $u(x)$ is the p.d.f. of the uniform distribution over I_n.

Proof. According to the definition of $\widetilde{\Pi}$ we have

$$\inf_{\widetilde{p}(\cdot) \in \widetilde{\Pi}} P_G(\widetilde{p}(\cdot)) = (1 - \varepsilon_1) \inf_{p(\cdot) \in \Pi} P_G(p(\cdot)) + \varepsilon_1 \inf_{q(\cdot) \in \Delta\Pi} P_G(q(\cdot)). \tag{15}$$

As well as in the proof of lemma 3 we can show that

$$\delta_3 \leq P_G(u(\cdot)) - (1 - \varepsilon_1) \inf_{p(\cdot) \in \Pi} P_G(p(\cdot)).$$

Since $P_G(p(\cdot))$ is bounded by 1, by lemma 4 we obtain

$$\delta_3 \leq P_G(u(\cdot)) - \inf_{p(\cdot) \in \Pi} P_G(p(\cdot)) + \varepsilon_1 = \delta_2 + \varepsilon_1 \leq \varepsilon_2 + \varepsilon_1.$$

The theorem is proved.

Lemma 5. *Let the following conditions hold:*

(i) *the support $A \subset I_n$ is convex with respect to each co-ordinate, $\mu_n(I_n \backslash A) \leq \varepsilon_3$ and*

$$0 \in A_M(x_1, \ldots, x_{j-1}, x_{j+1}, \ldots, x_n)$$

for each $j \in N$ and every $x \in S$, where $M = N \backslash \{j\}$;

(ii) *the family $\widetilde{\Pi}$ of densities satisfies condition* **(ii)** *of theorem 2 with the support A instead of the cube I_n;*

(iii) *the set $S \subset A$ satisfies condition* **(i)** *of theorem 2.*

Then

$$0 \le \delta_4 \stackrel{\text{def}}{=} P_S(\tilde{u}(\cdot)) - \inf_{p(\cdot)\in\widetilde{\Pi}} P_S(p(\cdot)) \le \frac{\varepsilon_3}{1-\varepsilon_3}\mu_n(S) \le \frac{\varepsilon_3}{1-\varepsilon_3}, \tag{16}$$

where $\tilde{u}(\cdot)$ *is the p.d.f. of the uniform distribution over* A.

Proof. Let $u(x)$ be the p.d.f. of the uniform distribution over I_n and Π satisfies condition **(ii)** of theorem 2 where the support is I_n. Since $A \subset I_n$ and every p.d.f. $p(\cdot) \in \widetilde{\Pi}$ satisfies condition **(ii)** of theorem 2, it follows that $p(\cdot) \in \Pi$. Hence $\widetilde{\Pi} \subset \Pi$. Therefore

$$\inf_{p(\cdot)\in\Pi} P_S(p(\cdot)) \le \inf_{p(\cdot)\in\widetilde{\Pi}} P_S(p(\cdot)).$$

By the virtue of theorem 2 we have

$$\inf_{p(\cdot)\in\Pi} P_S(p(\cdot)) = P_S(u(\cdot)).$$

Thus

$$\delta_4 \le P_S(\tilde{u}(\cdot)) - P_S(u(\cdot)) = \frac{\mu_n(S)}{\mu_n(A)} - \mu_n(S) = \frac{1-\mu_n(A)}{\mu_n(A)}\mu_n(S) = \frac{\mu_n(I_n\backslash A)}{1-\mu_n(I_n\backslash A)}\mu_n(S).$$

Since according to condition **(i)** $\mu_n(I_n\backslash A) < \varepsilon_3$ and $\mu_n(S) \le 1$, we have

$$\delta_4 \le \frac{\varepsilon_3}{1-\varepsilon_3}\mu_n(S) \le \frac{\varepsilon_3}{1-\varepsilon_3}.$$

Note that $\tilde{u}(\cdot)$ satisfies condition **(ii)**. For this reason $\tilde{u}(\cdot) \in \widetilde{\Pi}$, hence $\delta_4 \ge 0$. The lemma is proved.

Theorem 5. *Let the following conditions hold:*

(i) *the support* $A \subset I_n$ *satisfies condition* **(i)** *of lemma 5;*

(ii) *the family* Π_1 *of densities satisfies condition* **(ii)** *of theorem 2 with the support* A *instead of the cube* I_n;

(iii) *the family* $\widetilde{\Pi}$ *satisfies condition* **(ii)** *of lemma 3 with the support* A *instead of* I_n, *and* $p(\cdot) \in \Pi_1$;

(iv) *the set* $S \subset A$ *satisfies condition* **(i)** *of theorem 2.*

(v) *the set* G *satisfies condition* **(i)** *of lemma 4 with the support* A *instead of* I_n.

Then

$$\delta_5 \overset{\text{def}}{=} P_G(\tilde{u}(\cdot)) - \inf_{\tilde{p}(\cdot)\in\tilde{\Pi}} P_G(\tilde{p}(\cdot)) \le \varepsilon_1 + \frac{\varepsilon_2+\varepsilon_3}{1-\varepsilon_3}, \tag{17}$$

where $\tilde{u}(x)$ is the p.d.f. of the uniform distribution over the support A.

Proof. Since $\tilde{u}(x)$ is the p.d.f. of the uniform distribution over $A \subset I_n$ and $S \subset G$, we have

$$P_G(\tilde{u}(\cdot)) = \frac{\mu_n(G)}{\mu_n(A)} = \frac{\mu_n(S)+\mu_n(G\backslash S)}{1-\mu_n(I_n\backslash A)}.$$

By the virtue of theorem conditions the following bounds are valid: $\mu_n(G\backslash S) < \varepsilon_2$ and $\mu_n(I_n\backslash A) < \varepsilon_3$. Therefore

$$P_G(\tilde{u}(\cdot)) \le \frac{\mu_n(S)+\varepsilon_2}{1-\varepsilon_3}.$$

Taking into account that every p.d.f. $\tilde{p}(\cdot) \in \tilde{\Pi}$ has the structure (11) for all $x \in A \subset I_n$, where $p(\cdot) \in \Pi_1$ and $q(\cdot)$ is an arbitrary p.d.f. with the support A, in the same way as in the proof of lemma 3 we obtain

$$\inf_{\tilde{p}(\cdot)\in\tilde{\Pi}} P_G(\tilde{p}(\cdot)) \ge (1-\varepsilon_1) \inf_{p(\cdot)\in\Pi_1} P_G(p(\cdot)).$$

From $A \subset I_n$ it follows that $\Pi_1 \subset \Pi$, where Π satisfies condition **(ii)** of theorem 2. For this reason and taking into account $S \subset G$ we obtain

$$\inf_{p(\cdot)\in\Pi_1} P_G(p(\cdot)) \ge \inf_{p(\cdot)\in\Pi} P_S(p(\cdot)).$$

According to theorem 2 the inequality $P_S(u(\cdot)) \le P_S(p(\cdot))$ holds for every p.d.f. $p(\cdot) \in \Pi$, where $u(x)$ is the p.d.f. of the uniform distribution over I_n. Hence

$$\inf_{\tilde{p}(\cdot)\in\tilde{\Pi}} P_G(\tilde{p}(\cdot)) \ge (1-\varepsilon_1)P_S(u(\cdot)) = (1-\varepsilon_1)\mu_n(S).$$

Combining the obtained inequalities we obtain

$$\delta_6 = P_G(\tilde{u}(\cdot)) - \inf_{\tilde{p}(\cdot)\in\tilde{\Pi}} P_G(\tilde{p}(\cdot)) \le \frac{\varepsilon_3}{1-\varepsilon_3}\mu_n(S) + \varepsilon_1\mu_n(S) + \frac{\varepsilon_2}{1-\varepsilon_3}.$$

Recall that $\mu_n(S) \le 1$. Hence inequality (17) is valid.
The theorem is proved.

Lemma 6. *Let the following conditions hold:*

(i) *the support $A \subset I_n$ satisfies condition* **(i)** *of lemma 5;*

(ii) *the family $\tilde{\Pi}$ satisfies condition* **(ii)** *of theorem 3 with the support A instead of the cube I_n;*

(iii) *the set* $S \subset A$ *satisfies condition* **(i)** *of theorem 3.*

(iii) *the set* $S \subset A$ *satisfies condition* **(i)** *of theorem 3.*

Then inequality (16) holds.

Remark 18. The proof of lemma 6 is analogous to the proof of lemma 5 with application of theorem 3 instead of theorem 2.

Theorem 6. *Let the following conditions hold:*

(i) *the support* $A \subset I_n$ *satisfies condition* **(i)** *of lemma 5;*

(ii) *the family* Π_1 *of densities satisfies condition* **(ii)** *of theorem 3 with the support* A *instead of the cube* I_n*;*

(iii) *the family* $\widetilde{\Pi}$ *satisfies condition* **(ii)** *of lemma 3 where the support is* A *and* $p(\cdot) \in \Pi_1$*;*

(iv) *the set* $S \subset A$ *satisfies condition* **(i)** *of theorem 3.*

(v) *the set* G *satisfies condition* **(i)** *of lemma 4 with the support* A *instead of* I_n*.*

Then the inequality (17) holds.

Remark 19. The proof of theorem 6 is analogous to the proof of Theorem 5 with application of theorem 3 instead of theorem 2.

Remark 20. Note that the support in theorem 4 is not disturbed, since the product structure (2) leads automatically to a support of the rectangular type and we can use remark 14.

5 Conclusion

The results of section 3 confirm that practioners were almost correct when assuming that the worst-case distribution for probabilistic optimization models and for the statistical simulation is uniform. Indeed, in practical problems the density is usually unimodal (typical for the Gaussian distribution); it is quite justified in practice to adopt $p^*(x) \equiv const$ as the worst-case distribution, provided that the set, of which the probability measure is sought, is symmetric. The positive solution obtained in section 4 for the sensitivity problem makes such an aproach more reasonable, since insignificant deviations of the conditions of theorems 1–3 lead to a slight change of the worst-case probabilistic criterion value calculated according to the uniformity principle.

References

[1] ANDERSON, T.W. (1955) The Integral of a Symmetric Unimodal Function over a Symmetric Convex Set and Some Probability Inequalities. *Proc. Am. Math. Soc.,* 6, pp.170–176.

[2] BARMISH, B.R. AND C.M.LAGOA (1997) The Uniform Distribution: a Rigorous Justification for its Use in Robustness Analysis. *Math. Control, Signals, Systems.*, v.10., pp.203–222.

[3] BORELL, C. (1975) Convex Set Functions in *d*-Space. *Period. Math. Hung.*, v.6. No.2. pp.111–136.

[4] GILLILAND, D.C. (1968) On Maximization of the Integral of a Bell-Shaped Function over a Symmetric Set. *Naval Research Logistics Quarterly,* v.15, pp.507–517.

[5] KAN, YU.S. (2000) On Justification of the Uniformity Principle in the Optimization Problem for the Probabilistic Performance Index. *Avtomatika i telemekhanika,* No.1, pp.54–70 [in Russian].

[6] KIBZUN, A.I. (1998) On the Worst-Case Distribution in Stochastic Optimization Problems with Probability Function. *Automation and Remote Control,* v.59, No.11, pp.1587–1597.

[7] KIBZUN, A.I. AND YU.S.KAN (1996) *Stochastic Programming Problems with Probability and Quantile Functions.* Wiley, Chichester.

[8] KOLMOGOROV, A.N. AND S.V.FOMIN (1972) *Elements of the Functions Theory and Functional Analysis.* Nauka, Moscow [in Russian].

[9] LEHTVACE, K. (1980) *Convex Sets.* Nauka, Moscow [in Russian].

[10] MUDHOLKAR, G.S. (1966) The Integral of an Invariant Unimodal Function over an Invariant Convex Set – An Inequality and Applications. *Proc. Am. Math. Soc.,* 17, pp.1327–1333.

[11] PRÉKOPA, A. (1980) Logarithmic Concave Measures and Related Topics. *Stochastic Programming,* ed. M.A.H.Dempster. Academic Press, London, pp.63–82.

[12] ROCKAFELLAR, R.T. (1970) *Convex Analysis.* Princeton Univ. Press, Princeton (N.J.).

[13] WALD, A. (1950) *Statistical Decision Functions.* Wiley, New York.

On Maximum Reliability Problem in Parallel-Series Systems with Two Failure Modes

Vladimir Kirilyuk (kirilyukl@d130.icyb.kiev.ua)
V.M. Glushkov Institute of Cybernetics
Prospect Glushkova,40, Kiev 252022, Ukraine

Abstract

A problem of maximum reliability of parallel-series systems with both "open" and "shorted" failure modes is studied. In this paper we try to extend the results obtained for systems of nesting depth two to systems with a more complicated configuration. In particular, necessary conditions of optimality of systems containing subsystems with nesting depth two are proposed. Then these optimality conditions are specified for separate subproblems of the problem, and methods applicable for finding of optimal solutions are discussed. A possibility of designing of an optimal system configuration composed of not a maximum number of available components is considered.

Keywords: parallel-series systems, open and shorted failure modes, maximum reliability problem

1 Introduction

In the reliability optimization of parallel-series (PS) systems, the following problem occurs. Suppose that there are n identical components, each of which can be in two failure modes: "open" and "shorted". The problem is to compose a maximum reliable PS system among all possible systems which consist of these components.

Some papers, for instance [1]-[3], consider a general statement of this problem when a PS system and an optimized failure probability function are defined by recursion. In such case, 1) a simple component is treated as a PS system, and 2) any parallel or series design of PS systems is also regarded as a PS system. Besides this,

S.P. Uryasev (ed.), Probabilistic Constrained Optimization, 148–159.
© 2000 *Kluwer Academic Publishers.*

a failure probability function of a PS system is recursively calculated, depending on a system configuration.

For this problem, an efficient algorithm is not known, which would search for an optimal system configuration even when a number of components n is not large. This is because of a huge number of possible system configurations and due to complexity of a failure probability function which changes under recursion recalculation. For instance, the same problem is solved by Discrete Optimization Methods in [2] and [3] when n is only, respectively, 9 and 15.

In other papers, e.g. in [4]-[8], a similar problem is considered under an essential restriction on a PS system configuration. These systems are only composed of a collection of series subsystems connected in parallel, or of a collection of parallel subsystems connected in series. In terms of the general problem statement, such system configuration was called Nesting Depth Two (NDT). Besides this, unlike the general problem, its optimized failure probability function has a concrete form which allows to use its specific features. Therefore, a specific algorithm, applicable even for very large n, is developed in [6],[8]. For instance, the problem for $n = 10^6$ was solved by this algorithm in 1.2 min.

This paper tries to develop the results of the above mentioned papers for those PS system configurations which are more complicated, than the NDT systems. This point allows to slightly decrease the difference between these problem statements.

In brief, the main result may be formulated as follows. Within the general problem statement, if a system configuration is optimal, then any of its NDT subsystems should be composed of "almost equal" subsystems of nesting depth (ND) one, i.e. numbers of their components differ by not more than 1. Unfortunately, for PS systems with larger nesting depths, the similar symmetry property is true in very rare cases only: if a PS system of ND k is composed of identical PS subsystems of ND $k-2$, then the necessary condition of its optimality is "almost equal" to the subsystems of ND $k-1$.

Certainly, these optimality conditions do not allow developing of such efficient algorithms for general PS systems, that were elaborated for PS systems of NDT, but these conditions decrease a number of possible system configurations. In addition, such conditions can be taken into consideration in Discrete Optimization Methods (Branches and Boundaries, Improved Enumeration, etc.) when they are used for the general problem.

In this paper, optimal solutions of separate subproblems for NDT PS subsystems are characterized by necessary and sufficient optimality conditions. Furthermore, a bisection algorithm applicable for finding a solution is proposed. This algorithm makes it possible to accelerate the algorithm developed for NDT systems in [6],[8].

A possibility of designing of an optimal PS system configuration composed of not a maximum number of available components is discussed.

2 Problem statement and main results

Following [3], define a PS system as follows.

A system of components is a PS system if:
1) a simple component, denoted by $\square$, is a PS system;
2) when $S_1, ..., S_k (k \geq 2)$ are PS systems, their arrangement in parallel, denoted by $P(S_1, ..., S_k)$, and their arrangement in series, denoted by $S(S_1, ..., S_k)$, are also PS systems.

ND of brackets is a system ND. For instance, $P(S(\square, \square, \square), \square, S(\square, \square))$ is the PS system of NDT which is composed of 7 components.

Define an optimized failure probability function for a PS system as

$$f(S) = g(S) + h(S), \tag{1}$$

where $g(\cdot)$ and $h(\cdot)$ are calculated by recursion:

$g(\square) = p,$
$g(S(S_1, ..., S_k)) = 1 - \prod_{i=1}^{k} (1 - g(S_i)),$
$g(P(S_1, ..., S_k)) = \prod_{i=1}^{k} g(S_i);$
and
$h(\square) = q,$
$h(S(S_1, ..., S_k)) = \prod_{i=1}^{k} h(S_i),$
$h(P(S_1, ..., S_k)) = 1 - \prod_{i=1}^{k} (1 - h(S_i)),$

and $p > 0, q > 0$, where $p + q < 1$, are fixed values of "open" and "shorted" failure probabilities.

If a PS system of NDT is considered, then $f(S)$ from equation (1) is reduced to the following form

$$f(S) = \prod_{i=1}^{m} (1 - a^{x_i}) + 1 - \prod_{i=1}^{m} (1 - b^{x_i}), \tag{2}$$

where $a = 1 - p, b = q$ or $a = 1 - q, b = p$, depending on a system type.

Let there be n identical components. The problem is to find such configuration of PS system composed of these components that minimizes function $f(.)$.

The papers [1]-[4] consider PS systems consisting of exactly n components. Another problem is studied in [5]-[8], where PS systems have not more than n components. The statement of the latter problem differs from the statement of the former one not formally, since these papers describe some cases, where an optimal PS system configuration of NDT does not use all the available components n and contains a lower number of them. It seems rather interesting to know, whether this point is true for general PS systems.

Denote by G_n a set of possible configurations of PS systems consisting of exactly n components. Then the general problem statement is

$$f(S) \to \min, S \in \bigcup_{k=1}^{n} G_k. \tag{3}$$

A similar problem for PS systems of NDT is

$$\min_{m \in \{1,...,n\}} \begin{cases} \min & \prod_{i=1}^{m}(1-a^{x_i}) + 1 - \prod_{i=1}^{m}(1-b^{x_i}) \\ s.t. & \sum_{i=1}^{m} x_i \le n, x_i \ge 0, x_i \in N. \end{cases} \quad (4)$$

Its continuos relaxation is formulated as

$$\min_{m \in \{1,...,n\}} \begin{cases} \min & \prod_{i=1}^{m}(1-a^{x_i}) + 1 - \prod_{i=1}^{m}(1-b^{x_i}) \\ s.t. & \sum_{i=1}^{m} x_i \le n, x_i \in R^+. \end{cases} \quad (5)$$

The following necessary conditions of optimality for problems (4) and (5), described in [5] and [6], are known to be respectively held: any optimal solution of a continuos relaxation has equal non-zero components, and any optimal solution of an integer problem has "almost equal" non-zero components, i.e. these components differ by not more than 1.

Now consider the following problem statement. Let a PS system of NDT S_* be a subsystem of a general PS system S consisting of not more than n components. If a general system S is optimal, and if any subsystem of its configuration is fixed except for S_*, then the latter itself must be optimal as well. Find necessary condition of optimality of S_* subsystem.

This problem is formulated as follows:

$$\min_{m \in \{1,...,n\}} \begin{cases} \min & f(S(S_*)) \\ s.t. & \sum_{i=1}^{m} x_i \le n, x_i \ge 0, x_i \in N, \end{cases} \quad (6)$$

where function f is recursively calculated by equation (1).

Its continuos relaxation is

$$\min_{m \in \{1,...,n\}} \begin{cases} \min & f(S(S_*)) \\ s.t. & \sum_{i=1}^{m} x_i \le n, x_i \in R^+, \end{cases} \quad (7)$$

where function f is present in equation (1).

Now consider the following function and problems associated with this function

$$F_m(x_1, ..., x_m) = \gamma_1 \prod_{i=1}^{m}(1-a^{x_i}) - \gamma_2 \prod_{i=1}^{m}(1-b^{x_i}) + \gamma_0, \quad (8)$$

where $\overline{x} = (x_1, ..., x_m), 0 < b < a < 1$ and $\gamma_1 > 0, \gamma_2 > 0, \gamma_0$ are constants.

$$\min_{m \in \{1,...,n\}} \begin{cases} \min & F_m(x_1, ..., x_m) \\ s.t. & \sum_{i=1}^{m} x_i \le n, x_i \ge 0, x_i \in N, \end{cases} \quad (9)$$

$$\min_{m \in \{1,...,n\}} \begin{cases} \min & F_m(x_1, ..., x_m) \\ s.t. & \sum_{i=1}^{m} x_i \le n, x_i \in R^+. \end{cases} \quad (10)$$

Then the following Theorems are true.

Theorem 2.1 *Problems (6) and (7) are reduced, respectively, to problems (9) and (10), where optimized function $F_m(\cdot)$ has form (8).*

Theorem 2.2 *If a subsystem S_* in problem (7) is optimal, then non-zero components of an appropriate vector $\overline{x} = (x_1, ..., x_m)$ are equal.*

Theorem 2.3 *If a subsystem S_* in problem (6) is optimal, then non-zero components of a solution vector $\overline{x} = (x_1, ..., x_m)$ are "almost equal", i.e. such components differ by at most 1.*

To prove Theorems, the specific properties of the function $F_m(\cdot)$ from equation (8) should be studied. The proofs of these Theorems and some accompanying Lemmas can be found in Appendix.

Corollary 1 *Every PS subsystem of NDT, contained in an optimal PS system of problem (3), has "almost equal" subsystems of ND 1.*

Corollary 2 *If an optimal PS system of ND k is designed of identical subsystems ND $k-2$, then it has "almost equal" subsystems of ND $k-1$.*

Evidently, subsystems of ND $k-2$ can be considered in the problem as simple components.

Now consider the subproblem of problem (9) for fixed m

$$\begin{array}{c} F_m(x_1, ..., x_m) \rightarrow min \\ \sum_{i=1}^{m} x_i \leq n, x_i \geq 0, x_i \in N, \end{array} \tag{11}$$

and its $F_m(\cdot)$ function over set of vectors which satisfy the necessary optimality conditions from Theorem 2.3. To remove those vectors from consideration which are equivalent with respect to coordinate rearrangement, their coordinates are supposed to be ordered (more exactly, these coordinates are $x_i \geq x_j$ if $i \leq j$)).

Then the behavior of $F_m(\cdot)$ over the set $X_0 = \{\overline{x} = (x_1, x_2, ..., x_m) : |x_i - x_j| \leq 1, x_i \geq x_j, \text{ if } i \leq j, x_i \in N^+, i = 1, ..., m\}$ is the same as the behavior of a unimodal function on the line: its value monotonically decreases up to a minimum, and thereafter it monotonically increases.

This feature may be formulated as follows. Introduce the following order relation $\prec$ on the set X_0

$$\bar{x}_1 = (x_1^1, ..., x_m^1) \prec \bar{x}_2 = (x_1^2, ..., x_m^2) \text{ if}$$
$$x_i^1 \leq x_i^2, i = 1, ..., m \text{ and } \exists j : x_j^1 < x_j^2.$$

It is obvious that the order is complete, i.e. any two vectors in X_0 are comparable under $\prec$.

Vectors $\bar{x}_1$ and $\bar{x}_2$ are nearest as for the order if they differ in only one their coordinate by 1.

The following Theorems, the proofs of which are given in Appendix, are true.

Theorem 2.4 *A solution of subproblem (11) is attained at not more than two vectors $\bar{x}_*$, $\bar{x}^* \in X_0 : \bar{x}_* \prec \bar{x}^*$, which are nearest in this order if both of them exist. Also,*

$$F_m(\bar{x}_1) > F_m(\bar{x}_2) \text{ for } \bar{x}_1 \prec \bar{x}_2 \prec \bar{x}_* \text{ and} \tag{12}$$

$$F_m(\bar{x}_3) < F_m(\bar{x}_4) \text{ for } \bar{x}^* \prec \bar{x}_3 \prec \bar{x}_4 \tag{13}$$

Theorem 2.5 *Necessary and sufficient conditions, that $\bar{x}_* = (x_1, ..., x_m) \in X_0$ is a solution of subproblem (11), are as follows. If $\bar{x}_*$ is interior to the admissible set (i.e. if $\sum_{i=1}^m x_i < n$), then*

$$x \le \frac{1}{\ln(a/b)} \left(\ln \frac{1-b}{1-a} + (k-1) \ln \frac{1-b^{x+1}}{1-a^{x+1}} + (m-k) \ln \frac{1-b^x}{1-a^x} + \ln \frac{\gamma_2}{\gamma_1} \right), \tag{14}$$

and

$$x \ge \frac{1}{\ln(a/b)} \left(\ln \frac{1-b}{1-a} + k \ln \frac{1-b^{x+1}}{1-a^{x+1}} + (m-k-1) \ln \frac{1-b^x}{1-a^x} + \ln \frac{\gamma_2}{\gamma_1} \right), \tag{15}$$

hold, where components are not equal, i.e. $x_i = x+1, i = 1, ..., k$, $x_i = x, i = k+1, ..., m$ for $1 \le k \le m-1$, and

$$x \le \frac{1}{\ln(a/b)} \left(\ln \frac{1-b}{1-a} + (m-1) \ln \frac{1-b^x}{1-a^x} + \ln \frac{\gamma_2}{\gamma_1} \right) + 1, \tag{16}$$

and

$$x \ge \frac{1}{\ln(a/b)} \left(\ln \frac{1-b}{1-a} + (m-1) \ln \frac{1-b^x}{1-a^x} + \ln \frac{\gamma_2}{\gamma_1} \right), \tag{17}$$

hold for equal components $x_i = x, i = 1, ..., m$, respectively

If $\bar{x}_*$ is on a boundary of the admissible set, i.e. if $\sum_{i=1}^m x_i = n$, then only (14) or (16) holds.

The obtained results allow to decrease a number of possible system configurations in problem (3) and to improve the algorithms of Discrete Optimization Methods which are used to search for a problem solution.

In addition, these results can be used to calculate variants of a general PS system configuration optimized on a subsystem of NDT. In this case, concrete values of $\gamma_1, \gamma_2 > 0$ and γ_0 from formula (8) are computed, and then problem (9) with an obtained function $F_m(\cdot)$ is optimized on an appropriate subsystem of NDT. To solve problem (9), the algorithm developed in paper [6],[8] can be used.

Moreover, Theorems 2.4 and 2.5 allow to essentially accelerate the algorithm in the following meaning. At Step 5 of the algorithm, which is the most laborious one, a number of subproblems such as subproblem (12), should be solved. In this case, Theorems 2.4, 2.5 make it possible to use a simple bisection algorithm for every such subproblem over set X_0 in the following way.

Bisection algorithm

Let I_1 be a full chain of ordered vectors of the set X_0, where the leftmost vector is $\bar{x}_0 = (1, ..., 1)$ and the rightmost vector is $\bar{x}_1 = (x_1, ..., x_m) : x_i = \lceil n/m \rceil, i = 1, ..., l, x_i = \lfloor n/m \rfloor$, in which $l = n - m\lfloor n/m \rfloor$, i.e. $I_1 = [\bar{x}_0, \bar{x}_1]$ is an appropriate chain.

Step 1. Verify inequality (14) (or (16)) at rightmost vector of I_1. If it holds, then $\bar{x}_1$ is subproblem solution. Else go to Step 2.

Step k. Bisect I_{k-1} (or as nearly as possible, if there are an odd number of elements), and verify inequality (14) (or (16)) at a middle point. If it holds, reject the left set part, else reject the right part. Denote a set so obtained by I_k and go to Step $k + 1$.

The solution is obtained in $l = \lceil \log_2 N \rceil$ steps, where N is a number of I_1 elements.

In papers [5],[6], and [8] studying optimal PS systems of NDT, the cases when optimal configurations of such systems do not use all the available components n are described. These cases are explained by a behavior of a continuos relaxation problem.

If the result of Theorem 2.5 is taken into account, such cases can be explained in the following way. For a subproblem where minimum of problem (11) is attained, inequalities (14) and (15) hold simultaneously for non-equal components, or inequalities (16) and (17) hold for equal components. It is easily seen that fulfillment of this condition is rare enough, because the integer value x should get into a very narrow interval.

If general problem (3) is considered, the design of an optimal system configuration containing not the whole number of available components n is extremely improbable. Therefore, a number of possible PS system configurations is increased substantially compared to PS systems of NDT. Moreover, if such an optimal system consists of a less number of components, then every its subsystem of NDT should satisfy inequalities (14) and (15) (or inequalities (16) and (17) for equal components). Note, that, depending on a PS system of NDT type, the appropriate γ_1, γ_2, m, and k and the parameters $a = 1 - p, b = q$ or $a = 1 - q, b = p$, should be included into these conditions.

3 Appendix: Lemmas. Proofs of theorems

Lemma 1 *[6] Function* $\frac{1-b^x}{1-a^x}$ *for* $0 < b < a < 1$ *monotonically decreases on* $x > 0$.

Lemma 2 *[7] For* $x > 0$ *and* $\beta > 0$, *inequalities*

$$\frac{1-b^x}{1-a^x} \times \frac{b^\beta}{a^\beta} < \frac{1-b^{\beta+x}}{1-a^{\beta+x}} < \frac{1-b^\beta}{1-a^\beta}$$

are true.

Lemma 3 *If* $\bar{x}_1 = (x, x+\alpha, x_3, ..., x_m), \bar{x}_2 = (x+\beta, x+\alpha-\beta, x_3, ..., x_m)$ *and* $\bar{x}_3 = (x, x+\beta, x_3, ..., x_m)$ *for* $0 < \beta < \alpha$, *then at least one of the following inequalities holds:*

$$F_m(\bar{x}_1) > F_m(\bar{x}_2), F_m(\bar{x}_1) > F_m(\bar{x}_3).$$

Proof. Compare the function values $F_m(\cdot)$ at the vectors $\bar{x}_1$ and $\bar{x}_2$:

$$F_m(\bar{x}_1) - F_m(\bar{x}_2) = \gamma_1(1-a^x)(1-a^{x+\alpha})\prod_{i=3}^{m}(1-a^{x_i}) - \gamma_2(1-b^x)(1-b^{x+\alpha})\times$$

$$\prod_{i=3}^{m}(1-b^{x_i}) - \gamma_1(1-a^{x+\beta})(1-a^{x+\alpha-\beta})\prod_{i=3}^{m}(1-a^{x_i}) + \gamma_2(1-b^{x+\beta})(1-b^{x+\alpha-\beta})\times$$

$$\prod_{i=3}^{m}(1-b^{x_i}) = -\gamma_1[a^x(1-a^\beta) - a^{x+\alpha-\beta}(1-a^\beta)]\prod_{i=3}^{m}(1-a^{x_i}) +$$

$$\gamma_2[b^x(1-b^\beta) - b^{x+\alpha-\beta}(1-b^\beta)]\prod_{i=3}^{m}(1-b^{x_i}) = -\gamma_1 a^x(1-a^\beta)a^{\alpha-\beta}\prod_{i=3}^{m}(1-a^{x_i}) +$$

$$\gamma_2 b^x(1-b^\beta)b^{\alpha-\beta}\prod_{i=3}^{m}(1-b^{x_i}).$$

Consequently, $F_m(\bar{x}_1) - F_m(\bar{x}_2) > 0$ iff

$$x < \frac{1}{\ln(a/b)}\left(\ln\frac{1-b^\beta}{1-a^\beta} + \ln\frac{1-b^{\alpha-\beta}}{1-a^{\alpha-\beta}} + \sum_{i=3}^{m}\ln\frac{1-b^{x_i}}{1-a^{x_i}} + \ln\frac{\gamma_2}{\gamma_1}\right) \tag{18}$$

Perform the similar transformations with the difference of the function values at the vectors $\bar{x}_1$ and $\bar{x}_3$, then $F_m(\bar{x}_1) - F_m(\bar{x}_3) > 0$ is obtained iff

$$x > \frac{1}{\ln(a/b)}\left(\ln\frac{1-b^x}{1-a^x} + \ln\frac{b^\beta}{a^\beta} + \ln\frac{1-b^{\alpha-\beta}}{1-a^{\alpha-\beta}} + \sum_{i=3}^{m}\ln\frac{1-b^{x_i}}{1-a^{x_i}} + \ln\frac{\gamma_2}{\gamma_1}\right) \tag{19}$$

Compare inequalities (18) and (19). As it follows from Lemma 2, the right side of inequality (19) is lower than the respective side of inequality (18).

Therefore, if inequality (18) does not hold, then inequality (19) really holds. Thus, at least one of these inequalities really holds. Lemma is proved.

Corollary 3 *If* $\alpha > 1, \beta = 1$, *then at least one of the following inequalities holds:*

$$F_m(x, x+\alpha, x_3, ..., x_m) - F_m(x+1, x+\alpha-1, x_3, ..., x_m) > 0,$$

$$F_m(x, x+\alpha, x_3, ..., x_m) - F_m(x, x+1, x_3, ..., x_m) > 0.$$

Corollary 4 *If $\alpha > 0, \beta = \alpha/2$, then at least one of the following inequalities holds:*

$$F_m(x, x+\alpha, x_3, ..., x_m) - F_m(x+\alpha/2, x+\alpha/2, x_3, ..., x_m) > 0,$$

$$F_m(x, x+\alpha, x_3, ..., x_m) - F_m(x, x+\alpha/2, x_3, ..., x_m) > 0.$$

Lemma 4 *If $\bar{x}_1 = (x, x+1, x_3, ..., x_m), \bar{x}_2 = (x, x, x_3, ..., x_m)$ and $\bar{x}_3 = (x+1, x+1, x_3, ..., x_m)$, then at least one of the following inequalities holds:*

$$F_m(\bar{x}_1) > F_m(\bar{x}_2), F_m(\bar{x}_1) > F_m(\bar{x}_3).$$

Proof. Compare the function values $F_m(\cdot)$ at the vectors $\bar{x}_1$ and $\bar{x}_2$:

$$F_m(\bar{x}_1) - F_m(\bar{x}_2) = \gamma_1(1-a^x)(1-a^{x+1})\prod_{i=3}^{m}(1-a^{x_i}) - \gamma_2(1-b^x)(1-b^{x+1})\times$$

$$\prod_{i=3}^{m}(1-b^{x_i}) - \gamma_1(1-a^x)^2\prod_{i=3}^{m}(1-a^{x_i}) + \gamma_2(1-b^x)^2\prod_{i=3}^{m}(1-b^{x_i}) =$$

$$-\gamma_1(1-a^x)(1-a)a^x\prod_{i=3}^{m}(1-a^{x_i}) + \gamma_2(1-b^x)(1-b)b^x\prod_{i=3}^{m}(1-b^{x_i})$$

Consequently, $F_m(\bar{x}_1) - F_m(\bar{x}_2) > 0$ iff

$$x < \frac{1}{\ln(a/b)}\left(\ln\frac{1-b^x}{1-a^x} + \ln\frac{1-b}{1-a} + \sum_{i=3}^{m}\ln\frac{1-b^{x_i}}{1-a^{x_i}} + \ln\frac{\gamma_2}{\gamma_1}\right) \tag{20}$$

Perform the similar transformations with the difference of the function values at the vectors $\bar{x}_1$ and $\bar{x}_3$, then $F_m(\bar{x}_1) - F_m(\bar{x}_3) > 0$ is obtained iff

$$x > \frac{1}{\ln(a/b)}\left(\ln\frac{1-b^{x+1}}{1-a^{x+1}} + \ln\frac{1-b}{1-a} + \sum_{i=3}^{m}\ln\frac{1-b^{x_i}}{1-a^{x_i}} + \ln\frac{\gamma_2}{\gamma_1}\right) \tag{21}$$

Compare inequalities (20) and (21). As it follows from Lemma 1, the right side of inequality (21) is lower than the respective side of inequality (20).

Therefore, if inequality (20) does not hold, then inequality (21) really holds. Thus, at least one of these inequalities really holds. Lemma is proved.

Corollary 5 *If $F_m(x, x+1, x_3, ..., x_m) \geq F_m(x, x, x_3, ..., x_m)$ is met, then $F_m(x, x+1, x_3, ..., x_m) < F_m(x+1, x+1, x_3, ..., x_m)$ holds, and, vise versa,*
if $F_m(x, x+1, x_3, ..., x_m) \geq F_m(x+1, x+1, x_3, ..., x_m)$ is met, then
$F_m(x, x+1, x_3, ..., x_m) < F_m(x, x, x_3, ..., x_m)$ holds.

Proof of Theorem 2.1. ND of a general PS system increases if parallel and series design types alternate. Then, generally speaking, there are only two types of PS systems (either $\cdots SPS$, or $\cdots PSP$) which are defined by the first design type (S or P).

Consider a PS NDT system, for instance, of $\cdots SPS$ type:

$S_*^2 = P(S_1^1(\square, ...), ..., S_{m_1}^1(\square, ...))$.

Its failure probability is described according to equation (1) by the following function

$$f(S_*^2) = g(S_*^2) + h(S_*^2) = \prod_{i=1}^{m_1} (1 - (1-p)^{x_i}) + 1 - \prod_{i=1}^{m_1} (1 - q^{x_i}),$$

where p and q are, respectively, "open" and "shorted" failure probabilities.

If S_*^2 is included into a PS system of ND 3, i.e. $S_*^3 = S(S_1^2, ..., S_{m_2}^2, S_*^2)$, then its failure probability has the following form

$$g(S_*^3) + h(S_*^3) = 1 - \prod_{i=1}^{m_2} \left(1 - g(S_i^2)\right) \left(1 - \prod_{i=1}^{m_1}(1 - (1-p)^{x_i})\right) +$$

$$\prod_{i=1}^{m_2} h(S_i^2) \left(1 - \prod_{i=1}^{m_1}(1 - q^{x_i})\right).$$

If S_*^3 is included into a PS system of ND 4, i.e. $S_*^4 = P(S_1^3, ..., S_{m_3}^3, S_*^3)$, then its failure probability is

$$g(S_*^4) + h(S_*^4) = \prod_{i=1}^{m_3} g(S_i^3) \left[1 - \prod_{i=1}^{m_2}(1 - g(S_i^2)) \left(1 - \prod_{i=1}^{m_1}(1 - (1-p)^{x_i})\right)\right] +$$

$$1 - \prod_{i=1}^{m_3}(1 - h(S_i^3)) \left[1 - \prod_{i=1}^{m_2} h(S_i^2) \left(1 - \prod_{i=1}^{m_1}(1 - q^{x_i})\right)\right].$$

Similar reasonings may be continued. It is easily seen that the member $\prod_{i=1}^{m}(1 - (1 - p^{x_i}))$ is always introduced under recursive recalculation with sign (+), and the member $\prod_{i=1}^{m}(1 - q^{x_i})$ is introduced with sign (-). Thus, the failure probability in $\cdots SPS$ systems has the following form

$$f(S(S_*^2)) = \gamma_1 \prod_{i=1}^{m} (1 - (1-p)^{x_i}) - \gamma_2 \prod_{i=1}^{m} (1 - q^{x_i}) + \gamma_0,$$

where values $\gamma_1, \gamma_2, \gamma_0 > 0$ do not depend on the S_*^2 system configuration.

For the second PS system type, i.e. for $\cdots PSP$, the similar reasonings are applicable, and the failure probability function is

$$f(S(S_*^2)) = \gamma_1 \prod_{i=1}^{m} (1 - (1-q)^{x_i}) - \gamma_2 \prod_{i=1}^{m} (1 - p^{x_i}) + \gamma_0.$$

Evidently, the failure probability function has form of the function $F_m(\cdot)$ given by expression (8) in both cases, where either $a = 1 - p, b = q$, or $a = 1 - q, b = p$, and p and q are "open" and "shorted" failure probabilities. Theorem is proved.

Proof of Theorem 2.2. Let $\bar{x}_1 = (x_1, x_2, ..., x_m)$ be a solution of problem (5), in which at least two components differ. In order not to complicate the denotations, let these components be x_1 and x_2, $x_1 < x_2$. Fix the other components of $\bar{x}_1$ and consider the values of $F_m(\cdot)$ at the vectors $\bar{x}_1$, $\bar{x}_2 = (\frac{x_1+x_2}{2}, \frac{x_1+x_2}{2}, x_3, ..., x_m)$ and $\bar{x}_3 = (x_1, \frac{x_1+x_2}{2}, x_3, ..., x_m)$.

On the one hand, these vectors are admissible, because the set of constraints is convex and invariant with respect to rearrangement of the vector components. On the other hand, Corollary 4 implies that at least one of the following inequalities holds:

$$F_m(\bar{x}_1) > F_m(\bar{x}_2), F_m(\bar{x}_1) > F_m(\bar{x}_3).$$

Thus, $\bar{x}_1$ is not optimal, and the obtained contradiction proves Theorem.

Proof of Theorem 2.3. Let $\bar{x}_1 = (x_1, x_2, ..., x_m)$ be a solution of problem (4), in which at least two components differ at least by 2. In order not to complicate the denotations, let these components be x_1 and x_2, and $x_1 = x, x_2 = x + \alpha$, where $\alpha \geq 2$ is an integer number. Fix the other components of $\bar{x}_1$ and consider the values of $F_m(\cdot)$ at the vectors $\bar{x}_1 = (x, x + \alpha, x_3, ..., x_m)$, $\bar{x}_2 = (x + 1, x + \alpha - 1, x_3, ..., x_m)$ and $\bar{x}_3 = (x, x + 1, x_3, ..., x_m)$.

These vectors are admissible, because the set of constraints is convex and invariant with respect to rearrangement of the vector components. Besides this, Corollary 3 implies that at least of the following inequalities holds:

$$F_m(\bar{x}_1) > F_m(\bar{x}_2), F_m(\bar{x}_1) > F_m(\bar{x}_3).$$

Thus, $\bar{x}_1$ is not a solution, and this contradiction proves Theorem.

Proof of Theorem 2.4. Denote by $\bar{x}_*$ and $\bar{x}^*$ the minimal and maximal solutions of subproblem (11) in the sense of the introduced order relation.

Suppose the following: $\bar{x}_1 \prec \bar{x}_2 \prec \bar{x}_*$. Then, it is possible to design an ordered (in above sense) 'chain' of vectors from $\bar{x}_1$ up to $\bar{x}_*$, where the neighboring vectors differ in only one coordinate by 1. Choose such a vector from this chain that is the nearest one to the vector $\bar{x}_*$ on its left. Then, as it follows from Corollary 5, $F_m(\cdot)$ at this vector should be less than at the preceding one (the first part of the Corollary). Moving sequentially on the neighbor vectors to the left from $\bar{x}_*$ up to $\bar{x}_2$, follow these reasonings, and inequality (12) is obtained.

Inequality (13) is proved in the same way if the ordered chain of the vectors from $\bar{x}^*$ up to $\bar{x}_4$ is considered and the second part of Corollary 5 is applied.

Suppose that there exists such a vector $\bar{x}$, that $\bar{x}_* \prec \bar{x} \prec \bar{x}^*$. Then, by the similar reasoning and with the use of Corollary 5, the contradiction is obtained with under the assumption that $\bar{x}_*$ and $\bar{x}_*$ are the solutions of subproblem (11). Therefore, there are at most two solutions, and they are the nearest ones in the order if both of them exist.

Proof of Theorem 2.5. The proof easily follows from the line of the reasoning, similar to that is used in the proof of the preceding theorem. Suppose that $1 \leq k \leq m-1$. Then, if $\bar{x}_*$ is optimal and inside the admissible set, the value of function $F_m(\cdot)$ at this vector is less or equal than its values at the vectors, which are the left and the right neighbors of the vector under the order $\prec$. Writing the respective differences, inequalities (14) and (15) are obtained.

If $\bar{x}_*$ is on the boundary, then the value of $F_m(\cdot)$ at the right neighbor needs no consideration. Therefore, inequality (15) is absent.

Since inequalities (16) and (17) are equivalent to the inequalities for the differences of the function values in the respective vectors, they are sufficient in subproblem (11) as well.

In the case with equal coordinates, i.e. if $x_i = x, i = 1, ..., m$, then the proof is similar.

References

[1] B.W. Jenney and D.J. Sherwin (1986), Open and short circuit reliability of systems of identical components, IEEE Trans. Reliability, R-35, 532-538

[2] L.B. Page and J.E. Perry (1988), Optimal series-parallel networks of 3-state devices, IEEE Trans. Reliability, R-37, 388-394.

[3] W.J. Gutjahr (1994), A global optimization problem in series-parallel networks with maximum reliability, Global Optimization, 5, 403-404

[4] W.J. Gutjahr, G.Ch. Pflug, and A. Ruszscinsky (1996), Configuration of series-parallel networks with maximum reliability, Microelectron. Reliab., Vol. 36, 2, 247-253

[5] V. Kirilyuk (1995), On maximum reliable configuration for parallel-series scheme containing elements with failure states of two types, Kibernetica i Systemnyi Analiz, 1, 34-46 (in Russian).

[6] V. Kirilyuk (1997), On optimal configuration of parallel-series systems with two failure modes, Dopovidi NANU, Ser. A, 10, 109-113

[7] V. Kirilyuk (1999), On maximum reliable configuration of parallel-series systems with two failure modes, Kibernetica i Systemnyi Analiz, 2, 102-111 (in Russian).

[8] V. Kirilyuk and J.E. Falk, Minimizing the mean damage for parallel-series systems with two failure modes, to be published in Computational Optimization and Application.

Robust Monte Carlo Simulation for Approximate Covariance Matrices and VaR Analyses[1]

Alexander Kreinin (alex@algorithmics.com)
Algorithmics Inc.,
185 Spadina Ave., Toronto, Ontario, Canada, M5T 2C6

Alexander Levin (alevin@bmo.com)
Bank of Montreal, Global Treasury Group,
First Canadian Place,
Toronto, Ontario, M5X 1A1, Canada

Abstract

Value at Risk (VaR) analysis plays very important role in modern Financial Risk Management. The are two very popular approaches to portfolio VaR estimation: approximate analytical approach and Monte Carlo simulation. Both of them face some technical difficulties steaming from statistical estimation of covariance matrix decribing the distribution of the risk factors. In this paper we develop a new robust method of generating scenarios in a space of risk factors consistent with a given matrix of correlations containing possible small negative eigenvalues, and find an estimate for a change in VaR. Namely, we prove that the modified VaR of a portfolio $\mathbf{VaR}'$ satisfies the inequality $|\widetilde{\mathbf{VaR}}^2 - \mathbf{VaR}^2| \leq K \cdot \mu$, where μ is the maximum of the absolute values of negative eigenvalues of the approximate covariance matrix and K is an explicitly expressed constant, closely related to the market value of the portfolio.

Keywords : Value–at–Risk, risk management

[1]This paper was prepared when Alexander Levin was with Risk Lab, University of Toronto.

S.P. Uryasev (ed.), Probabilistic Constrained Optimization, 160–172.
© 2000 *Kluwer Academic Publishers.*

1 Introduction

Accurate and efficient Value-at-Risk (VaR) analysis is an important part of modern risk management strategy. This strategy requires that financial institutions implement a risk management engine to compute market risk of their full portfolio.

Computation of the VaR of a large portfolio represents a serious numerical problem. There are three major methodologies currently implemented in the risk management systems. The first methodology is based on an approximate analytical approach using a linear approximation of the portfolio pricing function. The second approach is based on Monte Carlo simulation of the underlying risk factors. These methodologies are based on a log-normal model of risk factors joint behavior that requires estimation of the covariance matrix of the risk factors.

The third methodology is historical simulation. Historical simulation draws scenarios from the observable discrete historical changes in the risk factors during a specified period of time. This method can be combined with some randomization technique, but direct implementation does not require any model of risk factor changes.

Historical simulation is not considered in this paper.

The number of underlying risk factors affecting the portfolio value may reach several hundreds or even thousands. In this case the estimation problem becomes ill–posed since the rounding procedure may affect computation of the elements of the covariance matrix and, in particular, the sign of the eigenvalues of the matrix. If an eigenvalue of the covariance matrix is negative, direct application of the Cholesky decomposition required for Monte Carlo simulation is impossible.

In this paper we consider two questions related to the Value at Risk analysis and Monte Carlo scenario generation based on Risk Metrics methodology:

- The algorithm of modification of an arbitrary non–positive definite approximate covariance matrix;
- Bounds on VaR if the covariance matrix is modified to be positive semi–definite.

Let the present value of the portfolio at time $t = 0$ be $V(0)$. The portfolio present value depends on risk factors, that are random variables, and therefore, at time $t = T$ the value will be different because of random changes of the risk factors. We say that the value at risk at time t of the portfolio is V_α, if

$$\mathbf{Pr}\left\{V(0) - V(t) < V_\alpha\right\} = \alpha, \quad 0 < \alpha < 1.$$

The parameter α is called *confidence probability* corresponding to the value at risk V_α.

The VaR analysis is based on the two major issues:

- the model of the market risk factor behavior.
- the statistical data to define the parameters of the model.

In [2] the following model was suggested to describe the movements of market parameters.

The short term dynamics of the risk factors is driven by the equation

$$R_t = R_{t-1} \cdot e^{\xi},$$

where the random vector of logreturns ξ has a joint normal distribution.

Based on this model, the following approximate formula to compute VaR was suggested [2]:

$$\text{VaR}(\alpha) = \sqrt{W^t \cdot \mathbf{C} \cdot W} \cdot x_\alpha, \tag{1}$$

where W is a vector of so–called *Daily Earning at Risk*, $\mathbf{C}$ is a correlation matrix of the logreturns of risk factors, and x_α is a quantile of a standard normal distribution. The notation W^t is always used for transposed vector.

The problem of VaR computation is related both to analytical approximation and to the Monte Carlo simulation. The Monte Carlo scenario generation algorithm, suggested in [2] uses Cholesky decomposition [1]. This transformation is applicable only to the positive definite covariance matrix $\mathbf{C}$. In theory, if a proper statistical estimation is applied to obtain the elements of the covariance matrix, then $\mathbf{C}$ must be symmetric positive definite. But in practice, because of computational and rounding errors, the eigenvalues of the covariance matrix might become zero or negative.

Another important issue is the statistical procedure itself. If the number of observations used for estimation of $\mathbf{C}$ is small enough with respect to the size of the matrix, then the result will be singular or almost singular matrix $\widetilde{\mathbf{C}}$, with the eigenvalues very close to zero. Again, rounding of the estimation of the elements of the matrix $\mathbf{C}$ most likely, might lead to the negative eigenvalues. In this case, Cholesky decomposition can not be performed.

This problem has been discussed in the literature. In [8] it is suggested to adjust the user–defined matrix so that it is positive semi–definite and remains as close as possible to the original matrix. The idea developed in [8] is to find an lower–triangular matrix A to minimize the weighted quadratic distance

$$\min_A \|C - A \cdot A^T\|,$$

between the original matrix C and $A \cdot A^T$. It is proved that the Gauss–Newton method converges to the solution under some regularity conditions.

Another approach to the problem was suggested by Davenport and Iman in [4]. They perform a spectral decomposition of the correlation matrix and replace negative eigenvalues with a small positive numbers. This method was also used in [3] to handle symmetric matrices with the negative eigenvalues.

In this paper we develop a new robust algorithm (the *Method of the Minimal Symmetric Pseudoinverse Operator*) which is stable with respect to perturbations of the covariance matrix. Our approach is close to that presented in [4], but differs in the way we replace the negative and small positive eigenvalues.

This paper is arranged as follows. In section 2 we show that the statistical VaR estimation is a stable problem in the following sense. If the portfolio value function $V(\cdot)$ is computed with an error not exceeding ε than the Value–at–Risk will be computed with the error bounded by the same quantity ε.

In section 3 the Method of the Minimal Symmetric Pseudoinverse Operator is described and the properties of the Pseudoinverse Operator are studied. It is shown that the modified matrix obtained after application of the Method of the Minimal Symmetric Pseudoinverse Operator has minimal deviation from the original matrix. The advantage of our approach is that we avoid to solve the minimization problem representing significant difficulties with respect to convergence properties.

Bounds on the Value–at–Risk based on the *"Minimal Symmetric Modification"* are considered in section 4. The bounds improve those obtained in [3]. The final section contains concluding remarks and some numerical results.

2 Stability of VaR computation

In this section we prove the following result about VaR. Suppose that the distribution of the portfolio value is to be estimated by generating n scenarios and by approximating the portfolio distribution with the empirical one. Let us assume, for simlicity, that the value of the portfolio W_i, $(i = 1, 2, \ldots, n)$ is positive on every scenario and that the estimator of the portfolio value admits a uniform error bound

$$\frac{|V_i - W_i|}{W_i} \leq \varepsilon \quad i = 1, 2, \ldots, n, \tag{2}$$

where V_i is the estimation of the portfolio value on ith scenario, W_i is the "true" value of the portfolio and ε is a small positive number. We introduce the vectors $\vec{V} = (V_1, \ldots, V_n)$ and $\vec{W} = (W_1, \ldots, W_n)$ and call them *the approximating vector* and the *vector of "true" values*, correspondingly.

Proposition 1 *The estimation of Value–at–Risk corresponding to probability $\alpha > 0$, $V(\alpha)$, admits the same error bound as the portfolio value. Namely, if $W(\alpha)$ is the VaR estimation based on "true" portfolio values W_i, $(i = 1, 2, \ldots, n)$, then*

$$\frac{|W(\alpha) - V(\alpha)|}{W(\alpha)} \leq \varepsilon. \tag{3}$$

Proof: The VaR estimations are obtained from the vector of observations $\vec{V}$ and $\vec{W}$ as follows. Consider first the vector $\vec{V}$. Let $\vec{V}^* = (V_{[1]}, \ldots, V_{[n]})$ be a new vector with the coordinates of $\vec{V}$ but reordered in ascending order:

$$V_{[1]} \leq V_{[2]} \leq V_{[k]} \leq \ldots \leq V_{[n]}.$$

Denote by $m(\alpha) = \lfloor \alpha \cdot n \rfloor$. Then

$$V(\alpha) = V_{[m(\alpha)]} \tag{4}$$

The same construction can be applied to the vector of "true" values $\vec{W}$. In this case the VaR corresponfing to the probability α is given by

$$W(\alpha) = W_{[m(\alpha)]}.$$

Let us prove that the approximate VaR estimation and the estimation based on the "true" portfolio values satisfy the inequality

$$\frac{|W(\alpha) - V(\alpha)|}{W(\alpha)} \leq \varepsilon \tag{5}$$

Inequality (5) is intuitively appealling but still need some justification because the permutation $\pi(V)$ of the vector V and the permutation $\pi(W)$ of the vector W resulting in monotonic ordering of the coordinates of V and W may be different.

The permutations $\pi(V)$ and $\pi(W)$ can be represented as a product of the transpositions of the adjacent components . Let us show that if two elements of the vector V, say, V_i and V_{i+1} are to be transposed but the corresponding elements of the true vector W follow in the right order, then after transposition the coordinates of the new vector satisfy inequality (2). Indeed, let us suppose that

$$W_i \leq W_{i+1},$$

but

$$V_i \geq V_{i+1}.$$

Then to find VaR the coordinates i and $i+1$ of the vector V will be transposed. We want to prove that inequality (2) implies that

$$\frac{|V_k - W_{k+1}|}{W_{k+1}} \leq \varepsilon,$$

and at the same time

$$\frac{|V_{k+1} - W_k|}{W_k} \leq \varepsilon.$$

We have from (2)

$$W_k \cdot (1 - \varepsilon) \leq V_k \leq W_k \cdot (1 + \varepsilon) \leq W_{k+1} \cdot (1 + \varepsilon).$$

At the same time,

$$W_{k+1} \cdot (1 - \varepsilon) \leq V_{k+1} \leq V_k \leq W_k \cdot (1 + \varepsilon) \leq W_{k+1} \cdot (1 + \varepsilon).$$

Thus, we find that V_k belongs to the interval $[W_{k+1}\cdot(1-\varepsilon), W_{k+1}\cdot(1+\varepsilon)]$. The same reasoning is applicable to V_{k+1} and leads to the conclusion that $V_{k+1} \in [W_k \cdot (1-\varepsilon), W_k\cdot(1+\varepsilon)]$. Therefore inequality (2) will be satisfied for the vectors V and W after the transposition was applied. Then we derive that inequality (2) is fulfilled after any finite number of transpositions, and, therefore after application of the permutations $\pi(V)$ and $\pi(W)$.

This proposition shows that the VaR estimation problem is well–posed problem.

3 The method of the minimal symmetric pseudoinverse operator

Before setting the numerical method, we will calculate the least estimation of the error of the covariance matrix $\mathbf{C}$ when the approximate covariance matrix $\tilde{\mathbf{C}}$ has small negative eigenvalues. We will measure perturbations of matrices using a spectral norm (see [7])

$$||A|| = \max_{x\neq 0} \frac{||Ax||}{||x||},$$

where $||\cdot||$ is the usual Euclidean norm in a vector space.

The spectral norm of a symmetric matrix A satisfies the relation

$$||A|| = \max_{x\neq 0} \frac{|(Ax,x)|}{(x,x)}, \tag{6}$$

where $(\cdot,\cdot)$ denotes inner product of vectors. Obviously, the spectral norm in this case is equal to the maximum of absolute values of the matrix eigenvalues.

Lemma 1 *The error h of an approximation of any covariance matrix $\mathbf{C}$ by the symmetric matrix $\tilde{\mathbf{C}}$ which has at least one negative eigenvalue, satisfies the inequality:*

$$h = ||\tilde{\mathbf{C}} - \mathbf{C}|| \geq \mu, \tag{7}$$

where μ is the maximum of absolute values of the negative eigenvalues of $\tilde{\mathbf{C}}$.

Proof. Any covariance matrix $\mathbf{C}$ is symmetric semi-positive definite: $(\mathbf{C}x,x)\geq 0$ for all vectors x. Let the eigenvector x_μ of the matrix $\tilde{\mathbf{C}}$ correspond to the negative eigenvalue $\lambda_- < 0, |\lambda_-| = \mu$. Then we have from (6):

$$\begin{aligned} h &= ||\mathbf{C}-\tilde{\mathbf{C}}|| = \max_{x\neq 0}\frac{|((\mathbf{C}-\tilde{\mathbf{C}})x,x)|}{(x,x)} \\ &\geq \frac{|((\mathbf{C}-\tilde{\mathbf{C}})x_\mu,x_\mu)|}{(x_\mu,x_\mu)} = \frac{(\mathbf{C}x_\mu,x_\mu)-(\lambda_- x_\mu,x_\mu)}{(x_\mu,x_\mu)} \geq \mu. \end{aligned}$$

In accordance with Lemma 1, if there is no any additional information about the error ε of the covariance matrix $\widetilde{\mathbf{C}}$, μ can be used as the least estimate of ε.

The proposed Method of the Minimal Symmetric Pseudoinverse Operator is a modification of the Method of the Minimal Pseudoinverse Operator [5] for the case of symmetric matrices. For the known approximate covariance matrix $\widetilde{\mathbf{C}}$ and the upper estimate $h \geq \mu$ of its accuracy let us consider the following "*Class of h–equivalent symmetric matrices*"

$$\Sigma_h(\widetilde{\mathbf{C}}) = \{B : \|B - \widetilde{\mathbf{C}}\| \leq h,\ B^t = B\}. \tag{8}$$

The unknown exact covariance matrix $\mathbf{C}$ belongs to the class $\Sigma_h(\widetilde{\mathbf{C}})$. For each matrix B, $B \in \Sigma_h(\widetilde{\mathbf{C}})$ we denote by B^+ the (unique) pseudoinverse matrix [6], [7]. Now we find the stable semi-positive definite approximation of the covariance matrix $\mathbf{C}$ and explicit formula for matrix A in the representation

$$C = A \cdot C^t \tag{9}$$

using the following

Theorem 1 *There exists a semi-positive definite matrix* $\mathbf{C}_h \in \Sigma_h(\widetilde{\mathbf{C}})$ *which has the minimal norm of its pseudoinverse matrix over all matrices from the class* $\Sigma_h(\widetilde{\mathbf{C}})$*:*

$$\mathbf{C}_h \in \Sigma_h(\widetilde{\mathbf{C}}) : \ \|\mathbf{C}_h{}^+\| = \min_{B \in \Sigma_h(\widetilde{\mathbf{C}})} \|B^+\|. \tag{10}$$

Proof. It is well known, that there exists the spectral factorization of the symmetric matrix $\widetilde{\mathbf{C}}$ of the form [7, 6]:

$$\widetilde{\mathbf{C}} = O \cdot \Lambda \cdot O^t,$$

where O is an orthogonal matrix $(O^{-1} = O^t)$, and Λ is a diagonal matrix with the eigenvalues $\lambda_1 \geq \lambda_2 \geq \ldots \geq \lambda_n$ of the matrix $\widetilde{\mathbf{C}}$ on the diagonal

$$\Lambda = \begin{pmatrix} \lambda_1 & & & \\ & \lambda_2 & & \\ & & \ldots & \\ & & & \lambda_n \end{pmatrix}$$

(under our assumptions, $\lambda_n = \lambda_- < 0$, $|\lambda_n| = \mu$). This factorization could be effectively done using Householder algorithm to tridiagonolize the matrix $\widetilde{\mathbf{C}}$ and then applying QR algorithm for diagonalization (for more details see [1, 7]). Then we define the matrix $\mathbf{C}_h$ as

$$\mathbf{C}_h = O \cdot \widetilde{\Lambda} \cdot O^t, \tag{11}$$

where

$$\widetilde{\Lambda} = \begin{pmatrix} \widetilde{\lambda_1} & & & \\ & \widetilde{\lambda_2} & & \\ & & \ldots & \\ & & & \widetilde{\lambda_n} \end{pmatrix}, \tag{12}$$

$$\widetilde{\lambda_i} = \omega(\lambda_i),\ i = 1, \dots, n;\ \ \omega(\lambda) = \begin{cases} \lambda + h & \text{if } |\lambda| > h \\ 0 & \text{if } |\lambda| \le h \end{cases}$$

In fact, we replace by zeros the eigenvalues whose absolute values are less or equal to h and add h to the other eigenvalues. It follows from (11) that $\mathbf{C}_h$ is symmetric and satisfies the relation

$$||\mathbf{C}_h - \mathbf{C}|| = h. \tag{13}$$

Therefore, $\mathbf{C}_h \in \Sigma_h(\widetilde{\mathbf{C}})$. The pseudoinverse matrix $\mathbf{C}_h^+$ is given by formula

$$\mathbf{C}_h^+ = O \cdot \widetilde{\Lambda}^+ \cdot O^t,\ \ \widetilde{\Lambda}^+ = diag(\rho(\widetilde{\lambda_1}), \dots, \rho(\widetilde{\lambda_n})),$$

$$\rho(\lambda) = \begin{cases} \lambda^{-1} & \text{if } \lambda \neq 0 \\ 0 & \text{if } \lambda = 0 \end{cases}$$

To finish the proof of Theorem we have to show, that $||B^+|| \geq ||\mathbf{C}_h^+||$ for all matrices $B \in \Sigma_h(\widetilde{\mathbf{C}})$. It follows from (11) - (13) and Wielandt-Hoffman inequality for symmetric perturbations [7], that the eigenvalues satisfy the inequality

$$\max_{0 \le j \le n} |\lambda_j(B) - \lambda_j(\widetilde{\mathbf{C}})| \le ||B - \widetilde{\mathbf{C}}|| \le h. \tag{14}$$

The triangle inequality for a spectral norm implies the bound on the error and prove the convergence of the method.

Corollary 1 *The upper bound on the error satisfies the relation*

$$||\mathbf{C}_h - \mathbf{C}|| \le ||\mathbf{C}_h - \widetilde{\mathbf{C}}|| + ||\widetilde{\mathbf{C}} - \mathbf{C}|| \le 2h. \tag{15}$$

Let us calculate the matrix A from (9) in the form

$$A_h = D \cdot O^t,\ \ A_h^t = O \cdot D, \tag{16}$$

$$D = \begin{pmatrix} \sqrt{\widetilde{\lambda_1}} & & & \\ & \sqrt{\widetilde{\lambda_2}} & & \\ & & \dots & \\ & & & \sqrt{\widetilde{\lambda_n}} \end{pmatrix}$$

Note that if the dimension of the problem is large, then many (positive or negative) of eigenvalues of $\widetilde{\mathbf{C}}$ are very close to zero. In the Method of the Least Symmetric Pseudoinverse Operator we assign them zero value. This transformation leads to a significant reduction of the dimension of the risk factors space. Moreover, this method gives the optimal compression of the dimension with respect to a given level of accuracy h of the approximate covariance matrix $\widetilde{\mathbf{C}}$:

Theorem 2 *The rank of the matrix $\mathbf{C}_h$ (and matrix A_h) is the least over ranks of all matrices B from the class $\Sigma_h(\widetilde{\mathbf{C}})$.*

Proof of the theorem immediately follows from the definition of the matrix $\mathbf{C}_h$ and Wielandt-Hoffman inequality (14) for the eigenvalues of the matrix $\widetilde{\mathbf{C}}$ which satisfy inequalities $\lambda_j > h$.

Importance of existense of optimal solution to the problem of regularization of empirical covariance matrix becomes clear if the scenario generation problem involves invertion of the matrix. Such a problem arises in the case of generation of conditional scenarios.

Suppose that we would like to generate a conditional distribution of the components of a vector given the values of some of the coordinates. We assume that this vector has a joint normal distribution. It is natural to represent the vector X in the form

$$X = (X^{(1)}, X^{(2)}),$$

where $X^{(1)}$ and $X^{(2)}$ are the random vectors of an arbitrary finite dimension. Let $\mu^{(i)} = \mathbf{E}X^{(i)}$, be the vector of expected values of the coordinates of $X^{(i)}$ and $\mathbf{Cov}X^{(i)} = \Sigma_{ii}$, be the covariance matrix of the coordinates of corresponding vector ($i = 1, 2$). Denote by

$$\Sigma_{12} = \mathbf{Cov}\left(X^{(1)}, X^{(2)}\right).$$

We assume that the vector $X = \left(X^{(1)}, X^{(2)}\right)$ has joint normal distribution $\mathcal{N}(\mu, \Sigma)$, where the vector

$$\mu = \left(\mu^{(1)}, \mu^{(2)}\right),$$

and the matrix

$$\Sigma = \begin{pmatrix} \Sigma_{11} & \Sigma_{12} \\ \Sigma_{21} & \Sigma_{22} \end{pmatrix}.$$

Obviously,

$$\Sigma_{21} = \Sigma_{12}^t,$$

where D^t denotes the transposed matrix D.

The problem of interest is to find the conditional distribution of the vector $X^{(2)}$ given the values of the coordinates of $X^{(1)}$. The solution to this problem (see [9]) is given by

Proposition 2 *The conditional distribution of the vector $X^{(2)}$ is a normal distribution*

$$\mathcal{L}\left(X^{(2)} | X^{(1)} = x^{(1)}\right) = \mathcal{N}(m, \mathbf{B}),$$

where the vector m and the matrices $\mathbf{A}$ and $\mathbf{B}$ are defined by the relations

$$m = \mu^{(2)} + \mathbf{A}(x^{(1)} - \mu^{(1)}), \quad \mathbf{A} = \Sigma_{21} \cdot \Sigma_{11}{}^{-1}, \tag{17}$$

and

$$\mathbf{B} = \Sigma_{22} - \Sigma_{21} \cdot \Sigma_{11}{}^{-1} \cdot \Sigma_{12}. \tag{18}$$

Obviously, computation of the parameters of the conditional distribution in (17) and (18) requires finding reciprocal matrix.

4 Bounds on the Value–at–Risk

The natural question to ask is how the changes in the covariance matrix affect the RiskMetrics **VaR**. It is not difficult to give such an estimate for a "linear" position. For any "linear" position X we have [2]

$$\mathbf{VaR}_X = MV_X \cdot \delta_X \cdot X_a$$

where MV is the market value, δ is the sensitivity and X_a is the maximal adverse movement per \$ of the market value with a given probability p (usually 0.95). For $p = 0.95$, X_a is approximately $1.65 * \sigma$, where σ is standard deviation of the position value. We will denote this coefficient by q. Thus

$$\mathbf{VaR}_X = MV_X \cdot \delta_X \cdot q \cdot \sigma \tag{19}$$

For a portfolio P comprised of n "linear" positions $X_1, \ldots, X_n$

$$\mathbf{VaR}_P = \sqrt{V \cdot C_X \cdot V^t},$$

where C_X is the correlation matrix and

$$V = (\mathbf{VaR}_{X_1}, \ldots, \mathbf{VaR}_{X_n})$$

However, our estimate is for the covariance matrix and therefore we have to use the following equality which follows from the associative property of matrix multiplication and (19):

$$\mathbf{VaR}_P^2 = V \cdot C_X \cdot V^t = \overline{V} \cdot \mathbf{C} \cdot \overline{V}^t,$$

where

$$\overline{V} = q \cdot (MV_{X_1} \cdot \delta_{X_1}, \ldots, MV_{X_n} \cdot \delta_{X_n})$$

Suppose now that the covariance matrix $\mathbf{C}$ has been modified and let μ be the estimate of its perturbation. Then for the modified matrix $\mathbf{C}_h$ we have

$$\mathbf{VaR}_P'^2 = \overline{V} \cdot \mathbf{C}_h \cdot \overline{V}^t$$

Therefore, from (15) and from (6) we obtain

$$|\mathbf{VaR}_P'^2 - \mathbf{VaR}_P^2| = |\overline{V} \cdot (\mathbf{C} - \mathbf{C}_h) \cdot \overline{V}^t| = |((\mathbf{C} - \mathbf{C}_h)\overline{V}, \overline{V})| \le 2\mu(\overline{V}, \overline{V}) \tag{20}$$

Finally,

$$|\mathbf{VaR}_P'^2 - \mathbf{VaR}_P^2| \le 2\mu \cdot q^2 \cdot (\sum_i MV_{X_i} \cdot \delta_{X_i})^2$$

Thus, we proved the following

Proposition 3 *For a portfolio P with assets $X_1, \ldots, X_n$ there is an estimate for a change in the* **VaR** *introduced by making the matrix of the correlations semi-positive definite:*

$$|\mathbf{VaR}_P'^2 - \mathbf{VaR}_P^2| \le 2\mu \cdot q^2 \cdot (\sum_i MV_{X_i} \cdot \delta_{X_i})^2$$

Note, that if $\mathbf{VaR}_P'$ and $\mathbf{VaR}_P$ are close, then $\mathbf{VaR}_P' + \mathbf{VaR}_P \approx 2\mathbf{VaR}_P'$ and we have that

$$|\mathbf{VaR}_P' - \mathbf{VaR}_P| \le \mu \cdot q^2 \cdot (\sum_i MV_{X_i} \cdot \delta_{X_i})^2 / \mathbf{VaR}_P'.$$

Also, if all the sensitivities are equal to one, then the inequality (20) simplifies to

$$|\mathbf{VaR}_P'^2 - \mathbf{VaR}_P^2| \le 2\mu \cdot q^2 \cdot MV_P^2$$

Therefore, we developed the stable numerical method of modification of the non-positive definite covariance matrix and calculated the explicit estimate of the error of VaR.

5 Examples

In this section we describe several examples of how the modification of the eigenvalues of the covariance matrix of the risk factors affect the changes in correlations and covariances.

In the first example, the risk factor space contains 41 risk factor. The key rates of the interest rate curves IRGBP, IRITL, IRUSD, FXITL and FXGBP comprise the risk factor space. The covariance matrix is read from the RM data files on 19 March 1998.

The covariance matrix is not positive–definite. It has 3 negative eigenvalues. The minimal eigenvalue is $-4.65 \cdot 10^{-8}$, the maximal eigenvalue is $8.0 \cdot 10^{-4}$. The modification of the eigenvalues leads to changes in the elements of the correlation matrix. The average change of the correlations is $1.1 \cdot 10^{-6}$; the maximal change in the correlations is $4.10 \cdot 10^{-4}$.

In the first example the number of negative eigenvalues did not exceed 10% of the total number of the eigenvalues. If the number of risk factors grow the number of negative eigenvalues may comprise up to 15 and even 20%.

The second example illustrates the modification of the covariance matrix when the number of risk factors is 83. The risk factors are the key rates of the interest rate curves DEM, FRF, ITL, USD, JPY, as well as foreign exchange rates between USD and DEM, FRF, ITL and JPY. In this case we have 13 negative eigenvalues, or 15.6%. The average change of correlations is $8.57 \cdot 10^{-5}$; the maximal change of correlations is $3.95 \cdot 10^{-5}$. The covariance matrix is read from the data set posted on 26 September 1996.

The method behaved well in all tests with the $RiskMetrics^{TM}$ data. In one of the tests, 50 eigenvalues were modified in the 190×190 covariance matrix. However, the maximal error in the matrix of correlations (it is more convenient to measure errors in the matrix of correlations as opposed to the matrix of covariations) was about 0.002 while the average error was 0.0001. The maximal relative error in the variations is 0.18% and the average about 0.03%. Smaller matrices usually produce even less error.

6 Conclusion

The statistical procedures applied to estimate the covariance structure of the risk factor space lead to unavoidable errors in computation of the eigenvalues. The regularization technique based on the method of the minimal symmetric pseudo–inverse operator allows to find optimal solution to the problem.

We believe that the method discussed above can be used by practitioners either for Monte Carlo simulation of the market risk scenarios or for quasi–analytical estimation of the Value–at–Risk when the information about covariance matrix is not precise.

The regularization algorithm admits small relative error in the Value–at–Risk estimation justified by the theoretical upper bound.

Acknowledgements

The authors are very grateful to Leonid Merkoulovitch, Dan Rosen and Michael Zerbs for fruitful discussions and suggestions.

References

[1] Press W., Vetterling W.,Teukolsky S. and Flannery B. *Numerical Recipes in C.* Cambridge University Press, 1992.

[2] *RiskMetrics^{TM} – Technical Document.* J.P.Morgan Guaranty Trust Company. Third edition, N.Y., May 26, 1995.

[3] Belkin M., Kreinin A. *Robust Monte Carlo Simulation for Variance/Covariance Matrices.* Algorithmics Inc. Technical paper No. 96-01, 1996.

[4] Davenport J.M., Iman R.L. *An Iterative Algorithm to Produce a Positive Definite Correlation Matrix from an Approximate Correlation Matrix.* Report SAND–81–1376, Sandia National Laboratories, Albuquerque, NM, 1981.

[5] Levin A.M. *Regularization of Unstable Criterion Problems and Solution of the Incompatible Operator Equations.* Dissertation of Doctor of Science (in Russian), Kiev State University, Ukraine, 1995.

[6] Fang Kai-Tai, Zhang Yao-Ting. *Generalized Multivariate Analysis.* Beijing: Science Press; Berlin: Springer-Verlag, 1990.

[7] Parlett B.N. *The Symmetric Eigenvalue Problem.* Englewood Cliffs, N.J.: Prentice-Hall, 1980.

[8] Lurie P.M., Goldberg M.S. *An Approximate Method for Sampling Correlated Random Variables from Partially Specified Distribution.* Management Science, 1998, vol. 44, 2, 203–218.

[9] Mood A., Graybill A. *Introduction to the Theory of Statistics.*, McGraw–Hill Inc., New York, 1963.

Structure of Optimal Stopping Strategies for American Type Options[1]

Alexander G. Kukush (kuog@mechmat.univ.kiev.ua)
Department of Mechanics and Mathematics, Kiev Taras Shevchenko University, 252601 Kiev, Ukraine

Dmitrii S. Silvestrov (dmitrii.silvestrov@mdh.se)
Department of Mathematics and Physics, Mälardalen University, SE-721 23 Västerås, Sweden

Abstract

The general pricing processes represented by an inhomogeneous vector Markov process with discrete time is considered. Its first component is interpreted as a price process and the second one as an index process controlling the price component. American type options with convex pay-off functions are studied. The structure optimal and ε-optimal buyer stopping strategies is investigated for various classes of convex pay-off functions.

Key words and phrases: Markov process, optimal stopping, convex pay-off function, American options.

AMS 1991 subject classification: Primary 62P05, 90C40, Secondary 60J25, 60J20.

1 Introduction

Traditional methods of option pricing are based on models of pricing processes which are various modifications of the classical model of geometrical Brownian motion. Stochastic differential equations can be written down for such pricing processes. Then

[1]This work has been supported by the EU Tempus-Tacis Programme (JEP-10353-97) and the Royal Swedish Academy of Sciences (Grant 1413).

S.P. Uryasev (ed.), Probabilistic Constrained Optimization, 173–185.
© 2000 *Kluwer Academic Publishers.*

partial differential equations and the corresponding variational problems can be derived for functions which represent optimal strategies, see for instance Øksendal (1992) and Duffie (1996) and Karatzas and Shreve (1998). Finally various numerical algorithmscan be applied to find optimal strategies for continuous time models and their discrete time approximations. The extended survey of latest results can be found in the book edited by Rogers and Talay (1998).

In the present paper an alternative approach is used for evaluation of optimal stopping buyer strategies for American type options. The structure of optimal stopping strategies is investigated by applying of the direct probabilistic analysis to discrete time models under general assumptions for underlying pricing processes. A model of pricing process under consideration is a two component inhomogeneous in time Markov process with a phase space $[0, \infty) \times X$ where the first component is the corresponding pricing process and the second component (with a general measurable phase space X) represents some stochastic index process controlling the pricing process. Pay-off functions under consideration are in sequel: (a) an inhomogeneous in time analogue of a standard one $g_n(x) = a_n[x - K_n]_+$; (b) piecewise linear convex functions, and finally (c) general convex functions.

We refer to the paper by Kukush and Silvestrov (1999) where similar results are presented for continuous time model and skeleton type approximations connecting continuous and discrete time models are given.

We think that the main advantage of direct probabilistic approach in structural studies of optimal stopping strategies is that this approach is much more flexible and less sensitive to the modifications of models of underlying pricing processes, pay-off functions and other characteristics of the models.

The knowledge of the explicit structure of optimal stopping strategies is the base for the creation of effective optimizing Monte Carlo pricing algorithms for numerical evaluation of the corresponding optimal strategies. Such algorithms and programs have been recently elaborated by Silvestrov, Galochkin and Sibirtsev (2000).

We would like to refer to the book by Pliska (1997) where one can find the basic theory concerning pricing processes with discrete time and the book by Shiryaev (1978) and the paper by Shiryaev, Kabanov, Kramkov, and Mel'nikov (1994) which stimulated the present research.

2 Optimal stopping in a discrete time model

Consider a two component inhomogeneous in time Markov process $Z_n = (S_n, I_n), n = 0, 1, \ldots$ with a phase space $Z = [0, \infty) \times Y$. Here $(Y, \mathcal{B}_Y)$ is a general measurable phase space and as usual we consider Z as a measurable space with the σ–field $\mathcal{B}_Z = \sigma(\mathcal{B}_+ \times \mathcal{B}_Y)$ where $\mathcal{B}_+$ a Borel σ–field on $R^+ = [0, \infty)$.

Let us denote $P^n_{x,y}$ a probability under the condition $S_n = x, I_n = y$ and as usual $P_n(x, y; A, B) = P^n_{x,y}\{(S_{n+1}, I_{n+1}) \in A \times B\}$ transition probabilities of Markov process

Z_n. Without loss of generality we assume that $Z_n = (S_0, I_0)$ is a non-random value in Z.

We interpret the first component S_n as a pricing process and the second component I_n as a stochastic index process controlling the pricing process.

Let $\{\mathcal{F}_n = \sigma(Z_0, \ldots Z_n),\ n = 0, 1, \ldots\}$ be a flow of σ–fields, associated with process Z_n. We shall consider Markov moments τ with respect to this flow. It means that τ is a random variable which can take values $0, 1, \ldots, +\infty$ and with the property $\{\tau(\omega) = n\} \in \mathcal{F}_n, n = 0, 1, \ldots$.

Introduce further pay-off functions $g_n(x), n = 0, 1, \ldots$. We assume that $g_n(x)$ is a nonnegative Borel measurable function on $R^+ = [0, \infty)$ for every $n = 0, 1, \ldots$. Let also $R_0 = 0, R_n = r_0 + r_1 + \ldots + r_{n-1}, n = 1, 2, \ldots$, where $r_m \geq 0$ is a riskless interest rate valid at the interval between the moments m and $m + 1$.

We fix parameter $N \in \mathcal{N}$ which we call an expiration date. Our goal is to maximise over all Markov moments $\tau \leq N$ the functional

$$\Phi_g(\tau) = Ee^{-R_\tau} g_\tau(S_\tau). \tag{1}$$

The Markov moment $\tau \leq N$, which delivers maximum to the functional $\Phi_g(\tau)$, is called the optimal stopping time $\check{\tau}_{opt}$. So, by definition

$$\Phi_g(\check{\tau}_{opt}) = \sup_{\tau \leq N} \Phi_g(\tau). \tag{2}$$

Consider typical examples of the process Z_n and pay-off functions.

Suppose that the pricing process and the index process are given in the dynamical form

$$S_n = A_n(S_{n-1}, I_n), I_n = B_n(I_{n-1}, \xi_n), n = 1, \ldots, \tag{3}$$

where $\xi_n, n = 1, 2, \ldots$ be a sequence of independent random elements which take values in a measurable space $(X, \mathcal{B}_X)$; $A_n : Z \to R^+$ and $B_n : Y \times X \to Y$ are measurable functions (with respect to natural σ-fields on the corresponding spaces). We suppose that (S_0, I_0) is a non-random value in Z.

Then the vector process $Z_n = (S_n, I_n), n = 0, 1, \ldots$ is an inhomogeneous Markov process with the flow of σ-fields $\mathcal{F}_0 = \{\emptyset, \Omega\}, \mathcal{F}_n = \sigma(\xi_1, \ldots, \xi_n), n = 1, 2, \ldots$.

A geometrical random walk can be considered as a particular case of the dynamical pricing process (3) with

$$S_n = S_{n-1} \cdot \xi_n,\ I_n = \xi_n, n = 1, 2, \ldots, \tag{4}$$

where $\xi_n, n = 1, 2, \ldots$ be a sequence of independent nonnegative random variables. Note that we consider inhomogeneous in time model. The standard model of geometrical random walk, widely used to model pricing processes, is the model (4), where $\xi_n, n = 1, 2, \ldots$ are i.i.d. nonnegative random variables.

The following pay-off function corresponds to the American call option

$$g_n(x) = a_n[x - K_n]_+ = \begin{cases} a_n(x - K_n), & \text{if } x > K_n, \\ 0, & \text{if } 0 \leq x \leq K_n, \end{cases} \tag{5}$$

where $a_n > 0$ and $K_n > 0, n = 0, 1, \ldots$ are, respectively, scale pricing coefficients and striking prices for moments $n = 0, 1, \ldots$. Note again that we consider inhomogeneous in time model. The standard model of American call option is the model where $a_n = a, K_n = K, n = 0, 1, \ldots$ are assumed to be independent on n.

To describe $\check{\tau}_{opt}$ in general situation, we use the classical results by Shiryaev (1978).

Define the operators T_n, which act on nonnegative measurable functions $f(x, y)$ acting from Z to the extended half line $[0, \infty]$ according to the equality

$$T_n f(x, y) = E^n_{x,y} f(S_{n+1}, I_{n+1}), \tag{6}$$

where $E^n_{x,y}$ means the expectation under the condition $S_n = x, I_n = y$ (as usual the expectation of a random variable which can take the value $+\infty$ with a positive probability is counted as equal to $+\infty$).

Define for $x \geq 0, y \in Y$:

$$w_0(x, y) = g_N(x), \tag{7}$$

and then for $k = 1, 2, \ldots, N$ by the recursion:

$$w_k(x, y) = \max\{g_{N-k}(x), e^{-r_{N-k}} T_{N-k} w_{k-1}(x, y)\}. \tag{8}$$

Introduce the set

$$\check{\Gamma}^N_n = \{(x, y) : g_n(x) = w_{N-n}(x, y)\} \tag{9}$$

and the cross-sections

$$\check{\Gamma}^N_n[y] = \{x : g_n(x) = w_{N-n}(x, y)\} \subseteq R^+, y \in Y. \tag{10}$$

Note that by (9) $\check{\Gamma}^N_N = R^+ \times Y$ and respectively $\check{\Gamma}^N_N[y] = R^+, y \in Y$.

The following statement is a variant of the classical result by Shiryaev (1978) modified for an inhomogeneous in time model. It can be obtained by the use of obvious transformation of the inhomogeneous in time Markov process Z_n to the homogeneous in time Markov process (n, Z_n).

THEOREM 1 (Shiryaev (1978)). *The optimal stopping time maximising the functional* (1) *is given by*

$$\check{\tau}_{opt} = \min\{0 \leq n \leq N : S_n \in \check{\Gamma}^N_n[I_n]\}$$

and

$$\Phi_g(\check{\tau}_{opt}) = w_N(S_0, I_0).$$

According to Theorem 1 we shall refer to the random sets $\check{\Gamma}^N_n[I_n]$ as to the optimal stopping domains. Our goal is to describe the structure of these domains for various types of pay-off functions in a more explicit form than those given by recursion formulas (7)-(8) and formulas (9)-(10).

It is useful to anticipate these studies by some general remark.

The first remark concerns the important particular case of the general model, where stopping domains do not depend on index component I_n. Let consider the model in which transition probabilities of Markov process Z_n can be represented in the form

$$P_n(x, y, A, B) = \int_B P_n(dz) P_n(x, z, A), A \in \mathcal{B}_+, B \in \mathcal{B}_Y. \qquad (11)$$

Condition (11) means that index component I_n is a sequence of independent random variables with the distributions $P_n(B)$. It is also conditionally independent of the first component S_n in the sense of relation (11). In this case the price process S_n itself is an inhomogeneous Markov process with transition probabilities

$$P_n(x, A) = \int_Y P_n(dz) P_n(x, z, A), A \in \mathcal{B}_+.$$

In the case, where the process Z_n is given in dynamical form (3), the analogue of (11) is condition that index process $I_n = B_n(\xi_n)$, i.e. it is a sequence of independent random variables which are transformations of random variables ξ_n. In this case

$$S_n = A_n(S_{n-1}, B_n(\xi_n)) = \tilde{A}_n(S_{n-1}, \xi_n), n = 1, 2, \dots. \qquad (12)$$

Relation (12) means that the process S_n itself is a inhomogeneous Markov process given in the dynamical form.

Conditions (11) or (12) imply that functions $w_n(x, y) = w_n(x)$ defined in (7)-(8) depend only upon x. This implies in sequel that domains $\check{\Gamma}_n^N = \tilde{\Gamma}_n^N \times Y$ and therefore cross-sections $\check{\Gamma}_n^N[y] = \tilde{\Gamma}_n^N, y \in Y$ do not depend on y. By definition $\tilde{\Gamma}_n^N = \{x \in R^+ : g_n(x) = w_{N-n}(x)\}$ and the optimal stopping time can be defined as

$$\check{\tau}_{opt} = \min\{0 \leq n \leq N : S_n \in \tilde{\Gamma}_n^N\}.$$

The second remark concerns the uniqueness of optimal stopping domains and the corresponding optimal stopping moments.

Let denote

$$\Pi_n = \{x \geq 0 : g_n(x) = 0\}, n = 0, 1, \dots$$

and define for $n = 0, 1, \dots, N-1$ the reduced domains

$$\hat{\Gamma}_n^N[y] = \check{\Gamma}_n^N[y] \setminus \Pi_n = \{x \geq 0 : g_n(x) = w_{N-n}(x, y) > 0\}, y \in Y.$$

It is obvious that if the option was not exercised before moment n and $S_n = x \in \Pi_n$, then there is no sense to exercise the option at the moment n either. To wait and to exercise option in some moment in the interval of time $(n, N]$ will be in any case not worse then to exercise the option at the moment n.

Using this remark it is easy to show that the stopping moment $\hat{\tau}_{opt} = \min\{0 \leq n \leq N : S_n \in \hat{\Gamma}_n^N[I_n]\} \geq \check{\tau}_{opt}$ (here we set $\hat{\Gamma}_N^N[y] = R^+$) can serve as an optimal stopping moment equally well with $\check{\tau}_{opt}$ because of $\Phi_g(\hat{\tau}_{opt}) = \Phi_g(\check{\tau}_{opt})$.

Moreover, let measurable domains $\Pi_n^* \subseteq \Pi_n, n = 0, 1, \ldots$ and $\Gamma_n^N[y] = \check{\Gamma}_n^N[y] \setminus \Pi_n^*, n = 0, 1, \ldots, N-1$, $\Gamma_N^N[y] = R^+$. Let now define

$$\tau_{opt} = \min\{0 \leq n \leq N : S_n \in \Gamma_n^N[I_n]\}. \tag{13}$$

By definition $\hat{\Gamma}_n^N[y] \subseteq \Gamma_n^N[y] \subseteq \check{\Gamma}_n^N[y], y \in Y$ and therefore $\check{\tau}_{opt} \leq \tau_{opt} \leq \hat{\tau}_{opt}$. However, $\Phi_g(\check{\tau}_{opt}) = \Phi_g(\tau_{opt}) = \Phi_g(\hat{\tau}_{opt})$.

In this paper we study the pay-off functions $g_n(x)$ such that $g_n(x) = 0$ for $x \leq K_n$ and $g_n(x) > 0$ for $x > K_n$. In this case the sets $\Pi_n = [0, K_n]$ and it is convenient to choose sets $\Pi_n^* = [0, K_n)$, i.e. to exclude from the optimal stopping domains the points $x < K_n$. In this case

$$\Gamma_n^N[y] = \check{\Gamma}_n^N[y] \setminus [0, K_n) = \{x \geq K_n : g_n(x) = w_{N-n}(x, y)\}, y \in Y. \tag{14}$$

3 The case of American call options

We specify the structure of optimal stopping domain for classical pay-off function given by formula (5). It was done in Shiryaev et al. (1994) for pricing processes represented by a homogeneous geometrical random walk on the lattice. Our setting of pricing processes is much more general.

We assume the following condition for transition kernel of Markov process Z_n:

A: For every $n = 0, 1 \ldots$ there exist transition kernels $P_n(y, B), y \in Y, B \in \mathcal{B}_Y$ and $P_n(x, y, z, A), x \in R^+, y, z \in Y, A \in \mathcal{B}_+$ such that for every $x \in R^+, y \in Y, A \in \mathcal{B}_+, B \in \mathcal{B}_Y$ the following representation holds:

$$P_n(x, y, A, B) = \int_B P_n(y, dz) P_n(x, y, z, A).$$

Under condition **A** the index component I_n itself is an inhomogeneous Markov process with transition kernel $P_n(y, B)$. As far as kernel $P_n(x, y, z, A)$ is concerned it is the conditional distribution of S_{n+1} with respect to random variables S_n, I_n and I_{n+1}.

In the case when the pricing process is given in dynamical form (3) condition **A** automatically holds. Note also that (11) is the particular case of condition **A**.

Let us introduce the condition which means that the conditional distributions of the pricing process $P_n(x, y, z, A)$ are stochastically monotonic with respect to the current price level:

B: For every $n = 0, 1 \ldots, y, z \in Y, s \geq 0$ and $0 \leq x' < x'' < \infty$:

$$P_n(x', y, z, [s, \infty)) \leq P_n(x'', y, z, [s, \infty)).$$

In the case when the pricing process is given in dynamical form (3) the following condition is analogue of **B**:

B$'$: For every $n = 0, 1, \ldots$ and $y \in Y$ function $A_n(x, y)$ is monotonically nondecreasing in $x \geq 0$.

It is obvious that **B**$'$ implies **B**.

Let now formulate the condition which means that the conditional distributions of the pricing process $P_n(x, y, z, A)$ possess some stochastic convexity property with respect to a current price level.

We introduce distribution functions $F_{n,x,y,z}(s) = P_n(x, y, z, [0, s]), s \geq 0$ and then define quantile functions $F^{-1}_{n,x,y,z}(u) = \sup\{s \geq 0 : F_{n,x,y,z}(s) \leq u\}, 0 \leq u \leq 1$. Let now ρ be a uniformly distributed on $[0, 1]$ random variable. It is obvious that $P\{F^{-1}_{n,x,y,z}(\rho) \leq s\} = F_{n,x,y,z}(s), s \geq 0$. Now, the convexity condition is:

C: For every $n = 0, \ldots, y \in Y, s \geq 0, B \in \mathcal{B}_Y$ and $0 \leq x' < x'' < \infty$:

$$P_n(\frac{x' + x''}{2}, y, z, [s, \infty)) \leq P\{\frac{F^{-1}_{n,x',y,z}(\rho) + F^{-1}_{n,x'',y,z}(\rho)}{2} \geq s\}.$$

In the case when the pricing process is given in dynamical form (3) condition **C** takes the following form:

C$'$: For every $n = 0, 1, \ldots$ and $y \in Y$ the function $A_n(x, y)$ is convex in $x \geq 0$.

Denote by V the space of functions $f(x, y) : Z \to [0, +\infty]$, which are: (a) measurable, (b) do not decrease in $x \geq 0$ for every $y \in Y$, and (c) are convex in $x \geq 0$ for every $y \in Y$.

Note that any function $f(x, y)$ from V, which takes values in $[0, \infty)$, is continuous in $x \geq 0$ for every $y \in Y$ due to its monotonicity and convexity in $x \geq 0$.

Lemma 1. *Assume the conditions* **A**–**C**. *If a function* $f(x, y)$ *belongs to* V *then for every* $n = 0, 1, \ldots$ *the function* $T_n f(x, y)$ *belongs to* V.

The following condition means that in the case of very large values of the pricing process it is better immediately to exercise the option:

D: For every $n = 0, \ldots$ and $y \in Y$:

$$\lim_{x \to \infty} \frac{1}{x} E^n_{x,y} S_{n+1} < \frac{a_n}{a_{n+1}} e^{r_n}.$$

Conditions **A**–**D** imply that $E^n_{x,y} S_{n+1} < \infty$ for every $n = 0, 1, \ldots$ and $x \geq 0, y \in Y$. Note that under **A**–**C** the limit in **D** exists because by Lemma 1 the function $h_n(x, y) = E^n_{x,y} S_{n+1}$ belongs to the space V for every $n = 0, 1, \ldots$.

In the case when the pricing process is given in dynamical form (3) the condition **D** takes the following form:

$\mathbf{D}'$: For every $n = 0, \dots$ and $y \in Y$:

$$\lim_{x \to \infty} \frac{1}{x} E A_{n+1}(x, B_{n+1}(y, \xi_{n+1})) < \frac{a_n}{a_{n+1}} e^{r_n}.$$

The following lemma gives important information about reward functions $w_n(x, y)$ defined in (7)–(8).

LEMMA 2. *Assume the conditions* **A**–**D**. *Then for every* $n = 0, 1, \dots, N$: (a) $w_n(x, y) < \infty, x \geq 0, y \in Y$, (b) $w_n(x, y)$ *is continuous in* $x \geq 0$ *for every* $y \in Y$, (c) $w_n(x, y) \in V$.

Lemma 2 implies also that under conditions **A**–**D**

$$\Phi_g(\tau_{opt}) = w_N(S_0, I_0) < \infty.$$

Now we are prepared to formulate the first main result.

THEOREM 2. *Suppose that the conditions* **A**–**D** *hold (in the case when the pricing process is given in dynamical form* (3) *conditions* **B**–**D** *can be replaced by conditions* **B**$'$-**D**$'$). *Then:*

(i) *for the pay-off function given by formula* (5) *there exists for every* $n = 0, 1, \dots,$ $N - 1$ *a unique root* $d_n^*(I_n)$ *of the equation*

$$g_n(x) = e^{-r_n} T_n w_{N-n-1}(x, I_n), \; x \geq K_n, \tag{15}$$

moreover $g_n(x) > e^{-r_n} T_n w_{N-n-1}(x, I_n), x > d_n^*(I_n)$ *and* $g_n(x) < e^{-r_n} T_n w_{N-n-1}(x, I_n)$, $K_n \leq x < d_n^*(I_n)$ *if* $K_n < d_n^*(I_n)$;

(ii) *the optimal stopping domains defined in* (14) *have the following form*

$$\Gamma_n^N[I_n] = [d_n^*(I_n), \infty) \subseteq [K_n, \infty)$$

and therefore the optimal stopping moment maximising the functional (1) *is given by* $\tau_{opt} = \min\{0 \leq n \leq N : S_n \geq d_n^*(I_n)\}$, *where we set* $d_N^*(I_N) = 0$.

It is not out of picture to note that $d_n^*(y)$ is a measurable function acting from Y to R^+. This function can be found by solving the equations $g_n(x) = e^{-r_n} T_n w_{N-n-1}(x, y)$, $x > K_n$ for $y \in Y$. Note also that $g_n(x) > e^{-r_n} T_n w_{N-n-1}(x, y)$ for $x > d_n^*(y)$ and $g_n(x) < e^{-r_n} T_n w_{N-n-1}(x, y)$ for $x < d_n^*(y)$.

In the case when assumptions (11) or (12) hold the equation (15) takes the form

$$g_n(x) = e^{-r_n} T_n w_{N-n-1}(x), \; x > K_n. \tag{16}$$

Under condition of Theorem 2 for every $n = 0, 1, \dots, N$ there exists a unique root d_n^* of equation (15) and the optimal stopping moment is given by $\tau_{opt} = \min\{0 \leq n \leq N : S_n \geq d_n\}$.

4 The case of piecewise linear convex pay-off function

Let us consider a more general than in (5) piecewise linear convex pay-off functions $g_n(x)$.

Let $p \geq 2$ (the case $p = 1$ was considered in Section 3) is fixed and independent from n. Let also for each $n = 0, 1 \ldots$ two sequences of positive numbers $0 < K_{n,1} < K_{n,2} < \ldots < K_{n,p} < \infty$ and $0 < a_{n,1} < a_{n,2} < \ldots < a_{n,p} < \infty$ are given. We assume that $g_n(x)$ be a continuous nonnegative function defined on $[0, \infty)$ such that $g_n(x) = 0, x \in [0, K_{n,1}]$ and $g_n(x)$ is linear on each of the intervals $[K_{n,1}, K_{n,2}], \ldots, [K_{n,p-1}, K_{n,p}], [K_{n,p}, \infty)$ with the corresponding slopes $a_{n,1}, \ldots, a_{n,p-1}, a_{n,p}$.

Conditions **B** (**B′**) and **C** (**C′**) do not require any changes since they do not depend of pay-off functions. In **D** (**D′**) the constants a_n must be replaced by $a_{n,p}$.

As it was in the case of simplest pay-off function (5) conditions **A**, **B** and **D** imply that $w_n(x, y) < \infty$ for all $x \geq 0, y \in Y$ and $n = 0, 1, \ldots, N$ and so $\Phi_g(\tau_{opt}) < \infty$.

THEOREM 3. *Suppose that $g_n(x)$ is a piecewise linear convex pay-off function and the conditions* **A**–**D** *hold (in the case when the pricing process is given in dynamical form* (3) *conditions* **B**–**D** *can be replaced by conditions* **B′**–**D′**). *Then:*

(i) *for every $n = 0, 1, \ldots, N-1$ and $i = 1, \ldots, p-1$ the inequality*

$$g_n(x) \geq e^{-r_n} T_n w_{N-n-1}(x, I_n), \; x \in [K_{n,i}, K_{n,i+1}] \tag{17}$$

has the set of solution $J_{n,i}$ which is either the empty set or an interval $[d^*_{n,i}(I_n), e^*_{n,i}(I_n)] \subseteq [K_{n,i}, K_{n,i+1}]$, moreover the left (right) point of interval $J_{n,i}$ either coincides with the corresponding end point of the interval $[K_{n,i}, K_{n,i+1}]$ or it is a solution of the equation

$$g_n(x) = e^{-r_n} T_n w_{N-n-1}(x, I_n), \; x \in [K_{n,i}, K_{n,i+1}]; \tag{18}$$

(ii) *there exists for every $n = 0, 1, \ldots, N-1$ a unique root $d^*_{n,p}(I_n)$ of the equation*

$$g_n(x) = e^{-r_n} T_n w_{N-n-1}(x, I_n), \; x \geq K_{n,p}, \tag{19}$$

moreover $g_n(x) > e^{-r_n} T_n w_{N-n-1}(x, I_n), x > d^*_{n,p}(I_n)$ and $g_n(x) < e^{-r_n} T_n w_{N-n-1}(x, I_n)$, $K_{n,p} \leq x < d^*_{n,p}(I_n)$ if $K_{n,p} < d^*_{n,p}(I_n)$;

(iii) *the optimal stopping domains have the following form*

$$\Gamma^N_n[I_n] = J_{n,1} \cup \ldots \cup J_{n,p-1} \cup [d^*_{n,p}(I_n), \infty), \tag{20}$$

and therefore the optimal stopping moment maximizing the functional (1) *is given by $\tau_{opt} = \min\{0 \leq n \leq N : S_n \in \Gamma^N_n[I_n]\}$, where we set $\Gamma^N_N[I_N] = R^+$.*

It is useful to note that $d^*_{ni}(y)$ and $e^*_{ni}(y)$ are measurable function acting from Y to R^+. This function can be found by solving the inequalities and equations which can be obtained from (17)-(19) by formal replacement of I_n by y.

In the case when assumptions (11) or (12) hold the end points of all intervals in (20) do not depend on I_n and therefore the optimal stopping domains also do not depend on I_n.

5 The case of an arbitrary convex pay-off function

Consider now the case where pay-off functions $g_n(x) : R^+ \to R^+$ satisfy the following condition:

E: for every $n = 0, 1, \ldots$: (a) $g_n(0) = 0$, (b) $g_n(x)$ is not identical zero, (c) $g_n(x)$ is convex in $x \geq 0$.

Condition **E** implies that there exist $K_n \geq 0$ such that $g_n(x) = 0, x \in [0, K_n]$, and $g_n(x)$ is continuous and strictly increasing for $x \geq K_n$.

Moreover, condition **E** implies that $g_n(x)$ is absolutely continuous for $x \geq 0$. Denote by $g'_n(x)$ the right-hand derivative of $g_n(x)$. It exists everywhere and $g'_n(x) = 0, x \in [0, K_n)$, $g'_n(x)$ is nondecreasing for $x \geq K_n$, $g'_n(K_n) \geq 0$ and $g'_n(x) > 0, x > K_n$.

Let first consider the case when the following condition holds:

F: for every $n = 0, 1, \ldots$: $G_n = \sup_{x \geq 0} g'_n(x) < \infty$.

We assume that conditions **E** and **F** hold and construct the approximation of $g_n(x)$ by piecewise linear convex functions.

Denote $G = \max_{1 \leq n \leq N} G_n$, and for every $p = 1, 2, \ldots$ introduce the sets $H_{n,m,p} = \{x \geq 0 : \frac{m}{p} G \leq g'_n(x) < \frac{m+1}{p} G\}$ for $0 \leq m \leq p-1$ and $H_{n,p,p} = \{x \geq 0 : g'_n(x) = G\}$. Due to monotonicity of derivatives $g'_n(x)$ these sets are intervals which have no intersections, are located on R^+ in the sequential order according to the index m (if $m' < m''$ then $H_{n,m',p}$ lies on the left side of $H_{n,m'',p}$) and their union covers the whole half line R^+.

We approximate derivatives $g'_n(x)$ by the functions $g'_{n,p}(x) = \frac{mG}{p}$ if $x \in H_{n,m,p}$, $m = 0, 1, \ldots p$, and the function $g_n(x)$ by the functions

$$g_{n,p}(x) = \int_0^x g'_{n,p}(y) dy, \ x \geq 0.$$

By definition $g_{n,p}(x)$ are piecewise linear convex functions. They are connected with $g_n(x)$ for every $n = 0, \ldots N$ and $p = 1, 2, \ldots$ by the following inequalities

$$0 \leq g_n(x) - g_{n,p}(x) \leq \frac{G}{p} [x - K_n]_+, \quad x \geq 0.$$

The structure of optimal stopping domains for general convex pay-off functions $g_n(x)$ can be very complicated. However, using approximation of these functions by piecewise linear convex functions we can describe the structure of ε-optimal stopping domains in such a way that the corresponding stopping times τ_ε deliver maximum to the goal functional $\Phi_g(\tau)$ with the deficiency less than or equal to a given small $\varepsilon > 0$.

Let us denote

$$\varepsilon_p = \sup_{\tau \le N} E\, e^{-R_\tau} \frac{G}{p} [S_\tau - K_\tau]_+. \tag{21}$$

Theorems 2 and 3 can be used to find the optimal stopping moment and the value of maximum on the right hand side in (21). Conditions **B** (**B**′) and **C** (**C**′) do not require any changes. In **D** (**D**′) the constants a_n must be replaced by G/p.

Under conditions **A**–**D** ε_p can be made as small as it is necessary by the appropriate choice of the parameter p.

Let us denote by $\tau[p]$ the optimal stopping time for the piecewise linear convex pay-off functions $g_{n,p}(x)$ which approximate functions $g_n(x)$ according the algorithm described above, i.e.

$$E\, e^{-R_{\tau[p]}} g_{\tau[p],p}(S_{\tau[p]}) = \sup_{\tau \le N} E\, e^{-R_\tau} g_{\tau,p}(S_\tau). \tag{22}$$

Theorems 2 and 3 can be used to find the optimal stopping times $\tau[p]$ and the optimal stopping domains and the value of maximum on the right hand side in (22). Conditions **B** (**B**′) and **C** (**C**′) do not require changes. In **D** (**D**′) the constants a_n must be replaced by G.

Let us summarize conditions, which have to be assumed in order to realize the approximation construction for given p. Conditions **B** (**B**′) and **C** (**C**′) do not require changes. In **D** (**D**′) the constants a_n can be replaced by 1. Really, if constants $a_n = a$ do not depend on n the expression on the right side of the inequality in **D** is equal to e^{r_n} and does not depend on a.

Conditions **A**, **B**, **D** and **F** imply that $w_n(x, y) < \infty$ for all $x \ge 0, y \in Y$ and $n = 0, 1, \ldots, N$ and so $\Phi_g(\tau_{opt}) < \infty$, moreover the same is valid for approximation functions $g_{n,p}(x)$.

THEOREM 4. *Let conditions* **A**-**F** *hold. Then the Markov moment* $\tau[p]$ *is an* ε_p*-optimal stopping moment for the optimization problem* (1)-(2) *in the sense that*

$$\Phi_g(\tau[p]) \ge \sup_{\tau \le N} \Phi_g(\tau) - \varepsilon_p.$$

To conclude the investigation let us consider also the case when pay-off functions $g_n(x)$ have unbounded derivatives, i.e. the following condition holds:

$\bar{\mathbf{F}}$: $\lim_{x \to +\infty} g_n'(x) = +\infty, n = 0, 1, \ldots.$

Let us choose $T > \max_{0\le n\le N} K_n$ and consider the function

$$g_n^T(x) = \begin{cases} g_n(x), & \text{if } 0 \le x \le T, \\ g_n(T) + g_n'(T)\cdot(x-T), & \text{if } x > T. \end{cases}$$

This functions satisfies **E** and **F**. Let us denote $g_{n,p}^T(x)$ the corresponding piecewise linear approximations constructed according the algorithm described above. In this case constant $G = \max_{0\le n\le N} g_n'(T)$.

Let denote

$$\varepsilon_{p,T} = \sum_{n=0}^{N} e^{-R_n} E\,(g_n - g_n^T)(S_n) + \sup_{\tau\le N} E\, e^{-R_\tau}\frac{G}{p}\,[S_\tau - K_\tau]_+. \tag{23}$$

Theorems 2 and 3 can be used to find the optimal stopping moment and the value of maximum of the second term on the right hand side in (23). Conditions **B** (**B′**) and **C** (**C′**) do not require any changes. In **D** (**D′**) the constants a_n must be replaced by G/p.

In order to provide the relations $\lim\limits_{T\to\infty} E(g_n - g_n^T)(S_n) = 0$ for all $n = 0, 1, \ldots N$ the following condition have to be assumed:

G: $E_{x,y}^0 g_n(S_n) < \infty$, $(x, y) \in Z, n = 0, 1, \ldots, N$.

Obviously $\varepsilon_{p,T}$ can be made as small as it is necessary by the appropriate choice of T and then p.

Let us denote by $\tau[p,T]$ the optimal stopping time for the piecewise linear convex pay-off functions $g_{n,p}^T(x)$ i.e.

$$Ee^{-R_{\tau[p,T]}} g_{\tau[p,T],p}^T(S_{\tau[p,T]}) = \sup_{\tau\le N} Ee^{-R_\tau} g_{\tau,p}^T(S_\tau).$$

Theorem 2 and 3 can be used to find the optimal stopping times $\tau[p,T]$ and the corresponding optimal stopping domains.

Let us again summarize conditions which have to be assumed in order to realize the approximation construction for given T and p. Conditions **B** (**B′**) and **C** (**C′**) do not require changes. In **D** (**D′**) constants a_n can be replaced by 1 as in Theorem 6.

Conditions **A**, **B**, **D** and **G** imply that $w_n(x,y) < \infty$ for all $x \ge 0, y \in Y$ and $n = 0, 1, \ldots, N$ and therefore $\Phi_g(\tau_{opt}) < \infty$, moreover the same is valid for approximation functions $g_{n,p}^T(x)$.

THEOREM 5. *Let conditions* **A**-**E** *and* $\bar{\mathbf{F}}$, **G** *hold. Then the Markov moment* $\tau[p,T]$ *is an* $\varepsilon_{p,T}$*-optimal stopping moment for the optimisation problem* (1)-(2) *in the sense that*

$$\Phi_g(\tau[p,T]) \ge \sup_{\tau\le N} \Phi_g(\tau) - \varepsilon_{p,T}.$$

In conclusion we would like to refer to the papers by Kukush and Silvestrov (2000) which is an extended precursor of the present paper and where one can find the more detailed presentation of results including proofs.

References

[1] Duffie, D. (1996). *Dynamical Asset Pricing Theory.* Princeton University Press.

[2] *Numerical Methods in Finance.* (1998). Ed. by L.G.G. Rogers and D.Talay. Cambridge University Press.

[3] Øksendal, B. (1992). *Stochastic Differential Equations: An Introduction with Applications.* Springer.

[4] Karatzas, I. and Shreve, S. E. (1998). *Methods of Mathematical Finance.* Springer Verlag.

[5] Kukush, A. G. and Silvestrov, D. S. (1999). Optimal stopping strategies for American type options with discrete and continuous time. *Theory Stoch. Proces.* **5(21)**, 1-2, 71-79. In *Proceedings of the Second International School on Actuarial and Financial Mathematics*, Kiev, 1999.

[6] Kukush, A.G. and Silvestrov, D.S. (2000). Optimal pricing of American type options with discrete time. Research Report 2000-1, Department of Mathematics and Physics, Mälardalen University.

[7] Pliska, S.R. (1997). *Introduction to Mathematical Finance.* Blackwell.

[8] Shiryaev, A.N. (1978).*Optimal Stopping Rules.* Springer.

[9] Shiryaev, A.N., Kabanov, Yu.M., Kramkov, D.O., and Mel'nikov, A.V. (1994). Toward a theory of pricing options of European and American types. I. Discrete time. *Theory Probab. Appl.* **39**, 14-60.

[10] Silvestrov, D. S., Galochkin, V. G. and Sibirtsev, V. G. (1999). Algorithms and programs for optimal Monte Carlo pricing of American type options. *Theory Stoch. Proces.* **5(21)**, 1-2, 175-187. In *Proceedings of the Second International School on Actuarial and Financial Mathematics*, Kiev, 1999.

Approximation of Value-at-Risk Problems with Decision Rules

Riho Lepp (lprh@ioc.ee)
Tallinn Technical University
Institute of Economics
Kopli 101, EE 11711 Tallinn, Estonia

Abstract

Probability function maximization and quantile function minimization problems are approximated starting from the weak convergence of discrete measures with increasing dimension. It is assumed that solutions of both problems depend on a random parameter, i.e., solutions are sought as decision rules from the class of bounded measurable functions L^∞. Both problems are approximated by sequences of finite dimensional extremum problems with discrete measures and with increasing dimensions. Convergence conditions for optimal values and solutions of both problems are presented.

Keywords Probability functional, quantile functional, discrete convergence and stability

Stochastic programs (SP) cover this part of the mathematical programming community where cost and/or constraint functions depend on a random parameter and are divided, largely speaking, into two classes: stochastic programs with recourse (or two-stage SP-s) and chance constrainsed programs. The latter ones were introduced into SP models in order to ensure a certain level of reliability of the solution of an extremum problem with random parameter.

Let $x \in R^r$ and let ξ be an m-dimensional random parameter with distribution $\sigma(\cdot)$. Starting from a $r \times m$-dimensional function $f(x,\xi)$, $f : R^r \times R^m \to R^1$, and from an one-dimensional parameter t, $t \in R^1$, one can define the probability and quantile functions of the form: for a fixed t the probability function $v_t(x)$ as

$$v_t(x) = \mathbf{P}\{\xi \mid f(x,\xi) \leq t\} \tag{1}$$

S.P. Uryasev (ed.), Probabilistic Constrained Optimization, 186–197.
© 2000 *Kluwer Academic Publishers.*

and for for a fixed probability level α, $0 < \alpha < 1$, the quantile function $w_\alpha(x)$ as

$$w_\alpha(x) = \min_t \{\mathbf{P}\{\xi \mid f(x,\xi) \leq t\} \geq \alpha\}. \tag{2}$$

Depending on the problem formulation in probabilistic optimization, we can look for its solution as for a deterministic vector $x \in R^r$ or as for a function which depends on the random parameter ξ, i.e., $x = x(\xi)$. The latter class of problems was introduced to stochastic programming community by Charnes and Cooper in early 60-ies [1], where solutions were looked from various classes of functions (linear, piecewise linear, e.t.c.) and were called "decision rules". In [2] solutions of problems with probability and quantile functionals $v_t(x)$ and $w_\alpha(x)$ were sought in Lebesgue L^p-spaces, $1 \leq p \leq \infty$, i.e., solution $x(\xi)$ was an integrable with p-th power function, $x(\cdot) \in L^p$.

In this paper we will approximate the following probabilistic programming problems: to maximize the probability functional $v_t(x)$,

$$\max_{x(\cdot)\in C} v_t(x) = \max_{x(\xi)\in C} \mathbf{P}\{\xi \mid f(x(\xi),\xi) \leq t\} \tag{3}$$

and to minimize its "inverse", the Value-at-Risk (or quantile) functional $w_\alpha(x)$,

$$\min_{x(\cdot)\in C} w_\alpha(x) = \min_{x(\xi)\in C} \min_t \{\mathbf{P}\{\xi \mid f(x(\xi),\xi) \leq t\} \geq \alpha\}, \tag{4}$$

where C is a certain bounded set and decision rule $x(\xi)$ should belong to the set C almost surely.

As an example of the Value-at-Risk minimization problem with decision rules consider the model of correction of a satellite orbit from [3], Ch. 1.8.

A geostationary satellite is put into the Clark's orbit in three stages. At the first stage the satellite is transferred to a near-circular equatorial orbit, at second it is removed to the desired longitude, where it should be "hung". At the third stage the satellite is kept in the neighbourhood of this point. Since transfers are not ideal, such as the thrust of the satellite engine, e.t.c., it should be dealt with a stochastic model of the transfer process. In their model the authors assumed that that the first two stages of the transfer have been carried out randomly. What is left is to compensate the rest of the drift speed. Only one correction was used for this purpose, and the mathematical model of the correction was given by the following recoursion relation

$$d = \xi_1 + x(1 + \xi_2), \tag{5}$$

where d is the drift speed after the correction, ξ_1 is the current drift speed at the apogee, x is the drift speed impulse and ξ_2 is the random coefficient characterizing the error of the correction impulse implementation. It was assumed that random variables ξ_1, ξ_2 are independent and normally distributed with means $E\xi_1 = m > 0$, $E\xi_2 = 0$ and variances $D\xi_1 = \sigma_1^2$, $D\xi_2 = \sigma_2^2$, respectively. Denote by $s = (s_1, s_2)$ realizations of the two-dimensional random vector $\xi = (\xi_1, \xi_2)$. Then the loss function

$$f(x,s) = \mid d(x,s) \mid = \mid s_1 + x(1 + s_2) \mid \tag{6}$$

characterizes the rest of the drift speed and it has to be minimized with respect to strategy x, which in the model of authors was considered as the feedback control, i.e., $x = x(s_1)$.

Fix a certain level t and define the probability function $v_t(x)$ as follows:

$$v_t(x) = \mathbf{P}\{s \mid \mid s_1 + x(s_1)(1+s_2) \mid \leq t\}. \tag{7}$$

Now fix a reliability (probability) level $\alpha > 0$ and define the quantile function $w_\alpha(x)$ of the form

$$w_\alpha(x) = \min_t \; \{\mathbf{P}\{s \mid \mid s_1 + x(s_1)(1+s_2) \mid \leq t\} \; \geq \alpha\} \tag{8}$$

In the satellite orbit correction problem function $w_\alpha(x)$ defines a level such that the correction error does not exceed this level with a given (high) probability $\alpha > 0$, and it has to be minimized with respect to strategy $x(s_1)$. Now the Value-at-Risk minimization problem looks as follows

$$\min_{x(\cdot)} w_\alpha(x) = \min_{x(s_1)} \min_t \{t \mid \; \mathbf{P}\{s \mid \mid s_1 + x(s_1)(1+s_2) \mid \leq t\} \geq \alpha\}. \tag{9}$$

To find a solution to the minimization problem (1.9) (in general, to problem (1.4)) is an extremely complicated optimization problem due to following reasons. First, solution $x(s_1)$ is a decision rule, i.e. it is sought as a function. Second, defining the probability functions in (1.3) and (1.4) via integral, we should introduce the Heaviside zero-one function as an integrand, which itself is a discontinuous function. Third, from which class of functions (continuous, discontinuous, etc.) we should look for the optimal solution? In [4] it was assumed that the correction strategy $x(s_1)$ was a Borel-measurable function. But the class of measurable functions is too wide for description and stability analysis of optimization problems in function spaces - it determines only a metric space. Assuming that admissible strategies (decision rules) are measurable and belong with probability one (almost everywhere (a.e.)) to a bounded set, we are able to look for optimal solutions $x(s_1)$ in the Banach space $L^\infty(\sigma)$ of essentially bounded measurable functions with *vraisup*-norm topology. Note that in L^p-spaces, $1 \leq p \leq \infty$, only in the L^∞-space sets of the type $\{x(s) \mid x(s) \in C \; a.s.\}$, have interior points, i.e. only in L^∞-space the Slater condition holds and therefore, the existence of Lagrange multipliers in L^p-spaces, $1 \leq p < \infty$, is under the doubt.

In [3] bounds (indirect estimates) such as mean squares and minimax ones (*min* over x, *max* over s) were introduced in order to estimate the optimal strategy $x_\alpha(s_1)$.

Here we will estimate the solutions of (1.3) and (1.4) directly, starting from the weak convergence of a sequence of discrete measures $\{(m_n, s_n)\}$ to the probability measure $\sigma(\cdot)$ of initial problems (1.3) and (1.4):

$$\sum_{i=1}^{n} h(s_{in}) m_{in} \to \int_S h(s)\sigma(ds), \; n \to \infty. \tag{10}$$

It is a well-known fact (see, e.g., [5]) that assuming boundedness of the support S of the measure $\sigma(\cdot)$ and its atomless,

$$\sigma\{s \mid \mid s-t \mid = const\} = 0 \;\; \forall t \in R^m,$$

the weak convergence (1.10) of a sequence of discrete measures $\{(m_n, s_n)\}$ is equivalent to the following partition $\{\mathcal{A}_n\}$, $\mathcal{A}_n = \{A_{1n}, A_{2n}, ..., A_{nn}\}$, $n \in N = \{1, 2, 3, ..., n, ...\}$ of S with properties A1) - A7):

A1) $\sigma(A_{in}) > 0$,

A2) $A_{in} \cap A_{jn} = \emptyset, \neq j$;

A3) $\cup_{i=1}^{n} A_{in} = S$;

A4) $\sum_{i=1}^{n} \mid m_{in} - \sigma(A_{in}) \mid \to 0, n \in N$;

A5) $\max_{1 \leq i \leq n} diam A_{in} \to 0, n \in N$;

A6) $s_{in} \in A_{in}$;

A7) $\sigma(int\ A_{in}) = \sigma(A_{in}) = \sigma(cl\ A_{in})$,

where $diam\ A = \max_{s,t \in A} \mid s-t \mid$ and $int\ A$ and $cl\ A$ denote interior and closure of a set A, respectively. Note that the collection of sets $\{\mathcal{A}_n\}$ with property A7) constitutes an algebra Σ_0 in the initial sigma-algebra Σ, $\Sigma_0 \subset \Sigma$ (see [5]), and if $S = [0, 1]$ and $\sigma(\cdot)$ is the Lebesgue measure on $[0, 1]$, then integrability relative to $\sigma \mid_{\Sigma_0}$ means the Riemann integrability.

We need the partition $\{\mathcal{A}_n\}$ in order to define the system of piecewise integral connection operators $\{p_n\}, n \in N$, from a Banach space $L^p(\sigma)$ to Euclidean space $R^{rn}, n \in N$, of the form:

$$(p_n x)_{in} = \sigma(A_{in})^{-1} \int_{A_{in}} x(s)\sigma(ds), \; i = 1, ..., n. \tag{11}$$

Denote this system of integral connection operators by $\mathcal{P}$, i.e., $\mathcal{P} = \{p_n\}$.

Remark 1 *Suppose, as in [6], that for any $x \in C$ and $\varepsilon > 0$ there exist a compact set S_ε such that*

$$\int_{R^m / S_\varepsilon} \mid f(x(s), s) \mid \sigma(ds) \; < \; \varepsilon.$$

Usually (see, e.g., [6]) in the partition of the domain S all sets from the initial sigma-algebra Σ are used, i.e., the probability measure $\sigma(\cdot)$ is discretized throughout conditional means [6, Ch. 4]. Let $\mathcal{S}_k, k = 1, 2, ...$, be a partition of S, $\mathcal{S}_k = \{S_{1k}, ..., S_{kk}\}$, with $S_{ik} \in \Sigma$, $i = 1, ...k$, $k \in N$, with properties S1) - S5):

S1) $S = \cup_{j=1}^{k} S_{jk}$;

S2) $S_{jk} \in \Sigma$;

S3) $S_{ik} \cap S_{jk} = \emptyset, i \neq j$;

S4) $\mathcal{S}_k \subset \mathcal{S}_{k+1}$ (the latter inclusion means that every set from $\mathcal{S}_k$ can be presented as a sum of sets from $\mathcal{S}_{k+1}$);

S5) $\max_{1 \leq j \leq k} \sigma(S_{jk}) \to 0, k \to \infty$.

Using the partition $\{\mathcal{S}_k\}$ one can define the system $\{\pi_k\}$ of projectors as conditional means from $L^\infty(\sigma)$ with functions $x(s)$, $x \in L^\infty(\sigma)$, to the subspace $l_k^\infty(\sigma)$ of k-valued Σ-measurable step functions:

$$(\pi_k x)(s) = \sum_{j=1}^{k} \sigma(S_{jk})^{-1} \int_{S_{jk}} x(s)\sigma(ds)\chi_{S_{jk}}(s), \tag{12}$$

where

$$\chi_{S_{jk}}(s) = \begin{cases} 1, & \text{if } s \in S_{jk}, \\ 0, & \text{if } s \notin S_{jk}. \end{cases}$$

Denote this system of integral connection operators $\{\pi_k\}$ by Π, $\Pi = \{\pi_k\}$. Clearly,

$$\| \pi_k x - x \|_\infty \ = \tag{13}$$

$$= \inf_N \sup_{s \in S/N} | \sum_{j=1}^{k} \sigma(S_{jk})^{-1} \int_{S_{jk}} x(s)\sigma(ds)\chi_{S_{jk}}(s) - x(s) | \to 0, k \to \infty$$

$\forall x \in L^\infty(\sigma)$ (see, e.g., [7], Ch. IV.8) where inf is taken over all sets $N \in \Sigma$ with σ-measure zero, $\sigma(N) = 0$, and if the convergence is uniform on a bounded set C, then this set C is compact in the strong (norm) topology in $L^\infty(\sigma)$, see [7], IV. Ch 8.18.

In our approximation of probability and quantile functionals (1.3) and (1.4) we will exploit the partition $\{\mathcal{A}_n\}$, i.e., we will use in the partition of the integration domain S only sets $\{A_{in}\}$ with σ-measure zero of their boundary. We are of the opinion that approximation means that we will replace a complicated problem with a simpler one. But using in the approximation of (1.3) and (1.4) all sets $\{S_{jk}\}$ from the initial sigma-algebra Σ, we will not simplify the approximation (in its essence the convergence (1.13) is equivalent to the definition of a L^∞-function).

To approximate probability and quantile optimization problems we should present probabilities $\mathbf{P}\{s \mid f(x(s), s) \leq t\}$ in (1.3) and (1.4) via integrals. In order to do it, let us present the constraint set
$\{s \mid f(x(s), s) \leq t\}$ via the Heaviside zero-one function $\chi(t - f(x(s), s))$:

$$\chi(t - f(x(s), s)) = \begin{cases} 1, & \text{if } f(x(s), s) \leq t, \\ 0, & \text{if } f(x(s), s) > t. \end{cases} \tag{14}$$

Now

$$v_t(x) = \int_S \chi(t - f(x(s), s)\sigma(ds), \tag{15}$$

and the approximate function $v_{nt}(x_n)$ to $v_t(x)$ (starting from the weak convergence (1.10) of a sequence of discrete measures $\{(m_n, s_n)\}$ to $\sigma(\cdot)$) looks as follows:

$$v_{nt}(x_n) = \sum_{i=1}^{n} \chi(t - f(x_{in}, s_{in}))m_{in}. \tag{16}$$

Formulas (1.15) and (1.16) clearly demonstrate us mathematical difficulties which arise in the approximate solution of problems (1.3) and (1.4): solutions of both problems are discontinuous functions, integrands as Heaviside functions are discontinuous too. Even more, differently from Lebesgue L^p-spaces, $1 \leq p < \infty$, in L^∞-space the space C of continuous functions is not dense there. This arises the question: how to understand the convergence the sequence $\{x_n\}$ of vectors as solutions of approximate maximization problems to an essentially bounded measurable function $x(s)$ as the solution of maximization of the probability functional $v_t(x)$ over the set C?

The answer to the last question is the following. In the approximate solution of operator equations (e.g., integral and differential ones) some projection-type method (e.g., Galerkin) is usually exploited. Galerkin method approach means that we define a certain (orthogonal) sequence of base functions, which is dense in our initial space, and then present the approximate solution of the initial problem as a linear but finite combination from these base functions. Unfortunately, the space L^∞ is nonseparable, so the Galerkin method approach does not work there. Nevertheless, there exists one possibility to approximate problems even in the space L^∞ of essentially bounded measurable functions. This approach is called in numerical analysis "discrete approximation", and differently from the projection methods approach, where the difference between initial and approximate solutions is estimated in the initial space metrics, in the discretization approach the solution of the initial problem is projected to another (in applications to the finite-dimensional) approximate space.

Formally, let X be a Banach space with norm $\| x \|$ and let $\{X_n\}$ be a sequence of Banach spaces with norms $\{\| x_n \|_n\}$. Let a system $\mathcal{P} = \{p_n\}$ of (linear) connection operators between spaces X and $\{X_n\}$, $p_n : X \to X_n$, $n \in N$, such that the norm consistency property,

$$\| (p_n x)_n \|_n \to \| x \|, \; n \to \infty, \tag{17}$$

holds, be defined.

Remark 2 *The consistency property (1.17) guarantees nondegeneracy of norms of projected elements $(p_n x)_n$ and thus, uniqueness of the limit process*

$$lim_{n \in N} \; \| x_n - (p_n x)_n \|_n = 0. \tag{18}$$

For L^∞-space of essentially bounded measurable functions the norm consistency property (1.17) was verified in [8] for integral connection operators p_n of the form:

$$(p_n x)_{in} = \sigma(A_{in})^{-1} \int_{A_{in}} x(s) \sigma(ds), \; i = 1, ..., n, \tag{19}$$

and for the partition $\{\mathcal{A}_n\}$ with properties A1) - A7). Therefore, the uniqueness of the limit $x(s)$ of a discretely convergent sequence $\{x_n\}$ of vectors $x_n = (x_{1n}, x_{2n}, ..., x_{nn})$,

$$\max_{1 \leq i \leq n} | x_{in} - (p_n x)_{in} | \to 0, \; n \to \infty, \tag{20}$$

to an essentially bounded measurable function $x(s)$ is guaranteed and thus, it is possible to approximate stocastic programming problems with recourse [9], optimal control problems [10] and integral equations [11] also in the Banach space $L^\infty(\sigma)$ of Σ-measurable essentially bounded functions.

Let us present conditions that will guarantee stability of approximations of probability and quantile problems (1.3) and (1.4), starting from the weak convergence (1.10) of a sequence of discrete measures $\{(m_n, x_n)\}$ to the measure $\sigma(\cdot)$ of initial problems (1.3) and (1.4). Since solutions $x(s)$ of both problems are discontinuous and the Heaviside zero-one function $\chi(\cdot)$ is discontinuous too, we are forced to introduce quite hard restrictions to the function $f(x, s)$.

Assumption 1 *Function $f(x, s)$ is continuous in $(x, s) \in R^r \times R^m$.*

Assumption 2 *To each bounded set $B \subset R^r$ there corresponds a bounded and Σ-measurable function $a(s)$, $a : S \to R^1$ such that*

$$| f(x, s) | \leq a(s) \quad \forall x \in B. \tag{21}$$

Remark 3 *Assumption 2 guarantees that the function $f(\cdot, \cdot)$ as a superposition operator maps an essentially bounded measurable function $x(s)$ to an integrable function $f(x(s), s)$.*

In order to avoid constant regions for the function $f(x, s)$ with positive measure, we will assume the so-called "platform" condition.

Assumption 3 *For each $x \in R^r$ let*

$$\sigma\{s \mid f(x, s) = const\} = 0. \tag{22}$$

Proposition 1 *Let function $f(x, s)$ be continuous in both variables (x, s) and let it satisfy growth and platform conditions (1.21) and (1.22). Then from the $\mathcal{P}$-convergence $\mathcal{P} - \lim x_n = x$, $n \in N$, of arguments it follows the convergence $v_{nt}(x_n) \to v_t(x), n \in N$ of functionals.*

Remark 4 *The convergence proposed above is the discrete analogue of the continuous convergence of functionals.*

Verification of the statement of the Proposition 1 is quite lenghty and technically sophisticated – we should first approximate the discontinuous function $\chi(t - f(x, s))$ by the following continuous function $\chi_c(t - f(x, s))$:

$$\chi_c(t - f(x, s)) = \begin{cases} 1, & \text{if } f(x, s) \leq t, \\ 1 - \delta^{-1}[f(x, s) - t], & \text{if } t < f(x, s) \leq t + \delta \\ 0, & \text{if } f(x, s) > t + \delta \end{cases}$$

for some (small) $\delta > 0$, and then in turn the discontinuous solution function $x(s)$ by a continuous one $x_c(s)$.

In order to analyze stability of solutions of problems (1.3) and (1.4), starting from weak convergence of a sequence of measures $\{\sigma_n\}$ to the measure σ

$$\int_S h(s)\sigma_n(ds) \to \int_S h(s)\sigma(ds), n \in N, \forall h \in C(S)$$

(in our case measures σ_n are discrete) we should also determine the convergence of the sequence of constraint sets $\{C_n\}$ of approximate problems to the constraint set C of initial problems (1.3) and (1.4).

In the following we will exploit the discrete analogue of the Painleve-Kuratowski convergence of the sequence of sets $\{C_n\}$ to the set C. The reason, why we prefer this convergence to the more spreaded, Mosco-convergence of sets, is that in the latter case also the weak convergence of elements is used. In the stability analysis of extremum problems with integral functionals the weak convergence of elements brings along the requirement about convexity of these functionals. Unfortunately, the probability functional $v_t(x)$ is never convex, since it is bounded from below and above, $0 \leq v_t(x) \leq 1$. Under very hard restrictions to $f(x, s)$, to $\sigma(\cdot)$ and to decision rule $x(s)$ the function $v_t(x)$ is quasiconcave. We will present these conditions later.

Let $\{C_n\}$ be a sequence of sets, $C_n \in R^{rn}$, that converge to the set C in the discrete Painleve-Kuratowski sense.

Definition 1 *Sequence of sets $\{C_n\}, C_n \subset R^{rn}, n \in N$, converges to the set C in the discrete Painleve-Kuratowski sense, if*

1) for any subsequence $\{x_n\}$, $n \in N' \subset N$, such that $x_n \in C_n$, from convergence $\mathcal{P} - lim\ x_n = x, n \in N'$, it follows that $x \in C$;

2) for any $x \in C$ there exists a sequence $\{x_n\}$, $x_n \in C_n$, which $\mathcal{P}$-converges to x, $\mathcal{P} - lim\ x_n = x, n \in N$.

Reformulate now both, probability function maximization problem (1.3) and quantile function minimization problem (1.4) via the Heaviside zero-one function (1.14):

$$\max_{x(\cdot)\in C} v_t(x) = \max_{x(s)\in C} \int_S \chi(t - f(x(s), s))\sigma(ds) \tag{23}$$

and

$$\min_{x(\cdot)\in C} w_\alpha(x) = \min_{x(s)\in C} \min_t \{\int_S \chi(t - f(x(s), s))\sigma(ds) \geq \alpha\}, \tag{24}$$

and their discretized finite-dimensional analogues:

$$\max_{x_n\in C_n} v_{nt}(x_n) = \max_{x_{in}\in C_n} \sum_{i=1}^{n} \chi(t - f(x_{in}, s_{in}))m_{in} \tag{25}$$

and

$$\min_{x_n\in C_n} w_{n\alpha}(x_n) = \min_{x_n\in C_n} \min_t \{\sum_{i=1}^{n} \chi(t - f(x_{in}, s_{in}))m_{in} \geq \alpha\}. \tag{26}$$

Now we can formulate the discrete stability conditions of the discrete approximation of the of the probability function maximization problem (1.3) (or equivalently, problem (1.23)) by the sequence of finite dimensional maximization problems (1.25). Denote by v^* and v_n^* the optimal values of maximization problems (1.23) and (1.25), respectively.

Theorem 1 *Let function $f(x,s)$ satisfy assumptions 1 – 3, closed bounded sets $\{C_n\}$ converge to the closed bounded set C in the discrete Painleve-Kuratowski sense and let the sequence of discrete measures $\{(m_n, s_n)\}$ converge weakly to the atomless measure $\sigma(\cdot)$. Then $v_n^* \to v^*$, $n \to \infty$, and sequence of solutions $\{x_n^*\}$ of approximate problems (1.25) has a subsequence which converges discretely to a solution x^* of the initial problem (1.23).*

As the second part of the paper, consider approximation of the quantile function $w_\alpha(x)$ minimization problem (1.4) by using the same, discrete approximation scheme.

Assumption 4 *Function $f(x,s)$ is convex in (x,s).*

Assumption 5 *Measure $\sigma(\cdot)$ is quasiconcave:*

$$\sigma(\lambda A + (1-\lambda)B) \ \geq \ min\{\sigma(A), \sigma(B)\} \ \ \forall A, B \in \Sigma.$$

It was verified in [12] that under these convexity and quasiconcavity assumptions 4 and 5 the quantile function minimization (1.2) is equivalent to the following Nash game:

$$\max_{x \in C} \ v_t(x) = \max_{x \in C} \int_S \chi(t - f(x,s))\sigma(ds), \tag{27}$$

$$\min_t \ [v_t(x) - \alpha]^2 = \min_t \ [\int_S \chi(t - f(x,s))\sigma(ds) - \alpha]^2. \tag{28}$$

Since we are dealing with the quantile functional minimization problem, we should assume also monotonicity of the solution $x(s)$ as a decision rule.

Assumption 6 *Decision rules $x(s)$ are monotone almost everywhere,*

$$(x(s) - x(t), s - t) \geq 0 \ \ a.s.$$

Under assumptions 1 - 6 the quantile functional minimization problem (1.4) is equivalent to the following Nash game:

$$\max_{x(\cdot) \in C} \ v_t(x) = \max_{x(s) \in C} \int_S \chi(t - f(x(s), s))\sigma(ds) \ = J_1^*, \tag{29}$$

$$\min_t \ [v_t(x) - \alpha]^2 = \min_{x(s) \in C} \ [\int_S \chi(t - f(x(s), s))\sigma(ds) - \alpha]^2 \ = J_*^2 \tag{30}$$

Proposition 2 *Let function $f(x,s)$ be convex and continuous in (x,s), satisfy growth and platform assumptions (1.21) and (1.22), respectively and let decision rule x(s) be monotone almost everywhere. Then from convergences $\mathcal{P}-lim\ x_n = x$ and $| t_n - t | \to 0, n \in N$, it follows the convergence $v_{t_n n}(x_n) \to v_t(x), n \in N$.*

Formulate the approximate to the Nash game (1.29),(1.30) finite dimensional game:

$$\max_{x_n \in C_n} v_{nt}(x_n) = \max_{x_{in} \in C_n} \sum_{i=1}^{n} \chi(t - f(x_{in}, s_{in}))m_{in} = J_{1n}^*, \tag{31}$$

$$\min_t [v_{nt}(x_n) - \alpha]^2 = \min_t [\sum_{i=1}^{n} \chi(t - f(x_{in}, s_{in}))m_{in} - \alpha]^2 = J_{2n}^*. \tag{32}$$

Theorem 2 *Let function $f(x,s)$ satisfy assumptions 1 - 4, decision rules $x(s)$ monotonicity assumption 6, let continuous measure $\sigma(\cdot)$ be quasiconcave, closed bounded sets $\{C_n\}$ converge to the closed bounded set C in the discrete Painleve-Kuratowski sense and let $t_n \to t$, $n \in N$. Then values (J_{1n}^*, J_{2n}^*) of approximate Nash games (1.31),(1.32) converge to the value (J_1^*, J_2^*) of the initial Nash game (1.29),(1.30) and there exists a subsequence $\{(x_n^*, t_n^*)\}$, $n \in N' \subset N$, such that $\mathcal{P} - lim\ x_n^* = x^*$, and $| t_n^* - t^* | \to 0$, $n \in N'$.*

As an example consider the satellite orbit correction problem (1.9) in its equivalent, minimax formulation:

$$\max_{x(\cdot) \in C} \mathbf{P}\{s \mid | s_1 + x(s_1)(1 + s_2) | \leq t\} = J_1^*, \tag{33}$$

$$\min_t [\mathbf{P}\{s \mid | s_1 + x(s_1)(1 + s_2) | \leq t) - \alpha]^2 = J_2^*, \tag{34}$$

where random parameters ξ_1 and ξ_2 were independent normally distributed variables with means $E\xi_1 = M > 0$, $E\xi_2 = 0$ and variances $D\xi_1 = \sigma_1^2$, $D\xi_2 = \sigma_2^2$, respectively.

Let a bounded domain $[-N, N] \times [-N, N]$ be given and let us divide the segment $[-N, N]$ into $2n$ parts $[s_{in} - s_{(i-1)n}]$, $i = -n, -n+1, ..., -1, 0, 1, ..., n$, so that for any continuous on $[-N, N]$ function $h(s)$ the convergence (1.10) holds:

$$1/2n \sum_{i=-n}^{n} h(s_{in})m_{in} \to \int_{-N}^{N} h(s)e^{-(s-m)^2/2\sigma_1^2} ds \tag{35}$$

with $m_{in} = e^{-(s_{in}-m)^2/2\sigma_1^2}(s_{in} - s_{(i-1)n})$ (weak convergence (1.10) of a sequence of discrete measures $\{(m_n, s_n)\}$ to the random Gauss variable X with $EX = m$ and $DX = \sigma_1^2$.

Replace now the satellite correction problem (1.33),(1.34) with the sequence of its finite dimensional approximations:

$$\max_{x_{in} \in C_n} \sum_{i=-n}^{n} \sum_{i=-n}^{n} \chi(t- \mid s_{1in} + x_{1in}(1 + s_{2in}) \mid) m_{1in} m_{2in} = J^*_{1n}, \quad (36)$$

$$\min_t [\sum_{i=-n}^{n} \sum_{i=-n}^{n} \chi(t- \mid s_{1in} + x_{1in}(1 + s_{2in}) \mid m_{1in} m_{2in} - \alpha]^2 = J^*_{2n}. \quad (37)$$

Function $f(x, s) = \mid s_1 + x(1 + s_2) \mid$ satisfies assumptions 1 - 4, two dimensional Gauss measure the quasiconcavity assumption 5. The solution $x(s_1)$ was sought in [3] as a piecewise continuous function. Here we should add the monotonicity assumption 6. Even a piecewise continuous function is Riemann integrable, we still will exploit the L^p-space approach, since the dual to the space of Riemann integrable functions is the space of finitely additive set functions (not measures). Of course, for piecewise continuous decision rules $x(s_1)$ we are able to appy the simpliest connection system $\mathcal{P}'$ of the form: $(p'_n x_1)_{in} = x_1(s_{in}),\ i = 1, ..., n,\ n \in N$.

Now, relying on the Theorem 2, we can conclude that a subsequence of solutions $\{(x^*_{1n}, t^*_n)\}$, $n \in N' \subset N$, of approximate problems (1.36),(1.37) converges to a solution $(x^*(s_1), t^*)$ of the initial satellite correction problem (1.33),(1.34).

References

[1] Charnes, A. and W.W. Cooper (1959). *Chance-constrained programming.* Management Science, **6**, No 1, 73 - 79.

[2] Raik, E. (1971) *On stochastic programming problems with probability and quantile functionals.* Proc. Acad. Sci. Estonian SSR. Phys.-Math., **21**, 142 - 148 (in Russian).

[3] Kibzun, A. and Y. Kan (1996) *Stochastic Programming Probles with Probability and Quantile Functions.* John Wiley and Sons, Chichester, New York, e.t.c., 1996, 300 p.

[4] Kan, Y. and A. Mistryukov (1998) *On the equivalence in stochastic programming with probability and quantile objectives.* In: "Stochastic Programming. Methods and Technical Applications." K. Marti and P. Kall, eds. Lecture Notes in Economic and Mathematical Systems, **458**, 145 - 153.

[5] Vainikko, G. (1971) *On the convergence of quadrature formulae method for integral equations with discontinuous kernels.* Siberian Math. Journal, **12,** 40 - 53.

[6] Birge, J. and R. Wets (1986) *Designing approximation schemes for stochastic optimization problems, in particular for stochastic programs with recourse.* Mathematical Programming Study, **27**, 54 - 102.

[7] Lepp, R. (1988) *Discrete approximation conditions of the space of essentially bounded measurable functions.* Proc. Acad. Sci. Estonian SSR. Phys.-Math. **38**, 204 - 208.

[8] Lepp, R. (1994) *Projection and discretization methods in stochastic programming.* J. Comput. Appl. Math., **56**, 55 - 64.

[9] Lepp, R. (1996) *On approximation of optimal controls with discontinuous strategies.* Proc. Estonian Acad. Sci. Phys.-Math., **45**, No-s 2/3, 193 - 200.

[10] Lepp, R. (1998) *Discrete approximation of Hammerstein integral equations with discontinuous kernels.* Numer. Funct. Anal. and Optim., **19**, 7/8, 835 - 848.

[11] Kan, Y. (1996) *A quasigradient algorithm for minimization of the quantile function.* Theory and Systems of Control, No 2, 81 - 86 (in Russian).

Managing Risk with Expected Shortfall

Helmut Mausser (hmausser@algorithmics.com)
Algorithmics Incorporated,
185 Spadina Avenue,
Toronto, Ontario, Canada, M5T 2C6

Dan Rosen (drosen@algorithmics.com)
Algorithmics Incorporated,
185 Spadina Avenue,
Toronto, Ontario, Canada, M5T 2C6

Abstract

This paper examines tools for managing a portfolio's risk as measured by expected shortfall, which has been proposed as an alternative to the more widely-used Value-at-Risk. These tools include the calculation of risk contributions, marginal risk, best hedge positions and trade risk profiles. We first derive the parametric, or delta-normal, versions of these tools and then extend them to the simulation-based, or non-parametric, case. We analyze two sample portfolios: one, consisting of foreign exchange contracts, is well-suited for parametric analysis while the other, which contains European options, is best addressed with simulation-based methods. While expected shortfall and Value-at-Risk are constant multiples of each other in the parametric case, expected shortfall tends to provide more robust estimates of relevant risk analytics under the simulation-based approach.

Keywords: risk management, expected shortfall

1 Introduction

Financial institutions worldwide have devoted much effort to developing enterprise-wide systems that integrate financial information across their organizations to *measure* their institution's risk. Probabilistic measures, most notably Value-at-Risk

S.P. Uryasev (ed.), Probabilistic Constrained Optimization, 198–219.
© 2000 *Kluwer Academic Publishers.*

(VaR), are now widely accepted by both financial institutions and regulators for assigning risk capital and monitoring risk. Since development efforts have been driven largely by regulatory and internal requirements to report risk numbers, tools needed to *understand* and *manage* risk across the enterprise have generally lagged behind those designed to measure it. Given the popularity of Value-at-Risk, research efforts in this area have focussed largely on creating VaR-based tools. As we will demonstrate, many of the concepts underlying these techniques extend naturally to expected shortfall (ES).

Unlike risk measurement, managing risk is a dynamic endeavor that requires tools to help identify and reduce the sources of risk. These tools should lead to an effective utilization of the wealth of financial products available in the markets to obtain the desired risk profiles. To achieve this, a comprehensive risk manager's toolkit includes the ability to

- decompose risk by asset and/or risk factor
- understand how new trades affect the portfolio risk
- understand the impact of instruments' non-linearities and of non-normal risk factor distributions to portfolio risks
- generate potential hedges and optimize portfolios.

Robert Litterman (1996a, 1996b, 1997a, 1997b) recently described a comprehensive set of analytical risk management tools for VaR, extending some of the insights originally developed by Markowitz (1952) and Sharpe (1964). Garman (1996, 1997) used a similar approach to decompose the risk of a portfolio and calculate the marginal VaR. These tools are based on a linear approximation of the portfolio to measure its risk and generally assume a joint (log)normal distribution of the underlying market risk factors, similar to the RiskMetrics^{TM} VaR methodology (J.P. Morgan 1996). While Litterman emphasized the dangers of managing risk using only such linear approximations, these tools often provide useful insights for understanding portfolio risk, in spite of their onerous assumptions.

Mausser and Rosen (1998) extended this analytic (parametric) approach to a simulation-based (non-parametric) environment. Simulation- based tools provide additional insights when the portfolio contains non-linear (i.e. derivative) securities, when the market distributions are not normal or when there are multiple horizons. While simulation-based methods to measure VaR (historical or Monte Carlo) are generally much more computationally intensive than parametric methods (such as the delta-normal method popularized by RiskMetrics), advances in computational simulation methods and hardware have rendered these methods practical for enterprise-wide risk measurement. Moreover, non-parametric risk analytics can typically be obtained with little or no additional simulation beyond that required to calculate VaR itself.

A non-parametric analysis can potentially expose several undesirable properties of VaR, such as its violation of sub-additivity, for example. The fact that VaR does not constitute a coherent risk measure has been demonstrated by Artzner et al. (1998), who propose expected shortfall (also known as tail conditional expectation) as a more coherent alternative to VaR. A further motivation for the use of expected shortfall is its tractability from an optimization perspective; as demonstrated recently by Uryasev and Rockafellar (1999), minimizing expected shortfall can be modeled as a linear program, unlike VaR, which requires an integer programming model.

In this paper, we extend VaR-based risk management tools to expected shortfall in both analytic and simulation-based environments. Specifically, we construct the trade risk profile and calculate the risk contribution, marginal risk and best hedge position for each holding in a portfolio. We also compare and contrast the properties of the VaR and the expected shortfall versions of these tools.

This paper is organized as follows. We first derive parametric tools for managing expected shortfall and use them to analyze a portfolio of foreign exchange forwards. We then develop the simulation-based counterparts. To demonstrate the methodology, we re-examine the foreign exchange portfolio (obtaining results consistent with the parametric version) and also consider a portfolio of stock options for which the parametric approach is inappropriate. In light of this option portfolio, we discuss the need for smoothing the non-parametric trade risk profile to remove sampling noise. Finally, we offer our concluding remarks.

2 Parametric approach

The parametric approach assumes that the change in the value of a portfolio is normally distributed with mean zero and volatility σ. Since we are interested specifically in potential losses, we refer to this as the portfolio's loss distribution (losses are positive while gains are negative under this scheme). Let us denote the size of the loss by L. The portfolio's $100(1-\alpha)\%$ Value-at-Risk is the size of the loss that will be exceeded with probability α:

$$\int_{VaR_\alpha}^{\infty} \frac{1}{\sqrt{2\pi}\sigma} e^{\frac{-L^2}{2\sigma^2}} dL = \alpha$$

Note that

$$VaR_\alpha = Z_\alpha \sigma \tag{1}$$

where Z_α is the standard normal z-value that delimits a probability of α in the right tail (e.g., $Z_{0.01} = 2.3263$).

The expected shortfall at the $100(1-\alpha)\%$ level is the conditional expectation of all losses that exceed VaR_α:

$$ES_\alpha = \frac{1}{\alpha} \int_{VaR_\alpha}^{\infty} L \frac{1}{\sqrt{2\pi}\sigma} e^{\frac{-L^2}{2\sigma^2}} dL \tag{2}$$

Substituting $L = \sigma y$ in Equation 2 and using Equation 1 yields

$$ES_\alpha = K_\alpha \sigma \tag{3}$$

where

$$K_\alpha = \frac{1}{\alpha} \int_{Z_\alpha}^{\infty} y \frac{1}{\sqrt{2\pi}} e^{\frac{-y^2}{2}} dy$$

is the conditional expectation of all standard normal variates exceeding Z_α (e.g., $K_{0.01} = 2.6652$).

From Equations 1 and 3, we see that both VaR and ES are linear in volatility and thus, for any α, they are constant multiples of each other, i.e.,

$$ES_\alpha = \frac{K_\alpha}{Z_\alpha} VaR_\alpha \tag{4}$$

Equation 4 implies that VaR and ES are effectively inter-changeable under the parametric approach. In particular, this means that the relevant risk analytics for expected shortfall follow directly from those of Value-at-Risk, which have been derived in detail elsewhere (see, for example, Litterman 1996a, Mausser and Rosen 1998). To interpret these risk analytics, it is necessary to first examine the parametric model in greater detail.

The parametric, or delta-normal, approach assumes the existence of a set of market risk factors whose log price changes are joint normally distributed with zero mean; that is, if r_k is the log return on risk factor k, then $\mathbf{r} \sim N(\mathbf{0}, \mathbf{Q}^*)$, where $\mathbf{Q}^*$ is the covariance matrix of risk factor returns. Consider a portfolio composed of positions $\mathbf{x}$ where x_i is the size of the holding in instrument i, for $i = 1, 2, ..., N$. The volatility of the portfolio is (note that we now explicitly identify volatility as a function of the positions)

$$\sigma(\mathbf{x}) = \sqrt{\mathbf{m}(\mathbf{x})^T \mathbf{Q}^* \mathbf{m}(\mathbf{x})}$$

where $\mathbf{m}(\mathbf{x})$ is a vector of the portfolio's exposure to the risk factors. Due to its role in the RiskMetrics methodology for measuring VaR, $\mathbf{m}(\mathbf{x})$ is known as the **VaR map** of the portfolio. The elements of $\mathbf{m}(\mathbf{x})$ equal the monetary value of the portfolio's position in each risk factor. Thus, the VaR map provides a reduced, or simplified, view of the portfolio from a risk management perspective. Note that

$$\mathbf{m}(\mathbf{x}) = \sum_{i=1}^{N} \mathbf{m}^i x_i \tag{5}$$

where $\mathbf{m}^i$ is the VaR map of one unit of the i-th instrument (i.e., m_k^i is the exposure to risk factor k that results from holding a single unit of instrument i) . Equation 5 shows that the portfolio's VaR map is the sum of the instruments' VaR maps, weighted by position. The portfolio's $100(1-\alpha)\%$ ES can be calculated as

$$ES_\alpha(\mathbf{x}) = \sqrt{\mathbf{m}(\mathbf{x})^T \mathbf{Q} \mathbf{m}(\mathbf{x})} \tag{6}$$

where $\mathbf{Q} = K_\alpha^2 \mathbf{Q}^*$ is a scaled covariance matrix (note that if K_α is replaced by Z_α, then one obtains the Value-at-Risk).

2.1 Trade risk profile and best hedge position

The **trade risk profile** (TRP) plots expected shortfall against the size of the position in a given instrument i (all other positions being fixed). It provides risk managers with a quick, visual summary of several key risk characteristics for each instrument in the portfolio. Under the parametric approach, all trade risk profiles have the same characteristic shape shown in Figure 1.

To construct the trade risk profile for instrument i, we fix the positions in all instruments other than i to their current values and consolidate them into a so-called **base portfolio**. Denote the volatilities of (one unit of) instrument i and the base portfolio by σ_i and σ_l, respectively, and their correlation by ρ_{il}. The volatility of the portfolio is

$$\sigma(\mathbf{x}) = \sqrt{(x_i\sigma_i)^2 + \sigma_l^2 + 2\rho_{il}(x_i\sigma_i)\sigma_l}$$

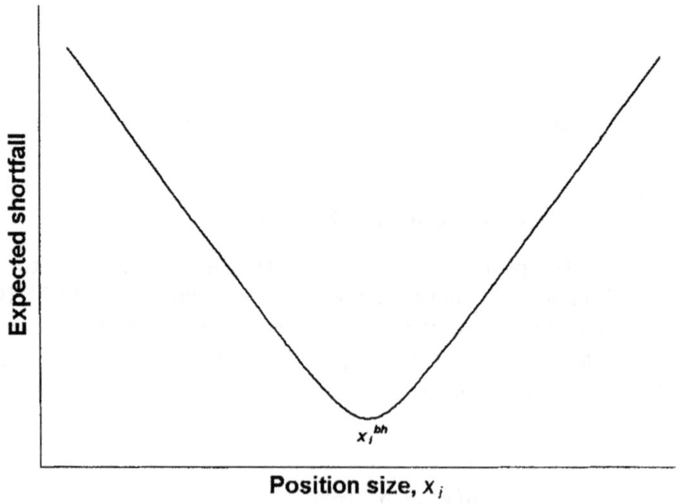

Figure 1. Trade risk profile

From Equation 3 it follows that the trade risk profile is a curve of the form

$$f(x_i) = \sqrt{ax_i^2 + bx_i + c} \tag{7}$$

where $a = (K_\alpha \sigma_i)^2, b = 2K_\alpha^2 \rho_{il} \sigma_i \sigma_l$ and $c = (K_\alpha \sigma_l)^2$.

The TRP has a unique minimum at the best hedge position, x_i^{bh}, which represents the most desirable (from the perspective of minimizing expected shortfall) position in instrument i, given that the remainder of the portfolio remains fixed. Differentiating Equation 7 with respect to x_i yields

$$\frac{df(x_i)}{dx_i} = \frac{2ax_i + b}{2f(x_i)} \tag{8}$$

Since $f(x_i)$ is strictly positive, the best hedge position occurs at

$$x_i^{bh} = -\frac{b}{2a} = -\frac{\rho_{il}\sigma_l}{\sigma_i}$$

2.2 Marginal ES

Managing risk requires an understanding of how new trades affect the portfolio risk. The marginal ES measures the impacts of small changes in risk factor exposures or instrument positions on the portfolio ES. From Equation 6, we find that the ES gradient with respect to the risk factor exposures is

$$\nabla_{\mathbf{m}} ES_\alpha(\mathbf{x}) = \frac{\mathbf{Qm}(\mathbf{x})}{ES_\alpha(\mathbf{x})}$$

The k -th element of $\nabla_{\mathbf{m}} ES_\alpha(\mathbf{x})$ is the change in ES that results from increasing the portfolio's exposure to the k-th risk factor (i.e., $m_k(\mathbf{x})$) by a single monetary unit.

Since the VaR map of the portfolio is the sum of the VaR maps for the positions (Equation 5), it follows that the derivative of ES with respect to the i-th position is

$$\frac{\partial ES_\alpha(\mathbf{x})}{\partial x_i} = (\mathbf{m}^i)^T (\nabla_{\mathbf{m}} ES_\alpha(\mathbf{x})) \tag{9}$$

Equation 9 indicates the change in ES due to adding one unit of instrument i to the portfolio (if $x_i < 0$, then this corresponds to reducing the short position). Note that this is simply the derivative of the trade risk profile for instrument i at the current position x_i (i.e., Equation 8). From Equation 4, it follows that marginal ES and the marginal VaR are in constant ratio:

$$\frac{\partial ES_\alpha(\mathbf{x})}{\partial x_i} = \frac{K_\alpha}{Z_\alpha} \frac{\partial VaR_\alpha(\mathbf{x})}{\partial x_i}$$

2.3 ES contribution

By decomposing ES, a risk manager is able to target the most significant sources of risk, or the portfolio's so-called "Hot Spots™". Like VaR, ES is a homogeneous function and so it admits a marginal decomposition:

$$\sum_{i=1}^{N} x_i \frac{\partial ES_\alpha(\mathbf{x})}{\partial x_i} = ES_\alpha(\mathbf{x}) \tag{10}$$

In Equation 10, each term in the summation is the product of position size and the rate of change of ES with respect to that position. This essentially represents the rate of change of ES with respect to a small *percentage* change in the size of the position. Let us define

$$C(x_i) = \frac{1}{ES_\alpha(\mathbf{x})} \times x_i \frac{\partial ES_\alpha(\mathbf{x})}{\partial x_i} \times 100\% \tag{11}$$

to be the percentage contribution to ES of the i-th position. Equation 11 must be interpreted on a marginal basis; it indicates the relative contributions to the change in ES that results if all positions are scaled by the same amount. Note that at the best hedge position, a position's marginal ES, and therefore also its ES contribution, is zero.

Similarly, we can define

$$C(m_k(\mathbf{x})) = \frac{1}{ES_\alpha(\mathbf{x})} \times m_k(\mathbf{x}) \frac{\partial ES_\alpha(\mathbf{x})}{\partial m_k(\mathbf{x})} \times 100\%$$

to be the percentage contribution to ES of the k-th risk factor. It follows from Equation 4 that the ES and VaR contributions are equal for the case of both positions and risk factors.

3 An example FX portfolio

Table 1 shows a portfolio of foreign exchange (FX) forward contracts as of July 1, 1997. The exchange rates, in USD, are 0.73 (CAD), 0.58 (DEM), 0.17 (FRF) and 0.0090 (JPY). The total value of the portfolio is 122,000 USD and its one-day 99% ES is 89,700 USD. For this example, we elect to use the RiskMetrics risk factor dataset for computing the parametric ES. We restrict our analysis to the position level; a similar analysis of this portfolio in terms of VaR, for both position and risk factor levels, can be found in Mausser and Rosen (1998).

Table 1. FX portfolio

Instrument	Currency	Days to Maturity	Strike Price (USD)	Position ($\times 10^6$)	Value ($\times 10^3$ USD)
CAD/USD .73 100d	CAD	100	0.73	0.5	2.5
CAD/USD .74 30d	CAD	30	0.74	1.0	-8.3
DEM/USD .57 60d	DEM	60	0.57	6.0	73.2
DEM/USD .59 120d	DEM	120	0.59	5.0	-28.2
FRF/USD .16 40d	FRF	40	0.16	8.0	83.3
JPY/USD .0091 11d	JPY	11	0.0091	10.0	-0.9

The analysis (Table 2) indicates that the DEM constracts are the major source of risk, contributing approximately 83% of the current portfolio risk. In contrast, the CAD contracts, with a total contribution of -0.70%, act as a hedge. The marginal ES values indicate that increasing the positions in the DEM, FRF and JPY contracts results in greater portfolio risk while a similar increase in the CAD contracts reduces risk. This is also reflected by the best hedge positions; for the DEM, FRF and JPY contracts, the best hedges are smaller than the current positions (and in fact suggest shorting contracts), but they are larger than the current positions in the case of the CAD contracts.

The impact of holding the best hedge position in a given instrument can be measured in terms of the percentage reduction in ES that can be achieved (i.e., the resulting decrease in ES expressed as a percentage of the current ES). At their best hedge positions, the DEM contracts each reduce the ES by almost 88% while each CAD contract offers a much smaller reduction of only 0.2%. In many cases, however, it may simply not be feasible to hold an instrument at its best hedge position. For example, being short seven million units of DEM/USD .57 60d contracts may well run counter to the underlying objectives of the portfolio. Thus, it is often useful to consult the trade risk profile (e.g., Figure 2) to determine the ES reduction that can be achieved within practical limitations. For example, eliminating the DEM/USD .57 60d position reduces the ES by almost half (to approximately 50,000 USD).

*Table 2.*Instrument data for the FX portfolio

Instrument	Marginal ES ($\times 10^{-4}$ USD)	ES Contribution (%)	Current Position ($\times 10^6$)	Best Hedge Position ($\times 10^6$)	ES Reduction (%)
CAD/USD .73 100d	67.87	45.4	6.0	-7.0	87.7
CAD/USD .74 30d	67.56	37.7	5.0	-8.0	87.6
DEM/USD .57 60d	18.55	16.5	8.0	-38.3	79.2
DEM/USD .59 120d	1.01	1.1	10.0	-209.2	13.2
FRF/USD .16 40d	-4.36	-0.2	0.5	1.4	0.2
JPY/USD .0091 11d	-4.41	-0.5	1.0	1.9	0.2

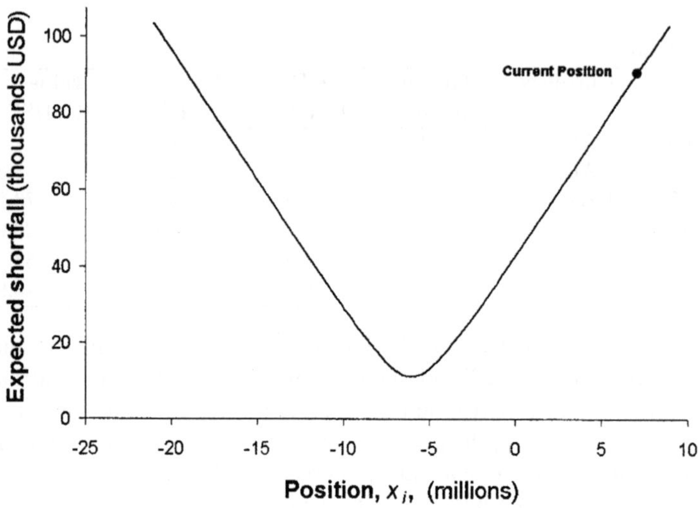

Figure 2. Trade risk profile for DEM/USD .57 60d

4 Simulation-based approach

The parametric approach makes several assumptions, such as the normality of the changes in portfolio value and the linearity of instrument value in the risk factors, that may be violated in practice. The latter assumption, for example, does not apply to derivative securities, whose prices are generally non-linear functions of the underlying asset price. These limitations can be overcome by using an alternative, non-parametric methodology that is based on simulation (such as the Mark-to-Future^{TM} framework described in Dembo et al. (2000)).

The simulation-based approach relies on a complete valuation of the portfolio under a set of scenarios, which may derive from historical data or a Monte Carlo simulation. Given a particular "base case" scenario (e.g., representative of current market conditions), it is straightforward to calculate the gain or loss in portfolio value in each scenario. Let ν_i^0 denote the unit value of instrument i in the base scenario and ν_{ij}^t denote its unit value in scenario j at some future time t. We refer to ν_{ij}^t as a **Mark-to-Future** value for instrument i. Since we assume exclusively a one-day time horizon, we will hereafter dispense with the t superscript to improve readability. Let us define

$$\Delta\nu_{ij} = \nu_i^0 - \nu_{ij}$$

to be the unit loss of instrument i in scenario j. If the current position in instrument i is x_i, then the loss incurred by the portfolio in scenario j is

$$L_j(\mathbf{x}) = \sum_{i=1}^{N} x_i \Delta\nu_{ij} \tag{12}$$

Suppose that the likelihood, or weight, of scenario j is p_j. If we order the losses from largest to smallest (since losses can be negative when the portfolio gains in value, "smallest" is taken here to mean "most negative") and calculate the cumulative scenario probability, then the non-parametric $100(1-\alpha)\%$ VaR, or $nVaR_\alpha$, equals the loss in that scenario for which the cumulative probability first meets or exceeds α. We refer to this scenario as the **threshold scenario**. To simplify the notation, we denote the threshold scenario as s_α, implicitly recognizing its dependence on the positions, $\mathbf{x}$. Thus, we obtain

$$nVaR_\alpha(\mathbf{x}) = L_{s_\alpha}(\mathbf{x})$$

Similarly, the non-parametric $100(1-\alpha)\%$ expected shortfall, or nES_α, is the conditional expectation of all losses that meet or exceed $nVaR_\alpha$:

$$nES_\alpha(\mathbf{x}) = E[L(\mathbf{x})|L(\mathbf{x}) \geq nVaR_\alpha(\mathbf{x})] \tag{13}$$

More formally, let γ denote a permutation of the scenarios so that $L_{\gamma(j)}(\mathbf{x}) \geq L_{\gamma(j+1)}(\mathbf{x})$ for $j = 1, 2, ..., M-1$. Thus, if the largest and smallest losses occur in scenarios g and h, respectively, then $\gamma(1) = g$ and $\gamma(M) = h$. We then have $s_\alpha = \gamma(j^\alpha)$ where

$$\sum_{j=1}^{j^\alpha - 1} p_{\gamma(j)} < \alpha \ \ and \ \ \sum_{j=1}^{j^\alpha} p_{\gamma(j)} \geq \alpha$$

Let

$$\xi_\alpha = \{\gamma(j)|j \leq j^\alpha\}$$

be the set of scenarios that comprise the "tail" (at level α) of the empirical loss distribution (again, we implicitly recognize that ξ_α depends on the positions, $\mathbf{x}$).We define

$$p^\alpha = \sum_{j=1}^{j^\alpha} p_{\gamma(j)}$$

to be the total probability of the tail scenarios and calculate the non-parametric expected shortfall as follows

$$nES_\alpha(\mathbf{x}) = \frac{\sum_{j=1}^{j^\alpha} p_{\gamma(j)} L_{\gamma(j)}(\mathbf{x})}{p^\alpha} \tag{14}$$

Note that Equations 13 and 14 are only consistent if $P(L(\mathbf{x}) \geq nVaR_\alpha(\mathbf{x})) = p^\alpha$. This relationship may fail to hold if multiple scenarios have losses equal to $nVaR$. For example, suppose that all scenarios have probability 0.01 and the four largest losses are 10, 5, 5 and 4. Then, if $\alpha = 0.02$, we have $p^\alpha = 0.02$ and $nVaR_\alpha(\mathbf{x}) = 5$, but $P(L(\mathbf{x}) \geq nVaR_\alpha(\mathbf{x})) = 0.03$. Given the large number of scenarios in a typical simulation, however, the effects of using Equation 14 in place of Equation 13 are negligible.

While valuing the portfolio under multiple scenarios can be a computationally intensive task, the non-parametric analysis has the desirable property of requiring only a single simulation of the portfolio. Once the instruments' Mark-to-Future values have been obtained, Equation 12 can be used to calculate losses for individual holdings (and hence, the portfolio) under subsequent changes in the positions.

4.1 Trade risk profile and best hedge position

Mausser and Rosen (1998) showed that that the non-parametric trade risk profile (nTRP) for VaR is piecewise linear, with each segment spanning a range of position sizes for which the threshold scenario remains constant. A similar argument shows that the expected shortfall nTRP is also piecewise linear. Using Equation 12, we can re-write Equation 14 as

$$nES_\alpha(\mathbf{x}) = \sum_{i=1}^{N} \left\{ \frac{\sum_{j=1}^{j^\alpha} p_{\gamma(j)} \Delta\nu_{i,\gamma(j)}}{p^\alpha} \right\} x_i \tag{15}$$

Thus, nES is linear in the size of each position and the nTRP has constant slope as long as the set of tail scenarios, ξ_α, remains unchanged. It follows that the nTRP is piecewise linear, with each segment corresponding to a ξ_α that is in effect for a given range of position sizes.

While the TRP and the best hedge position can be derived analytically in the parametric case, constructing the nTRP and finding its global minimum, x_i^{nbh}, requires numerical procedures. A precise nTRP can be "traced" by finding the position sizes at which ξ_α changes, calculating $nES_\alpha(\mathbf{x})$ at these points and then connecting them. Since ξ_α can change only when the threshold scenario changes, the algorithm described in Mausser and Rosen (1998) can be used to find the relevant points. An alternative method involves simply calculating $nES_\alpha(\mathbf{x})$ at a pre-defined set of points and then interpolating linearly between them. This "sampling" approach may produce a less-detailed nTRP than one that is traced (e.g., Figure 3) but it typically requires far less computational effort.

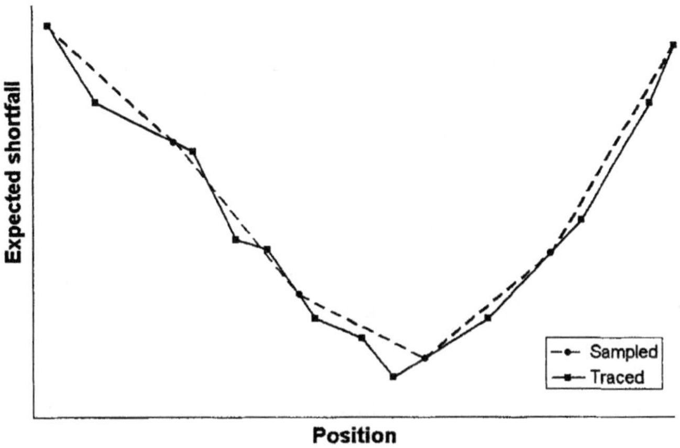

Figure 3. Non-parametric trade risk profile (nTRP)

4.2 Marginal nES

Since nES is a linear function of position size, it is a simple matter to obtain the marginal nES. Let us denote the expected unit loss of instrument i, conditional on ξ_α, as

$$\Delta\nu_i^\alpha = \frac{\sum_{j=1}^{j^\alpha} p_{\gamma(j)} \Delta\nu_{i,\gamma(j)}}{p^\alpha} \tag{16}$$

From Equations 15 and 16, it follows immediately that the marginal nES is

$$\frac{\partial nES_\alpha(\mathbf{x})}{\partial x_i} = \Delta\nu_i^\alpha$$

providing that ξ_α remains unchanged for small variations in x_i. Thus, the i-th component of the nES gradient is simply the conditional expected difference between the instrument's values in the base and tail scenarios. Note that the marginal nES is undefined at vertices of the nTRP; at such points, it is necessary to consider two one-sided sub-gradients. However, knowledge of the slopes and endpoints of the segments comprising the nTRP allows marginal nES, and the range for which it is valid, to be reported for any position.

4.3 nES contribution

Equation 15 immediately provides us with a risk decomposition - nES is the sum of the conditional expected differences between the positions' values in the base and tail scenarios. Thus, the percentage contribution to nES of the i-th position is simply

$$nC(x_i) = \frac{1}{nES_\alpha(\mathbf{x})} \times x_i \Delta\nu_i^\alpha \times 100\% \tag{17}$$

Note that Equation 17 is identical to Equation 11 in that the risk contribution is based on the product of the position size and the marginal nES. Thus, as in the parametric case, the above decomposition must be interpreted on a marginal basis. If we scale all positions by $(1+\varepsilon)$, where ε is a small constant, then nES increases by $\varepsilon \times nES_\alpha(\mathbf{x})$ and Equation 17 indicates the relative contribution of the i-th position to this increase.

5 The FX portfolio revisited

To compare the parametric and simulation-based VaR analyses, we simulated the FX portfolio over a set of 1,000 Monte Carlo scenarios. The one-day 99% nES is 89,000 USD, which differs from the parametric value by less than 1%. The resulting loss histogram is approximated well by a normal distribution (Figure 4) and the nES analysis (Table 3) is consistent with its parametric counterpart in Table 2. Note the close agreement between the parametric and simulation-based trade risk profiles for DEM/USD .57 60d (Figure 5).

Table 3. nES analytics for FX portfolio (ranked by nES contribution)

Instrument	Marginal nES ($\times 10^{-4}$ USD)	nES Contribution (%)	Current Position ($\times 10^6$)	Best Hedge Position ($\times 10^6$)	nES Reduction (%)
CAD/USD .73 100d	68.34	46.1	6.0	-7.2	87.4
CAD/USD .74 30d	68.13	38.3	5.0	-8.2	87.3
DEM/USD .57 60d	18.25	16.4	8.0	-38.0	78.9
DEM/USD .59 120d	1.15	1.3	10.0	-258.6	21.5
FRF/USD .16 40d	-12.47	-0.7	0.5	2.2	0.9
JPY/USD .0091 11d	-12.51	-1.4	1.0	2.7	1.0

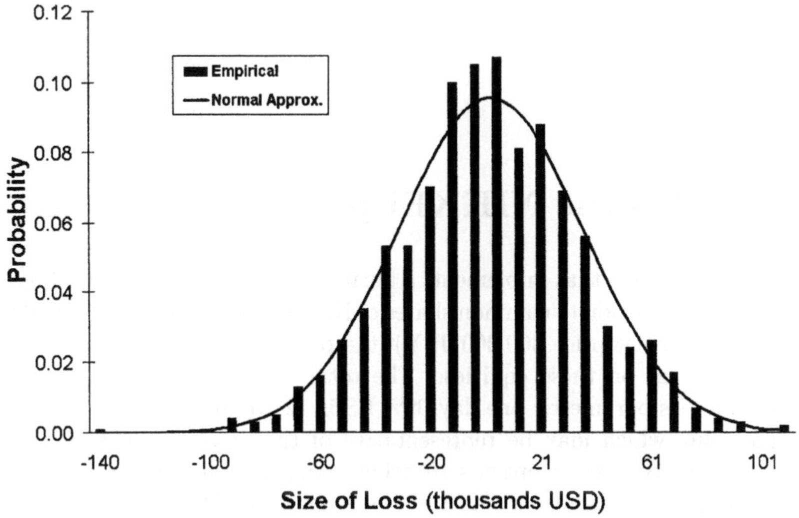

Figure 4. Distribution of losses for the FX portfolio with best normal approximation (1,000 scenarios).

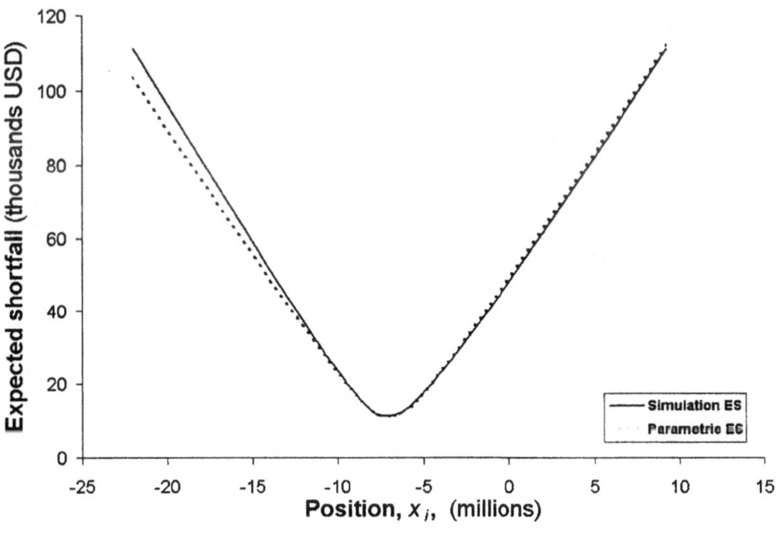

*Figure 5.*TRP and nTRP for DEM/USD .57 60d.

6 Example - the NIKKEI portfolio

Table 4 shows a portfolio that implements a butterfly spread on the NIKKEI index, as of July 1, 1997. In addition to common shares of Komatsu (current price 840,000 JPY) and Mitsubishi (current price 860,000 JPY), the portfolio includes several European call and put options on these equities. The total value of the portfolio is 12,493 million JPY and its parametric one-day 99% ES is 121 million JPY.

This portfolio, which may be representative of the positions held by a trading desk, makes extensive use of options to achieve the desired payoff profile. A histogram showing the distribution of losses over a set of 1,000 Monte Carlo scenarios (Figure 6) indicates that the normal distribution fits the data poorly, and that the parametric ES is likely to over-estimate the true expected shortfall. Indeed, simulating the portfolio over these 1,000 scenarios results in a one-day 99% nES of 3.6 million JPY, reflecting the fact that the portfolio is well-hedged. Because the parametric approach is inadequate in this case, we perform only a simulation-based risk analysis.

*Table 4.*NIKKEI portfolio

Instrument	Type	Days to Maturity	Strike Price (10^3 JPY)	Position ($\times 10^3$)	Value ($\times 10^3$ JPY)
Komatsu	Equity	N/A	N/A	2.5	2,100,000
Mitsubishi	Equity	N/A	N/A	2.0	1,720,000
Komatsu Cjul29 900	Call	7	900	-28.0	-11,593
Mitsubishi Cjul29 800	Call	7	800	-16.0	-967,280
Mitsubishi Csep30 836	Call	70	836	8.0	382,070
Mitsubishi EC 6mo 860	Call	184	860	11.5	563,340
Komatsu Cjun2 760	Call	316	760	7.5	1,020,110
Komatsu Cjun2 670	Call	316	670	22.5	5,150,461
Komatsu Paug31 760	Put	40	760	-10.0	-68,919
Komatsu Paug31 830	Put	40	830	10.0	187,167
Mitsubishi Psep30 800	Put	70	800	40.0	2,418,012

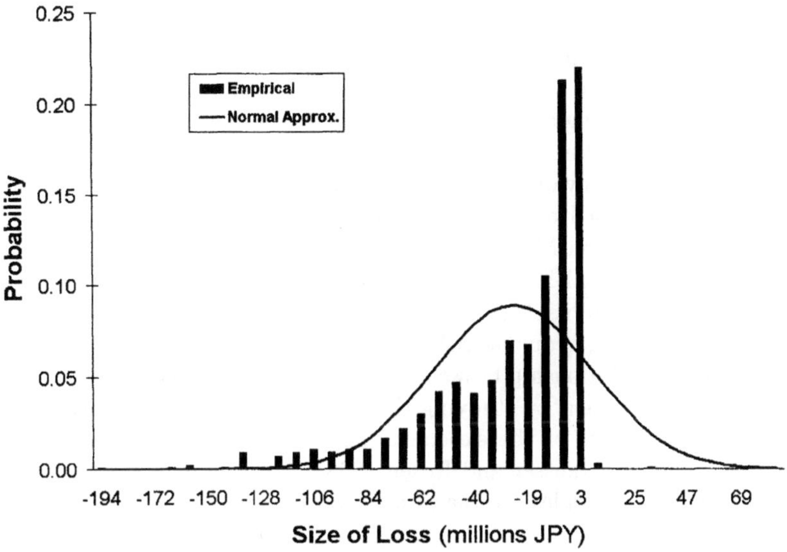

Figure 6. Distribution of losses for the NIKKEI portfolio with best normal approximation (1,000 scenarios)

*Table 5.*Analysis of the NIKKEI portfolio based on 1,000 scenarios

Instrument	Marginal nES (JPY)	nES Contribution (%)	Current Position ($\times 10^3$)	Best Hedge Position ($\times 10^3$)	nES Reduction (%)
Komatsu Cjun2 670	1846	1138	22.5	20.4	33.9
Komatsu Cjun2 760	1751	360	7.5	5.0	34.8
Mitsubishi Csep30 836	1113	244	8.0	5.3	28.7
Mitsubishi EC 6mo 860	375	118	11.5	6.1	15.0
Komatsu	1508	103	2.5	0.5	28.1
Mitsubishi	1390	76	2.0	-0.2	28.1
Komatsu Paug31 760	-99	27	-10.0	20.8	29.0
Komatsu Cjul29 900	34	-26	-28.0	-116.2	27.7
Komatsu Paug31 830	-441	-121	10.0	16.9	28.8
Mitsubishi Cjul29 800	1361	-597	-16.0	-18.2	28.2
Mitsubishi Psep30 800	-1116	-1223	40.0	42.7	28.9

Table 5 summarizes the relevant expected shortfall analytics for the NIKKEI portfolio (note that the tracing approach was used to construct the nTRP's). The magnitudes of the nES contributions are quite large, ranging from -1233% (Mitsubishi Psep30 800) to 1138% (Komatsu Cjun2 670). This is due to the fact that the portfolio is highly-leveraged and well-hedged, so that the risks incurred by individual positions tend to offset each other to a large extent. In particular, considering the relative sizes of the positions in the portfolio, note that the Komatsu Cjun2 670 position stands to gain considerably if the market appreciates while the Mitsubishi

Psep30 800 position acts in the opposite manner. More generally, as indicated by their negative contributions, the two short calls and the two long puts act as a hedge for the portfolio, protecting against drops in the Nikkei index.

Based on the marginal nES, the most attractive possibilities for lowering the overall portfolio risk include any one of the following trades: reducing the current holdings in Komatsu Cjun2 670 or Komatsu Cjun2 760, selling one of the common stocks, or shorting additional calls on Mitsubishi (i.e., Mitsubishi Cjul29 800). Purchasing additional units of Mitsubishi Psep30 800 is also a promising strategy. If it is feasible to hold an instrument at its best hedge position, then Komatsu Cjun2 760 offers the greatest potential for reducing risk (i.e., a reduction of 34.8%). Additional information about the relationship between position size and expected shortfall can be obtained for any instrument by studying the appropriate nTRP (e.g., Figure 7).

In general, the results of the expected shortfall analysis are consistent with a Value-at-Risk analysis (Mausser and Rosen 1998). For example, the sources of risk and the most attractive trades are the same in both cases.

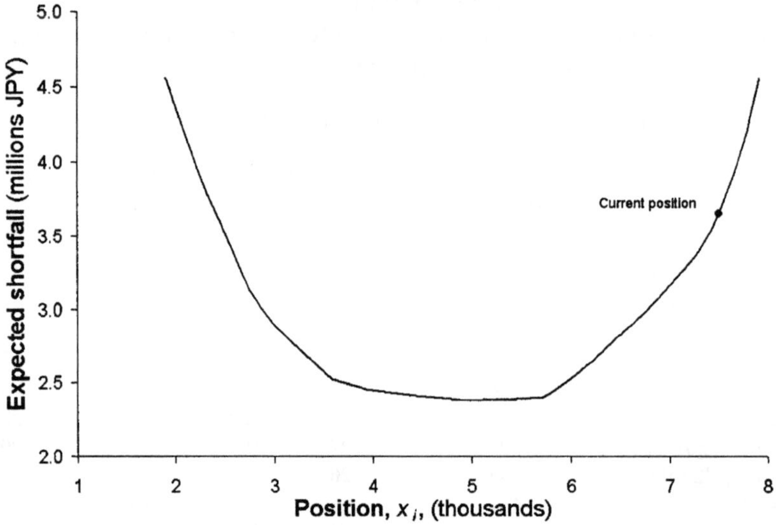

Figure 7. nTRP for Komatsu Cjun2 760

7 Smoothing the nTRP

Since it derives from a finite set of scenarios, the nTRP contains sampling "noise" that may distort the values of the calculated risk analytics. This problem is particularly relevant in the case of Value-at-Risk as demonstrated, for example, by the VaR nTRP for Komatsu Cjun2 760 (Figure 8). Adjacent segments often have markedly different slopes, giving the nTRP a jagged appearance and potentially resulting in erroneous estimates of the marginal VaR. Notice, for example, the downward-sloping segment at a position size of 7,300, which is clearly incompatible with the overall shape of the nTRP. For positions on this segment, the resulting negative marginal nVaR provides a poor estimate of the true implications of trading Komatsu Cjun2 760.

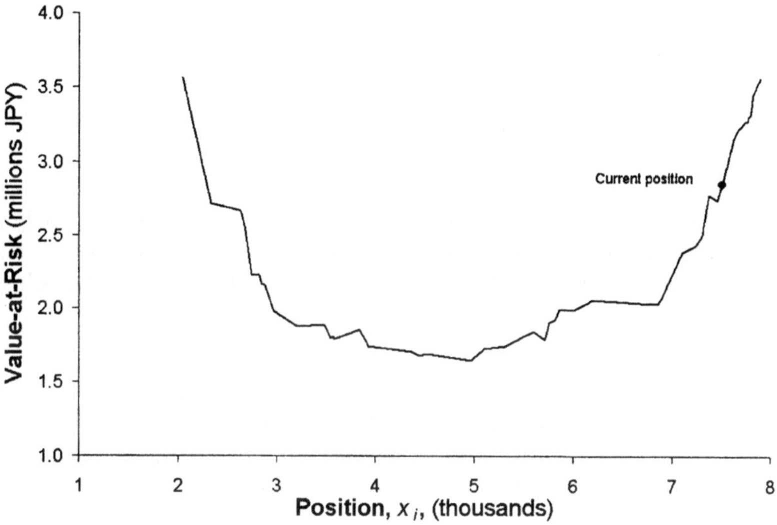

Figure 8. Value-at-Risk nTRP for Komatsu Cjun2 760.

In comparison to Figure 8, the expected shortfall nTRP for Komatsu Cjun2 760 (Figure 7) exhibits less apparent sampling noise. Essentially, this is because the slopes of the segments depend on an expected loss across all tail scenarios, rather than only on the threshold scenario as in the case of VaR (in a related manner, Hallerbach (1999) estimated the marginal VaR as a weighted average of an instrument's unit losses in the scenarios surrounding the threshold scenario). Furthermore, since the segments change only when ξ_α (rather than s_α) changes, the nTRP for expected shortfall has fewer segments than the corresponding nTRP for VaR. Thus, expected shortfall tends to allow for a more robust estimation of relevant risk analytics than VaR in the simulation-based approach.

To mitigate the effects of sampling noise in the case of VaR, Mausser and Rosen (1998) propose fitting a smooth approximation to the nTRP and then deriving the relevant risk analytics from this approximation. The same approach is also applicable to expected shortfall. Let $P_i(x_i)$ represent a smooth approximation to the nTRP of instrument i (as could be obtained from a least-squares fit to the endpoints of the segments, for example). From this approximation, one can then obtain the following estimates:

-marginal nES at position x_i

$$\frac{\partial nES_\alpha(\mathbf{x})}{\partial x_i} = P_i'(x_i)$$

-best hedge position

$$x_i^{nbh} = \{x_i^* | P_i(x_i^*) \leq P_i(x_i) \text{ for all } x_i\}$$

-nES contribution at position x_i

$$nC(x_i) = \frac{x_i P_i'(x_i)}{\sum_{l=1}^{N} x_l P_l'(x_l)} \times 100\%$$

To demonstrate the effects of smoothing, we obtain a fourth-order polynomial approximation to each nTRP using least-squares regression. The adjusted-R^2 statistic is 0.99 or better in all cases except Komatsu Cjul29 900 (0.8799) and Mitsubishi EC 6mo 860 (0.9889). The resulting risk analytics (Table 6) are generally consistent with the unsmoothed values in Table 5.

Table 6. Analysis of the NIKKEI portfolio based on smooth approximations

Instrument	Marginal nES (JPY)	nES Contribution (%)	Current Position ($\times 10^3$)	Best Hedge Position ($\times 10^3$)	nES Reduction (%)
Komatsu Cjun2 670	1606	1632	22.5	20.0	35.0
Komatsu Cjun2 760	1521	515	7.5	4.7	36.1
Mitsubishi Csep30 836	972	351	8.0	4.7	28.1
Mitsubishi EC 6mo 860	247	128	11.5	6.8	17.3
Komatsu	1320	149	2.5	0.2	27.7
Mitsubishi	1216	110	2.0	-0.5	27.7
Komatsu Paug31 760	-85	38	-10.0	26.6	28.8
Komatsu Cjul29 900	29	-37	-28.0	-136.5	31.9
Komatsu Paug31 830	-381	-172	10.0	18.0	28.4
Mitsubishi Cjul29 800	1191	-861	-16.0	-18.6	27.7
Mitsubishi Psep30 800	-971	-1755	40.0	43.3	28.4

The fit between the nTRP and its polynomial approximation for Komatsu Cjun2 760 (Figure 9) is typical of the instruments in this portfolio. A notable exception is the approximation for Komatsu Cjul29 900 (Figure 10), which fits well around the current position of -28,000 but poorly for positions below -110,000 (thus, the difference of approximately 20,000 in the best hedge positions). A better fit may be obtained by using a higher order polynomial or, alternatively, one might consider increasing the number of scenarios to establish whether or not the irregular shape of the nTRP is due to sampling error in this case.

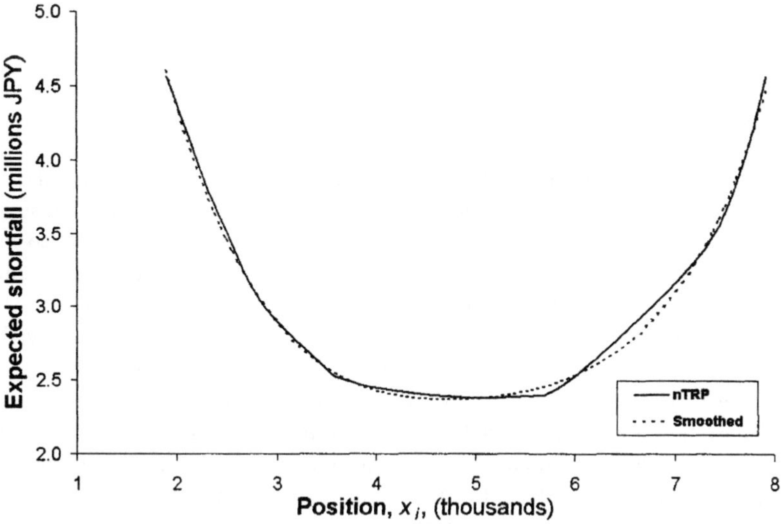

Figure 9. nTRP and smooth approximation for Komatsu Cjun2 760.

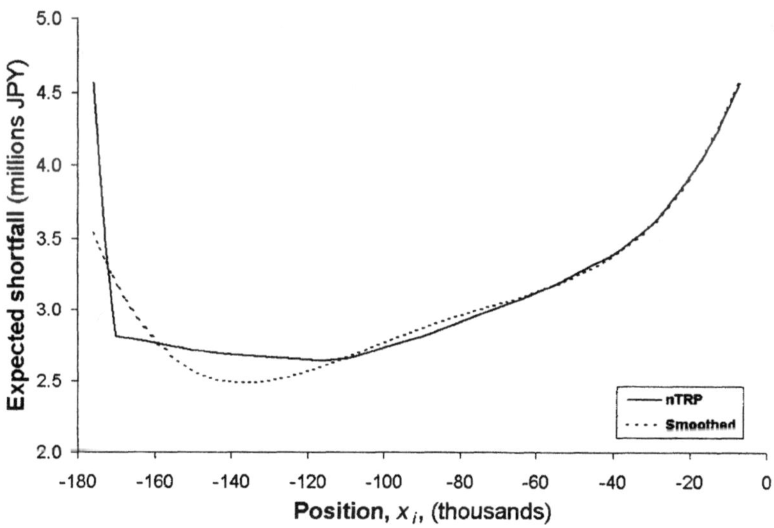

Figure 10. nTRP and smooth approximation for Komatsu Cjul29 900.

8 Conclusions

This paper has examined tools for managing risk as measured by expected shortfall. Tools for decomposing risk, assessing its marginal impacts and constructing best hedges, allow managers to understand the sources of risk better and to manipulate the portfolio to effect the desired risk characteristics. Analytical techniques are applicable to portfolios containing instruments that depend linearly on a set of market risk factors whose log price changes are joint normally distributed with zero mean. Simulation-based tools extend these capabilities to portfolios that contain non-linearities or that are subject to non-normal market distributions. An attractive feature of these methods is their need for only a single simulation to obtain the Mark-to-Future values of the instruments. Furthermore, it is straightforward to incorporate new instruments into the analysis by simulating them independently of the portfolio itself. Thus, while our analyses considered trading only those instruments currently held in the portfolio, it extends naturally to encompass so-called incremental risk.

In a simulation-based environment, expected shortfall is less sensitive than VaR to sampling noise and therefore, tends to provide more robust risk analytics. Nevertheless, it may be appropriate to use a smooth approximation to the non-parametric trade risk profile in some cases (a simple visual inspection of the nTRP will generally establish the need for such an approximation). To demonstrate this concept, we used regression to fit a fourth-order polynomial to the nTRP. The use of more powerful fitting techniques, including methods that smooth in both the position ($\mathbf{x}$) and probability (α) dimension (e.g., Mausser and Rosen 1999), presents an interesting research opportunity.

References

[1] Artzner, P., F. Delbaen, J.-M. Eber and D. Heath, 1998, "Coherent measures of risk," Working Paper, Institut de Recherche Mathmatique Avance, Universit Louis Pasteur et C.N.R.S.

[2] Dembo, R.S., A. Aziz, D. Rosen and M. Zerbs, 2000, *mark to future*™: A Framework for Measuring Risk and Reward, Toronto:Algorithmics Inc.

[3] Garman, M., 1996, "Improving on VaR," *Risk* 9(5): 61-63.

[4] Garman, M., 1997, "Taking VaR to pieces," *Risk* 10(10): 70-71.

[5] Hallerbach, W. G., 1999, "Decomposing Portfolio Value-at-Risk: A General Analysis", working paper, Tinbergen Institute Rotterdam.

[6] Litterman, R., 1996a, "Hot spots and hedges," Risk Management Series, Goldman Sachs.

[7] Litterman, R., 1996b, "Hot spots and hedges," *Journal of Portfolio Management* (Special Issue): 52-75.

[8] Litterman, R., 1997a, "Hot spots and hedges (I)," *Risk*10(3): 42-45.

[9] Litterman, R., 1997b, "Hot spots and hedges (II)," *Risk*10(5): 38-42.

[10] Markowitz, H., 1952, "Portfolio selection," *Journal of Finance* 7(1): 77-91.

[11] Mausser, H. and D. Rosen, 1998, "Beyond VaR: From Measuring Risk to Managing Risk," *Algo Research Quarterly* 1(2), 5-20.

[12] Mausser, H. and D. Rosen, 1999, "Managing Risk using the VaR Trade Risk Surface," Working Paper, Algorithmics Inc.

[13] RiskMetrics^{TM} Technical Document, 4th Edition, J.P. Morgan, December 1996.

[14] Sharpe, W., 1964, "Capital asset prices: a theory of market equilibrium under conditions of risk," *Journal of Finance* 19(3): 425-442.

[15] Uryasev, S. and R.T. Rockafellar, 1999, "Optimization of conditional Value-at-Risk," Research Report #99-4, Center for Applied Optimization, University of Florida.

On the Numerical Solution of Jointly Chance Constrained Problems

János Mayer(mayer@ior.unizh.ch)
IOR University of Zurich
Zurich, Switzerland

Abstract

This paper considers jointly chance constrained problems from the numerical point of view. The main numerical difficulties as well as techniques for overcoming these difficulties are discussed. The efficiency of the approach is illustrated by presenting computational results for large–scale jointly chance constrained test problems.

Keywords: Stochastic programming, joint chance constraints, probabilistic constraints, implementation of algorithms.

1 Introduction

In this paper we consider jointly chance constrained problems of the following type: only the right hand side is stochastic and it has a joint multinormal distribution. The point of view of the paper is numerical. After the problem formulation we give an overview of the main numerical difficulties associated with jointly chance constrained problems.

The algorithms for jointly chance constrained problems can roughly be subdivided into two classes: algorithms in the first class are based on some general nonlinear programming algorithmic framework which is adapted to the special structure. The second class consists of methods which directly attack the problem by utilizing probabilistic techniques. In this paper we consider the first approach and discuss techniques which are used to adapt general nonlinear programming methods to jointly chance constrained problems.

S.P. Uryasev (ed.), Probabilistic Constrained Optimization, 220–235.
© 2000 *Kluwer Academic Publishers.*

We also present numerical results which illustrate the efficiency of the approach. The test problems are large scale with respect to the dimension of the random right hand side. Numerical results concerning two test problem batteries are presented with 20– and 30–dimensional random right hand sides, respectively. For comparison let us remark that the largest jointly chance constrained problems for which numerical results have been published involve a 7–dimensional random right hand side.

For applications of chance constrained problems see [19]. Let us point out that a new application area of chance constrained problems is finance, more specifically Value–at–Risk (VaR), see [25] and the references therein. The technique introduced in [25], is also promising for solving general jointly chance constrained problems; further research is needed however for working out the details.

2 Problem formulation

We consider the following jointly chance constrained model (probabilistic constrained model):

$$
\begin{array}{rll}
\min & c^{\mathrm{T}}x & \\
\boldsymbol{P}(& Tx \geq \xi) & \geq \alpha \\
& Ax & = b \\
& x & \geq 0
\end{array}
\tag{1}
$$

where α is a prescribed (high) probability level and ξ is a random vector $\Omega \to \mathbb{R}^r$ on a probability space $(\Omega, \mathcal{F}, \boldsymbol{P})$. A is an $m \times n$ matrix; the other arrays have compatible dimensions.

We assume that ξ has a nondegenerate multinormal distribution with covariance matrix C and expectation vector μ. We denote the probability distribution function of ξ by $F(y)$ where we have omitted in the denotation the explicite reference to the parameters C and μ. By introducing the variable $y = Tx$ problem (1) can equivalently be written as

$$
\left.
\begin{array}{lcrl}
\min \; c^{\mathrm{T}}x & & & \\
& & F(y) & \geq \alpha \\
Tx & - & y & = 0 \\
Ax & & & = b \\
x & & & \geq 0
\end{array}
\right\}
\tag{2}
$$

Although problems (1) and (2) are trivially equivalent, the second formulation (2) has some advantages for implementing algorithms. To illustrate the point let us mention that for practical problems usually $r \ll n$ holds. Considering now e.g. a cutting plane method the cuts are performed w.r. to the nonlinear constraint which involves usually much fewer variables in (2) as in (1) thus allowing for reduced storage for the cuts.

From the geometrical point of view the feasible domain of (2) has the following structure: the convex polyhedron $\{x \mid Ax = b, x \geq 0\} \subset \mathbb{R}^n$ is mapped into $\mathbb{R}^r$ by the linear mapping $y = Tx$ and in the image space we have the constraint $F(y) \geq \alpha$.

Problem (2) is a nonlinear programming problem with a single nonlinear constraint. Both from the theoretical and from the practical point of view the most important question is whether (2) is a convex programming problem. For a broad class of probability distributions this question has been answered in the affirmative by the basic theorem of [17] on logconcave measures. For the details see also [19], [8] and [12]. This theorem in our case immediately implies that $F(y)$ is a logarithmically concave function. Having $F(y) > 0 \;\; \forall\, y \in \mathbb{R}^r$ logconcavity simply means that $\log F(y)$ is a concave function on $\mathbb{R}^r$. From this immediately follows that $F(y)$ is quasiconcave thus implying that the feasible domain of (2) is a convex set. Replacing the nonlinear constraint (2) by the equivalent constraint

$$\log F(y) \geq \log \alpha \tag{3}$$

results in a convex programming problem. Let us notice that we work with chance constrained problems in the form as formulated in (2). The reason is that carrying out the logarithmic transformation does not result in a numerically better behaving problem, this point will be discussed in the next section.

Let us notice that jointly chance constrained problems can be formulated in a more general form involving several joint probability constraints, see e.g. [13]. In this paper we consider a single joint constraint because our numerical experiences are for this type of problems.

The probability distribution function is in our case smooth; the gradient $\nabla F(y)$ can be computed according to

$$\frac{\partial F(y)}{\partial y_i} = F_i(y_j;\, j = 1, \ldots, r,\, j \neq i \,|y_i) \cdot f_i(y_i) \;\; \forall i \tag{4}$$

where the conditional distribution function $F_i(y_j;\, j = 1, \ldots, r,\, j \neq i \,|y_i)$ corresponds to a multivariate nondegenerate normal distribution with dimension $r - 1$ and $f_i(y_i)$ is the density function of a one–dimensional normal distribution. Formula (4) was proposed by Prékopa for computing the gradient of multivariate distribution functions in conjunction with chance constrained problems; for the details see [18]. The formula has the following obvious advantage: once we have an algorithm for computing $F(y)$, the same algorithm can be utilized for computing $\nabla F(y)$.

3 Numerical considerations

In this section we discuss some numerical difficulties associated with the problem (2).

The main difficulty is that the computation of $F(y)$ and $\nabla F(y)$ involves evaluating the multivariate normal distribution function. This means numerical integration in several variables which can in higher dimensions ($r > 4$) only be carried out by Monte–Carlo methods. In conjunction with the numerical solution of chance constrained problems such algorithms have been developed by [3], [4], [6], [23], [24]. An immediate consequence of this simulation approach is that

- the computation of $F(y)$ and $\nabla F(y)$ is time–consuming and
- the precision of the computed values is relatively low.

A first idea for solving (2) is the following: the problem is a nonlinear programming problem with nice theoretical properties therefore let us employ a general–purpose nonlinear programming solver for solving it. We have tried to solve the problem by employing Minos 5.4, [15]. According to our experiences Minos was unable to solve the problem even for low dimensional random vectors (e.g. $r = 3$) where $F(y)$ could be computed with high precision. In general the solver could not make improvements at the starting point and got stuck there. [5] report on a successful use of Minos for solving a water management problem formulated as a jointly chance constrained problem with $r = 3$. Let us mention however that their starting point was quite close to the optimal solution.

This indicates that beyond the difficulties associated with the computation of $F(y)$ there are some further intrinsic properties of jointly chance constrained problems (2) which make them difficult to solve.

The second difficulty we would like to point out is associated with $\nabla F(y)$. Figure 1 shows the bivariate standard normal distribution function.

The figure shows that outside of a roughly L-shaped region the graph of the function consists of nearly horizontal flat regions indicating that the components of $\nabla F(y)$ are small there.

To get a feeling on the magnitude of $\nabla F(y)$ let us consider the value of the partial derivatives of the $r-$ dimensional standard multivariate normal distribution at the

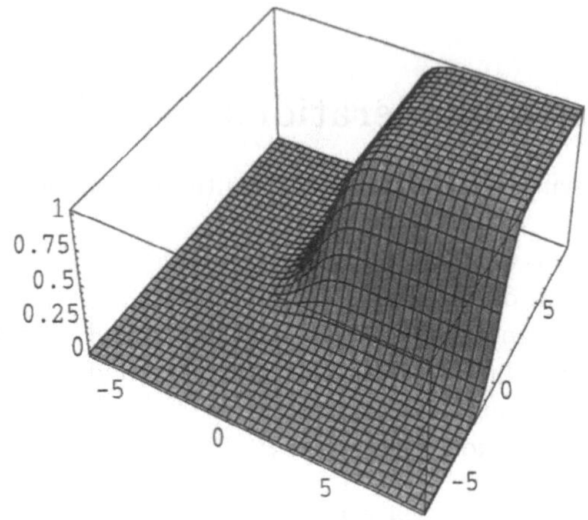

Figure 1: The distribution function of the bivariate standard normal distribution.

point $y_i = -5$, $i = 1, \ldots, r$, with varying r. For $r = 10, 20, 30$ the value of $\frac{\partial F(y)}{\partial y_i}$ is $1.9 \cdot 10^{-65}$, $7.2 \cdot 10^{-131}$ and $2.7 \cdot 10^{-196}$, respectively. For $y_i = -20$, $i = 1, \ldots, r$ and $r = 30$ the value of $\frac{\partial F(y)}{\partial y_i}$ is $3.2 \cdot 10^{-2656}$. Let us notice that the logarithmic transformation (3) of the probabilistic constraint does not help us in overcoming this difficulty because the Monte–Carlo methods compute $F(y)$; there is no known method for directly computing $\log F(y)$.

This behavior implies that $\nabla F(y)$ is practically zero outside of a narrow region. This means that outside of that region there is no gradient–based local information available concerning $F(y)$.

A further inconvenient feature is encountered when we consider problem (2) with increasing dimension r of the random right hand side. Figure 2 shows the graph of the multivariate standard normal distribution function along the line

$$y(\lambda) = \{y \mid y_i = \lambda \, \forall i, \; \lambda \in \mathbb{R}\}.$$

with increasing r.

The figure shows that in the nonflat region the graph of the distribution function becomes quite steep with increasing dimension. Formulated in another way this means that the region where information based on the gradient is available becomes narrower with increasing dimensions.

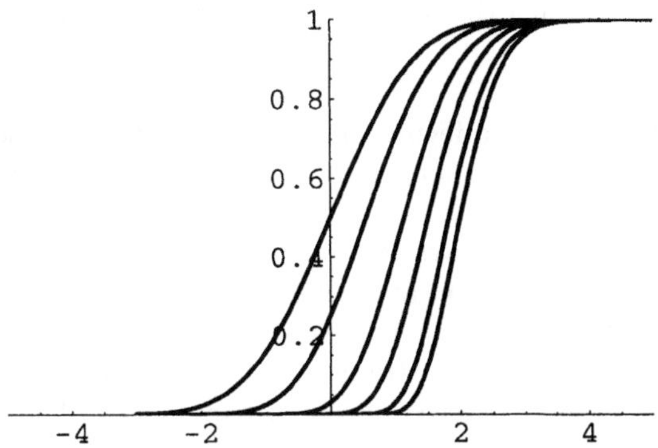

Figure 2: The distribution function of the multivariate standard normal distribution along the line $\{y \mid y_i = \lambda \; \forall i\}$. The curves from left to right correspond to dimensions $r = 1, 2, 5, 10, 20, 30$, respectively.

Summarizing the second difficulty: although Problem (2) is basically a convex programming problem it involves a hidden nonconvexity feature; local gradient based information is only available within a narrow nonconvex region.

4 Algorithms

The algorithmic approaches for the solution of jointly chance constrained problems can roughly be subdivided into two classes as follows:

- A general nonlinear programming solution scheme (e.g. a cutting plane method) is taken as a basis which is subsequently adapted to the special structure of the problem.
- The stochastic programming problem is directly attacked by probabilistically based methods.

Overviews and detailed descriptions of algorithms for jointly chance constrained problems can be found in [12], [13], [14], [18], [19] and the references therein. A recently proposed algorithm belonging to the first class is a central cutting plane method of [14] while the algorithms of [7], [16] and [20] belong to he second class.

In this section we discuss various techniques used for adapting general purpose nonlinear programming methods to jointly chance constrained problems. They have been designed for overcoming the numerical difficulties outlined in the previous section.

Let us begin with the first difficulty outlined in the previous section: computing $F(y)$ is time–consuming and the computed result has relatively low accuracy.

The following simple idea serves for overcoming this difficulty: instead of computing $F(y)$ utilize cheaply computable lower and upper bounds on $F(y)$ as far as possible. The key ingredient of all nonlinear programming approaches implemented for the solution of jointly chance constrained problems is the following linesearch subproblem: for a given feasible point $(\bar{x}, \bar{y})$ and direction (u, v) find the intersection of the straight line

$$\{(x, y) \mid x = \bar{x} + \lambda u, \; y = \bar{y} + \lambda v, \; \lambda \in \mathbb{R}\}$$

and the surface

$$\{y \mid F(y) = \alpha\}.$$

Considering a bisection–type procedure the interval reduction is carried out on the basis of bounds as far as possible. If this is no longer possible a switch to the computation of $F(y)$ occurs and the bisection proceeds using function values.

The usefulness of this idea depends heavily on the availability of appropriate bounds. The first type of bounds used in algorithms for jointly chance constrained problems were the Boole–Bonferroni bounds, see e.g. [22]. For computing the bounds only the one– and two-dimensional marginal distribution functions are to be computed. For a multinormal distribution the marginal distributions are normal as well and the bounds can be computed fastly an accurately.

Szántai used these bounds in a cutting plane method, see [24]. The author utilized the Boole–Bonferroni bounds in a reduced gradient framework as well as in a central cutting plane method, see [14].

Let us point out that the efficiency of the algorithms using bounds crucially depends on the quality (tightness) of the bounds. Tighter bounds have been found via discrete moment problems by Prékopa, Prékopa and Boros, see [19]. For recent developments see [1] and [2].

Let us address the second difficulty discussed in the previous section: $\nabla F(y)$ is practically zero outside of a narrow region.

Let us assume that our current iteration point (x^k, y^k) satisfies the linear constraints but it is infeasible w.r. to the nonlinear constraint and y^k lies in the lower flat region with no local gradient information available, see Figure 1. This is a typical situation encountered in cutting plane type algorithms. One idea would be to utilize the fact that for any direction $u > 0$ the relation $\lim_{\lambda \to \infty} F(y^k + \lambda u) = 1$ holds. Based on this fact search directions could be computed for finding feasible points.

The author has found the following approach more efficient in practice: Assume that at each iteration a Slater-point is also known, i.e. we have a feasible point $(\bar{x}^k, \bar{y}^k)$ with the property $F(\bar{y}^k) > \alpha$. This Slater point is used as a navigational aid: a feasible point on the boundary $\{y \mid F(y) = \alpha\}$ is located using linesearch along the line segment connecting (x^k, y^k) and $(\bar{x}^k, \bar{y}^k)$. For applying this technique we need the

Assumption: Problem (2) is Slater-regular, i.e. there exists a feasible point $(\bar{x}^0, \bar{y}^0)$ such that $F(\bar{y}^0) > \alpha$ holds.

A first Slater–point can be found by carrying out iterations of an optimization algorithm for solving the problem

$$\begin{array}{llll} \max & & F(y) & \\ & Tx & - \quad y & = 0 \\ & Ax & & = b \\ & x & & \geq 0 \end{array} \tag{5}$$

till a Slater point is found. For this purpose usually the same algorithm is used with obvious modifications as for solving the original problem (2).

The Slater point is moved during the iterative procedure for solving (2). In conjunction with cutting plane type algorithms heuristics for performing the move can be found in [24] and [14].

5 Implementation

The following two solvers have been implemented on the basis of the ideas discussed in the previous section

- *PCSPIOR* is an implementation of the supporting hyperplane method of [24],
- *PROBALL* is an implementation of the central cutting plane method of [14].

Both solvers have been implemented by the author; in implementing PCSPIOR the guidelines in [24] have been utilized.

For computing $F(y)$ and $\nabla F(y)$ as well as for computing the Boole–Bonferroni bounds the 1996 version of the subroutine MDMNOR of T. Szántai has been utilized, see [23]. This subroutine also computes the sharper Hunter–bound (see [19]).

For solving linear programming subproblems Minos 5.4, [15] has been used.

As a testing environment we have utilized SLP–IOR, our model management system for stochastic linear programming, see [9]; for details consult [14]; a list of solvers currently connected to SLP–IOR can be found in [10]. SLP–IOR served as a workbench for generating test problems and for carrying out the test runs.

The following features have been used:

- Randomly generating test problem batteries for jointly chance constrained problems with multinormal distribution and with known optimal objective value.
- Generating test problem batteries by imposing random perturbations on the data of a fixed test problem. This feature serves for generating variants of a fixed test problem.
- Performing test runs with a selected test problem battery by running selected solvers in turn on the test problems.

The hardware/software environment was the following:

- IBM PC/Pentium II, 400MHz, 128MB RAM;
- MSDOS operating system.
- SLP–IOR has been developed in Borland Pascal 7.0 in an object oriented style. The development of a Windows 32 version is in progress.
- The solvers have been developed in Fortran, the compiler LaheyFortran F77L–EM/32 has been used.

6 Computational results

Computational results for jointly chance constrained problems have recently been published by [7], [14] and [11]. Considering the dimension of the random right hand side the largest problems for which computational results have been published were those in [11], with $r = 7$. The purpose of this section is to illustrate that with present day hardware technology and with appropriate implementation of the methods much larger problems can be solved.

Table 1: Test problem batteries. s(A) and s(B) denote the sparsities of the corresponding matrices; r denotes the dimension of the random right hand side.

Name	# probs	A	s(A)	T	s(T)	r
STABIL	11	48×46	8%	4×46	5%	4
CAM	10	80×160	10%	20×160	5%	20
CBJ	10	50×120	5%	30×120	5%	30

The main characteristics of the test problem batteries are summarized in Table 1.

The test problem battery *STABIL* consists of the standard test problem STABIL of [21] and perturbed versions of it. It is the same battery as published in [14] (see Test #13 there) and has been included here for the purpose of comparing solution times. The test problems in this battery are maximization problems.

Test batteries *CAM* and *CBJ* are randomly generated with array entries between -10 and 10. The test problems have the following property: when choosing the probability level as $\alpha = 0.9$ then exactly 10 of the deterministic constraints are active at the generated optimal solution. Both batteries consist of minimization problems.

All test runs have been carried out with the default setting of the run–time parameters of the solvers.

Table 2: PCSPIOR: Test results for the battery STABIL. STABIL0 is the original STABIL testproblem; $\alpha = 0.9$ and the random RHS is 4–dimensional.

	elapsed time (sec)	*objective value*
STABIL0	0.38	4370.324016
STABIL01	0.39	4315.231023
STABIL02	0.50	4345.390472
STABIL03	0.44	4358.448581
STABIL04	0.38	4543.280717
STABIL05	0.44	4409.216199
STABIL06	0.39	4515.090907
STABIL07	0.39	3971.862817
STABIL08	0.44	4300.541908
STABIL09	0.44	4289.313896
STABIL10	0.44	4663.230828

Tables 2 and 3 show the results with the first test problem battery *STABIL*. Notice that the optimal objective values differ slightly from those published in [14]. This

Table 3: PROBALL: Test results for the battery STABIL. STABIL0 is the original STABIL testproblem; $\alpha = 0.9$ and the random RHS is 4–dimensional.

	elapsed time (sec)	*objective value*
STABIL0	0.49	4370.310724
STABIL01	0.55	4315.278862
STABIL02	0.55	4345.462722
STABIL03	0.55	4358.157237
STABIL04	0.49	4543.269744
STABIL05	0.39	4411.221964
STABIL06	0.44	4515.374999
STABIL07	0.39	3969.638259
STABIL08	0.61	4299.978985
STABIL09	0.55	4289.280101
STABIL10	0.60	4663.215792

is due to the fact that both solvers have been further–developed in the meantime. Considering solution times a similar behavior can be observed for the two solvers.

Table 4: PCSPIOR: Test results for the battery CAM; $\alpha = 0.9$ and the random RHS is 20–dimensional.

	elapsed time (sec)	*objective value*
CAM01	94.19	-2665.172342
CAM02	51.63	328.305673
CAM03	208.22	4372.883690
CAM04	340.43	4704.557191
CAM05	129.02	-168.757369
CAM06	228.6	4279.159963
CAM07	363.22	3042.091725
CAM08	175.99	3484.348103
CAM09	91.17	-719.559098
CAM10	313.08	4840.253687

Tables 4 and 5 display the results with the test problem battery *CAM* consisting of test problems with 20–dimensional random right hand sides. The results indicate that PROBALL (central cutting plane method) is in general faster than PCSPIOR (supporting hyperplane method).

The test problem battery *CBJ* contains test problems with 30–dimensional random right hand sides. For these large–scale test problems we have displayed more

Table 5: PROBALL: Test results for the battery CAM; $\alpha = 0.9$ and the random RHS is 20–dimensional.

	elapsed time (sec)	*objective value*
CAM01	50.37	-2665.188584
CAM02	40.53	328.300711
CAM03	141.05	4372.890811
CAM04	150.27	4704.561423
CAM05	221.68	-168.761995
CAM06	145.50	4279.159905
CAM07	289.13	3042.096881
CAM08	126.38	3484.347312
CAM09	58.55	-719.396123
CAM10	346.36	4840.262148

detailed information on the runs. Table 6 shows the results with the solver PCSPIOR for the first three test problems. The $*$ at CBJ3 indicates that the solver has prematurely terminated because the maximum number of iterations has been exceeded. Due to excessive computational times we have stopped the test run with the rest of the battery; these test problems seem to be too large for
PCSPIOR.

Table 6: PCSPIOR: Test results for the battery CBJ; $\alpha = 0.9$ and the random RHS is 30–dimensional. "Iter" denotes number of iterations, "F-eval" stands for the number of evaluations of F(y); time is measured in minutes.

	Iter	*F-eval*	*Time(min)*	*objective*
CBJ01	451	3097	56.7	1717.81
CBJ02	768	7700	117.5	2915.48
CBJ03	800*	7164	132.9	634.58

Table 7 displays the computational results for PROBALL (central cutting plane method) for the test problem battery *CBJ*. We have included the number of iterations and the number of functional evaluations for $F(y)$. By the latter we mean the number of occassions where the bounds were insufficient to proceed and the function value has been evaluated by Monte–Carlo integration. We have also included elapsed time. The elapsed time shown in the third column is measured in minutes and is mainly used for computing $F(y)$ and $\nabla F(y)$; for solving linear programming subproblems a small fraction of it is used, on the average just about 0.6 minutes per testproblem. Despite this fact the elapsed time can not directly be estimated on the basis of number of iterations and number of function evaluations. The reason is on the one hand that the computation of the gradient involves computation of r-1 dimensional distribution functions. On the other hand Monte–Carlo integration is carried out with varying samplesizes depending on the quality of the bounds. The second column shows the overall number of function evaluations regardless of the differences in computing time as outlined above.

We have also solved all problems of the battery *CBJ* with probability level $\alpha =$ 0.99. The results are shown in Table 8. For test problem CBJ05 the solver could not solve the problem with the default run–time parameter settings. This indicates that for large scale problems adjustment of the run–time parameters of the solver ("tuning") may become necessary. This is in accordance with computational experience from other fields of mathematical programming.

Table 7: PROBALL: Test results for the battery CBJ; $\alpha = 0.9$ and the random RHS is 30–dimensional. "Iter" denotes number of iterations, "F-eval" stands for the number of evaluations of F(y); time is measured in minutes.

	Iter	*F-eval*	*Time(min)*	*objective*
CBJ01	586	926	18.5	1717.82
CBJ02	434	2493	16.3	2915.48
CBJ03	782	2950	30.1	629.50
CBJ04	636	4010	28.3	4931.63
CBJ05	204	205	8.1	-1418.34
CBJ06	423	664	10.5	172.27
CBJ07	122	411	3.4	2007.77
CBJ08	655	1384	22.0	1796.93
CBJ09	529	1721	18.5	2612.00
CBJ10	705	5418	35.6	943.38

Table 8: PROBALL: Test results for the battery CBJ; $\alpha = 0.99$ and the random RHS is 30–dimensional. "Iter" denotes number of iterations, "F-eval" stands for the number of evaluations of F(y); time is measured in minutes.

	Iter	*F-eval*	*Time(min)*	*objective*
CBJ01	342	343	7.1	1718.14
CBJ02	648	2159	13.9	2916.81
CBJ03	385	386	8.0	629.73
CBJ04	347	848	8.0	4934.48
CBJ05	—	—	—	—
CBJ06	209	210	4.1	172.32
CBJ07	310	314	6.2	2008.18
CBJ08	265	266	5.2	1797.31
CBJ09	575	577	12.0	2612.20
CBJ10	251	1014	5.5	943.79

References

[1] Bukszár, J. (1999). Probability bounds with multitrees. Research Report RRR 5-99, RUTCOR.

[2] Bukszár, J. and Prékopa, A. (1999). Probability bounds with cherry–trees. Research Report RRR 33-98, RUTCOR.

[3] Deák, I. (1980). Three digit accurate multiple normal probabilities. *Numerische Mathematik*, 35:369–380.

[4] Deák, I. (1990). *Random number generators and simulation.* Akadémiai Kiadó, Budapest.

[5] Dupačová, J., Gaivoronski, A., Kos, Z., and Szántai, T. (1991). Stochastic programming in water management: A case study and a comparison of solution techniques. *European Journal of Operational Research*, 52:28–44.

[6] Gassmann, H. I. (1988). Conditional probability and conditional expectation of a random vector. In Ermoliev, Y. and Wets, R.-B., editors, *Numerical Techniques for Stochastic Optimization*, pages 237–254. Springer Verlag.

[7] Gröwe, N. (1997). Estimated stochastic programs with chance constraints. *European Journal of Operational Research*, 101:285–305.

[8] Kall, P. (1976). *Stochastic linear programming.* Springer Verlag.

[9] Kall, P. and Mayer, J. (1996). SLP–IOR: An interactive model management system for stochastic linear programs. *Mathematical Programming*, 75:221–240.

[10] Kall, P. and Mayer, J. (1998a). On solving stochastic linear programming problems. In Marti, K. and Kall, P., editors, *Stochastic Programming Methods and Technical Applications*, pages 329–344. Springer Verlag.

[11] Kall, P. and Mayer, J. (1998b). On testing SLP codes with SLP–IOR. In Giannessi, F., Rapcsák, T., and Komlósi, S., editors, *New trends in mathematical programming*, pages 115–135. Kluwer Academic Publishers.

[12] Kall, P. and Wallace, S. W. (1994). *Stochastic programming.* John Wiley & Sons.

[13] Mayer, J. (1992). Computational techniques for probabilistic constrained optimization problems. In Marti, K., editor, *Stochastic Optimization: Numerical Methods and Technical Applications*, pages 141–164. Springer.

[14] Mayer, J. (1998). *Stochastic Linear Programming Algorithms: A Comparison Based on a Model Management System.* Gordon and Breach.

[15] Murtagh, B. A. and Saunders, M. A. (1995). Minos 5.4 user's guide. Technical Report SOL 83-20R, Department of Operations Research, Stanford University.

[16] Norkin, V. I., Ermoliev, Y. M., and Ruszczyński, A. (1998). On optimal allocation of indivisibles under uncertainty. *Operations Research*, 46:381–395.

[17] Prékopa, A. (1971). Logarithmic concave measures with applications to stochastic programming. *Acta. Sci. Math*, 32:301–316.

[18] Prékopa, A. (1988). Numerical solution of probabilistic constrained programming problems. In Ermoliev, Y. and Wets, R.-B., editors, *Numerical Techniques for Stochastic Optimization*, pages 123–139. Springer Verlag.

[19] Prékopa, A. (1995). *Stochastic programming.* Kluwer Academic Publishers.

[20] Prékopa, A. (1999). The use of discrete moment bounds in probabilistic constrained stochastic programming models. *Annals of Operations Research*, 85:21–38.

[21] Prékopa, A., Ganczer, S., Deák, I., and Patyi, K. (1980). The STABIL stochastic programming model and its experimental application to the electricity production in Hungary. In Dempster, M. A. H., editor, *Stochastic Programming*, pages 369–385. Academic Press.

[22] Szántai, T. (1986). Evaluation of a special multivariate gamma distribution. *Mathematical Programming Study*, 27:1–16.

[23] Szántai, T. (1987). Calculation of the multivariate probability distribution function values and their gradient vectors. Working Paper WP-87-82, IIASA.

[24] Szántai, T. (1988). A computer code for solution of probabilistic-constrained stochastic programming problems. In Ermoliev, Y. and Wets, R.-B., editors, *Numerical Techniques for Stochastic Optimization*, pages 229–235. Springer Verlag.

[25] Uryasev, S. and Rockafellar, R. T. (1999). Optimization of conditional Value–at–Risk. Research Report #99-4, Center for Applied Optimization, University of Florida.

Management of Quality of Service through Chance-constraints in Multimedia Networks

E.A. Medova (e.medova@jims.cam.ac.uk)
The Judge Institute of Management Studies,
University of Cambridge,
England CB2 1AG.

J.E. Scott (j.scott@jims.cam.ac.uk)
The Judge Institute of Management Studies,
University of Cambridge,
England CB2 1AG.

Abstract

Recently stochastic programming with recourse has been used in telecommunication models with multiple traffic scenarios to result in large scale linear programmes for network design. In this paper we use large deviation theory to invert chance constraints involving simultaneous quality-of-service blocking probabilities at several network time scale layers to result in node and arc-path incidence formulations of continuous multicommodity network flow problems involving deterministic effective bandwidths of origin-destination stochastic traffic flows in the network. These compact deterministic linear programmes are used in the paper to size network resource capacities, route traffic in a proportionally fair manner, price resources and design peak resource pricing customer tariffs. Our Integrated Network Design System software, which incorporates more network resource and traffic types than are treated here in detail, is also described briefly.

S.P. Uryasev (ed.), Probabilistic Constrained Optimization, 236–251.
© 2000 *Kluwer Academic Publishers.*

1 Introduction

The last few years have demonstrated overwhelming acceptance of new internet-based businesses. With the increasing variety of ways in which networks are being used, coupled with growing customer expectations of network performance, the modelling of telecommunications is becoming ever more important. Because of the wide variety of traffic requirements, the design, dimensioning and traffic management of *Broadband Integrated Services Digital Networks* (B-ISDN) involve the application of diverse methods and theories. In the early 1970s optimization models were applied to the design of the data networks (ARPANET, BITNET, TYMNET) which started the 'internet revolution' ([7]). However the performance of public telephone networks was mainly analyzed using results from queueing theory and only recently have stochastic programming methods been applied to the design of telecommunication networks ([1, 4, 19]).

In this paper we summarize results from our collaborative work with British Telecom, previously reported in ([5, 16, 17]), and extend our *Path and Capacity Allocation* (PCA) model to the problem of infrastructure costing and customer tariffing. Our formulation of constraints on quality of service allows us to invert chance-constraints involving random *decisions* (traffic flows) – rather than random parameters – using large deviation theory to obtain their deterministic equivalents.

The remainder of this introduction will give a brief account of network technology and our modelling assumptions. The derivation of an effective bandwidth of a stochastic multi-service traffic is presented in Section 2 The stochastic optimization traffic management model and its deterministic equivalent is given in Section 3 Section 4 outlines our proposal for tariffing. Our implementation of the PCA model and its extensions—the *Integrated Network Design* System (INDS) providing optimal capacity allocation and traffic routing for guaranteed quality of service—is discussed in Section 5 Conclusions and future research directions are summarized in Section 6

B-ISDN is designed to carry voice, data and video over the same physical infrastructure with a standardised multiplexing and switching digital format called *Asynchronous Transfer Mode* (ATM). The notion of a committed quality or *Grade of Service* (GoS) per connection corresponding to a negotiated contract is particular to ATM networks. In ATM networks all information coming from different traffic sources is transmitted by means of short fixed length cells comprising a 5 byte header and a 48 byte information payload. With routing based on information in the header of a packet, an ATM network behaves like a hybrid of a circuit-switched and a packet network. The packets belonging to different *virtual paths* (VPs) are merged and transmitted as a single stream via *statistical multiplexing*. Statistical multiplexing leads to shorter average delay per packet, particularly for 'irregular' traffic, but requires a complex *effective bandwidth* allocation. For example, data transmissions are highly bursty (burstiness = peak bit rate / average bit rate) and bits must not be lost. On the other hand, high definition television (HDTV) distribution has low burstiness but will consume a bit rate above 50 Mb/s per HDTV channel and is insensitive to bit losses. In our work we consider *constant bit rate* (CBR) service classes such as telephony, N-ISDN video and TV, and *variable bit rate* (VBR) service classes such as VBR video retrieval, Ethernet and local area network (LAN) data. In high speed optical networks the ATM packets are further framed, according to the *Synchronous Digital Hierarchy* (SDH) protocol with standard rate of 155Mb/s (STM-1) and its multiples (STM-N) ([18]).

2 Effective bandwidth as a deterministic measure of stochastic multiservice traffic

Here the term *call* means a predefined mixture of multimedia traffic. The call may be viewed alternatively at the levels of connection request, packet bursts, or transmitted cells. GoS is measured by *blocking probability* or probability of cell loss at the different time scales. For example, call, burst or cell grades of service may be 10^{-3}, 10^{-5} and 10^{-7} respectively. The definition of effective bandwidth varys depending on the network link model used (unbuffered or buffered), see e.g. [6, 9, 15, 3]. We adopt Hui's *principle of layered switching*, which considers traffic entities at call, burst and cell levels simultaneously ([11, 12]). The effective bandwidth of a specific traffic flow is defined by an *admission policy* at the call level ([13]), but *offered* traffic to the network at the call level should not disturb currently *carried* traffic at the cell and burst level. Our algorithm recursively solves for the effective bandwidth of a given traffic flow. Due to very small probabilities of call rejection/blocking the results are based on large deviation techniques ([20]).

Consider, for each service type, independent stationary Poisson offered call traffic and l-state Markov modulated fluid bursty calls with constant cell flow rates in each burst state ([11]). We assume sufficient separation of the call, burst and cell time-scale layers that each lower layer sees the parameters of higher layers as quasi-static. At the call time-scale layer the offered call process for each independent source is a stationary Poisson process with intensity N and state given by the number $\mathbf{N}$ of calls offered per unit time in statistical equilibrium, so that $\mathbb{E}\mathbf{N} = \mathbf{N}$. We wish to set N so as to limit the call blocking probability, i.e. call GoS, of carried traffic from this source to g_{call}.

For a potential call in progress, bursts are modelled by an l-state stationary Markov modulated fluid transmitting cells at constant rate a_r in state r, which obtains with probability $p_r := \tau_r/(\tau_1 + \cdots + \tau_l)$, where τ_r denotes the expected holding time in the state $r = 1, \ldots, l$.

We are assuming that cell loss is absent (i.e. set $g_{cell} := 0$) and wish to set the *total* end-to-end carried call mix so as to meet the call GoS and simultaneously to limit burst blocking probability, i.e. burst GoS, to g_{burst}.

Consider for each O-D pair of the network a mixture of constant bit rate (CBR) and multiple service type, $j = 1, \ldots, k$, variable bit rate (VBR) traffic $\mathbf{W}$. We assume that

$$\mathbf{W} = \sum_{\mathbf{j=1}}^{\mathbf{k}} \mathbf{N_j W_j} \tag{1}$$

is composed of independent stationary call streams $\mathbf{N_j}$ with independent identically distributed stationary call bandwidth requirements $\mathbf{W_j}$.

Let us denote by $\mathbf{N}$ the vector of numbers of calls in progress of each service type. Alternatively—for example, for the purposes of stationary state individually offered call routing—we may think of a number $\boldsymbol{\nu} := \sum_{j=1}^{k} \mathbf{N_j} \pi_\mathbf{j}$ of representative compound calls each of which is of service type j with required bandwidth $\mathbf{W_j}$ with probability $\pi_j := N_j/(N_1 + \cdots + N_k)$.

Defining the logarithmic generating functions of the compound traffic process $\mathbf{W}$ as

$$\mu_{\mathbf{W}}(s) := \ln \mathbb{E}(^{\sim \mathbf{W}}) \tag{2}$$

and of individual service type processes as

$$\mu_j(s) := \ln \mathbb{E}(^{\sim \mathbf{W_j}}) \quad j = 1, \ldots, k, \tag{3}$$

it follows from our assumptions of independence that

$$\mu_{\mathbf{W}|\mathbf{N}}(s) = \sum_{j=1}^{k} N_j \mu_j(s), \tag{4}$$

where at burst layer for the l_j-state Markov modulated fluid VBR source j

$$\mu_j(s) := \ln \left\{ \sum_{r=1}^{l_j} p_{jr} e^{s a_{jr}} \right\}, \tag{5}$$

with a_r, r and $p_r := \tau_r/(\tau_1 + \cdots + \tau_l)$ defined as above and $\mathbf{W}|\mathbf{N}$ denoting conditioning the total VBR traffic $\mathbf{W}$ on a *fixed* call in progress profile N.

Admission policy at each level satisfies the GoS agreed between user and network. With respect to network provisioning at the first stage of network planning, these can be specified in terms of common probabilities of call / burst / cell blocking resulting from the lack of transmission capacity. Assuming zero blocking at cell level, these conditions are:

$$P \left\{ \sum_{j=1}^{k} \mu_j \mathbf{N_j} \geq \mathbf{C} \right\} = g_{call}, \tag{6}$$

and simultaneously

$$P\{\mathbf{W} > \mathbf{C}\} = g_{burst}. \tag{7}$$

Thus, the traffic carried over the virtual path is bounded by maximum capacity on an end-to-end basis and is considered to be stationary with certain statistical measures (GoS) controlled at each considered time-scale level.

Each multiservice source is characterized by its effective bandwidth and therefore the virtual path capacity C can be approximated in terms of the aggregate effective bandwidth D_{N^*} by setting the average call profile $N := N^* = (N_1^*, N_2^*, \ldots, N_k^*)'$ such that 6 holds and 7 becomes

$$P\{\mathbf{W} > \mathbf{D}_{\mathbf{N}^*}\} = g_{burst}. \tag{8}$$

To see this in more detail, let us consider first the burst layer effective bandwidth corresponding to a given (not necessarily optimal) multiservice average call profile $N := (N_1, N_2, \ldots, N_k)'$. The *Chernoff bound* of large deviation theory is given by

$$P\{\mathbf{W} \geq \mathbf{D}\} \leq \exp\{-max_s[sD - \mu_{\mathbf{W}}(s)]\}, \tag{9}$$

or rearranging

$$-\ln P\{\mathbf{W} \geq \mathbf{D}\} \geq max_s[sD - \mu_{\mathbf{W}}(s)]\}. \tag{10}$$

Consequently, conditional on $\mathbf{N} = \mathbf{N}$, we wish to find s_N^* to solve the equation

$$max_s[sD - \mu_{\mathbf{W}|\mathbf{N}}(s)] = -\ln g_{burst} \tag{11}$$

in order to determine the burst layer equivalent bandwidth

$$D_N^* = \mu'_{\mathbf{W}|\mathbf{N}}(s_N^*) \tag{12}$$

(in cells/s or Mb/s) corresponding to the offered multiservice call profile N. This may be accomplished numerically using a Newton-Raphson iteration since the left hand side of 11 concerns the maximization of a strictly concave function of s for fixed D.

Now, at the call layer, we may define the convex boundary of the admissible call profile region in terms of acceptable burst layer performance as

$$\{N \in \mathbb{R}^{\urcorner} | \sim_{\mathbb{N}}^* \mathbb{D}_{\mathbb{N}}^* - \mu(\sim_{\mathbb{N}}^*)'\mathbb{N} = -\ln \eth_{burst}\} \tag{13}$$

with tangent plane at a specific call profile N_0 (to be defined below) given by

$$\mu(s_{N_0}^*)'N = \mu(s_{N_0}^*)'N_0 \left(\approx s_{N_0}^* D_{N_0}^*\right), \tag{14}$$

where

$$\mu(s)' := (\mu_1(s), \mu_2(s), \ldots, \mu_k(s)). \tag{15}$$

Any call profile N on the linearized admissible call profile region boundary defined by 13 will have effective bandwidth $D_{N_0}^*$.

At the call level the total offered VBR source traffic mix is appropriately modelled as independent stochastic processes with offered call rates λ_j and mean holding times μ_j^{-1} for service types j, $j = 1, \ldots, k$, to result in a steady state number of *offered calls in progress* vector $\mathbf{N} := (\mathbf{N_1}, \ldots, \mathbf{N_k})'$, whose coordinates are independent Poisson random variables with *means* $\rho_{call,j} := \lambda_j/\mu_j := N_j^0, j = 1, ..., k$. However we wish to adjust the mean carried VBR call mix downwards from N_0 to solve 6 (approximately) as

$$P\left\{\mu(s^*_{N_*})'\mathbf{N} \geq \mu(\mathbf{s}^*_{\mathbf{N}_*})'\mathbf{N}\right\} = g_{call}. \tag{16}$$

Once again applying the Chernoff bound to the probability of the offered call mix $\mathbf{N}$ lying outside the admissible call profile region linearized about $N := (N_1, N_2, \ldots, N_k)'$ we obtain

$$P\left\{\mu(s^*_N)'\mathbf{N} \geq \mu(\mathbf{s}^*_{\mathbf{N}})'\mathbf{N}\right\} \leq \exp\left\{-max_s[s\mu(s^*_N)'N - \mu_{\mu(s^*_N)'\mathbf{N}}(s)]\right\} \tag{17}$$

where

$$\mu_{\mu(s^*_N)'\mathbf{N}}(s) = \sum_{j=1}^{k} N_j(e^{s\mu_j(s^*_N)} - 1). \tag{18}$$

We can use simultaneous coordinate bisection search to adjust N with coordinates $N_j \in [b_j^{-1}\rho_{call,j}, \rho_{call,j}], j = 1, ..., k$, ranging between mean and peak call rate (b_j is the burstiness parameter for source j) until a second Newton-Raphson process achieves the maximum in the right-hand side of 17 at $s^*_{N_*}$ for $N_* := (N_1^*, \ldots, N_k^*)'$ satisfying

$$s^*_{N_*}\mu(s^*_{N_*})'N_* - \mu_{\mu(s^*_{N_*})'N_*}(s^*_{N_*}) = -\ln g_{call} \tag{19}$$

with

$$\frac{d}{ds}\mu_{\mu(s^*_{N_*})'N_*}(s^*_{N_*}) = D_{N_*}, \tag{20}$$

the required burst level effective bandwidth for the stationary offered multiservice traffic $\mathbf{W}$ with call profile having mean N_*. This level of offered traffic $\mathbf{W}$ meets GoS requirements at *both* call and burst levels (with no cell loss) and results in *carried* traffic with maximum bandwidth requirement within GoS specifications $C := D_{N_*}$—a conservative approximation (overestimated) determined by total effective bandwidth.

Examples of the effective bandwidth calcultions for different scenarios of traffic demands have been reported in [16]. In what follows we use these effective bandwidth calculations for specific random traffic flows in a stochastic programming model of the entire network.

3 The chance-constrained stochastic programme

The optimal traffic management policy includes:-

- allocation of capacity for each virtual connection, i.e. *origin-destination* (O-D) pair in the network, subject to a negotiated acceptance policy or quality of service contract.
- routing of stochastic traffic flows over the network with designated capacities.

At the network level the probabilistic distribution of offered traffic is conditioned on the random number of calls accepted and the composition of the random mix of traffic types. The functional form of this distribution depends on the nature of the burst and cell level models. In the effective bandwidth calculations we assume dependence between time-scale layers (call, burst and cell), but, for example, independence of an individual call's burst statistics from the number of calls in progress at the call layer. The level of call acceptance is fixed by the linearized call acceptance region defined by 16 for $N := N_*$ and the inverse problem of finding the number of calls by service type corresponding to the specified loss probabilities at the call layer is solved numerically as described above.

We are now in a position to formulate our traffic management *path and capacity allocation* (PCA) problem for a network.

For a fixed network topology represented by a directed graph $G(N, A)$, with node set $n \in N$ and directed arc (link) set $a \in A$ consider *stationary* stochastic *cell flow* traffic $\mathbf{f_w}$ (in cells/s or Mb/s) between each OD (node) pair $w \in W$ in the network during a specified traffic regime. The network dimensioning and traffic management problem is to allocate *link capacity* $C_a, a \in A$ and route traffic $\mathbf{f_p}$ on *paths* $p \in P_w$, a set of specified routes through the network from origin to destination node of w, so as to optimize the expected net revenue resulting from carrying the stochastic traffic in terms of total OD pair cell flows

$$\mathbf{f_w} = \textstyle\sum_{\mathbf{p}\in\mathbf{P_w}} \mathbf{f_p} \quad w \in W \tag{21}$$

subject to GoS constraints at call, burst and cell time-scale levels. Consider the *chance constrained stochastic programme*

$$\textbf{(PCA)} \qquad \min[\sum_{\mathrm{a}\in\mathrm{A}} \mathrm{b_a C_a} - \sum_{\mathrm{w}\in\mathrm{W}} \mathrm{r_w} \sum_{\mathrm{p}\in\mathrm{P_w}} \textit{ess}\ \sup\{\mathbf{f_p} \mid \mathbf{f_w} \leq \mathbf{D_w}\}] \tag{22}$$

s.t.

$$P\{\textstyle\sum_{p\in P_w} \mathbf{f_p} \geq D_w\} \leq g_{burst} \qquad w \in W \tag{23}$$

$$P\{\textstyle\sum_{p\in Q_a} \mathbf{f_p} \geq C_a\} \leq g_{burst} \qquad a \in A \tag{24}$$

$$\mathbf{f_p} \geq 0 \qquad p \in P_w \qquad w \in W. \tag{25}$$

The decision variables of this problem are stochastic OD pair traffic flows and the deterministic capacities of the network links implicitly defined by the optimum routes chosen in the solution of the flow optimization problem. The objective 22 of this optimization problem contains coefficients representing the unit *cost of link capacity provision* $b_a, a \in A$, and unit *OD pair net revenue* $r_w, w \in W$, resulting from carrying traffic between OD pairs. Here *ess* $\sup\{\cdot \mid \mathbf{f_w} \leq \mathbf{D_w}\}$ denotes the supremum with probability one of carried stochastic path traffic conditional on the stationary stochastic OD pair offered traffic flow $\mathbf{f_w}$ not exceeding the capacity D_w which is consistent with burst GoS requirements for each $w \in W$. The first chance, or *probabilistic* constraint 23 states that the probability of offered stochastic OD pair traffic flow $\mathbf{f_w}$ exceeding a specified *demanded capacity* level D_w must not exceed the GoS for *burst refusal probability* g_{burst}. In the implementation the demands D_w will be set so as to *simultaneously* maintain the call GoS for the corresponding calls in progress profile $\mathbf{N}$. The probabilistic constraints 24 state that the burst GoS g_{burst} must also be maintained for the sum of offered traffic flows on the set Q_a of *all paths* passing *through* each (directed) *link* $a \in A$. The constraints 25 require the stochastic path traffic flows to be nonnegative (with probability one).

Stochastic optimization problems of the form (PCA) have a *deterministic equivalent* problem which may be obtained by replacing the probabilistic constraints involving random variables 23, 24 and 25 with equivalent constraints written in terms of deterministic variables which meet the probability restriction as nearly as possible. Therefore the maximum OD pair demanded bandwidths D_w are the effective bandwidth requirements of OD pair stochastic offered traffic $\mathbf{f_w}, \mathbf{w} \in \mathbf{W}$, for statistically multiplexed multimedia traffic obtained numerically by procedure given in Section 2

We may thus replace the stochastic offered traffic flows $\mathbf{f_p}$ and their carried values determined by the conditional expectations in 22 by *deterministic* fluid flows $f_p, p \in P_w, w \in W$, which represent flows at the rate just above which fluctuations of the stochastic offered flows $\mathbf{f_p}$ would begin to exceed GoS requirements.

The deterministic equivalent problem of **PCA** is given by:

$$\textbf{(PCA)} \qquad \min[\sum_{a\in A} \mathrm{b_a C_a} - \sum_{w\in W} \mathrm{r_w} \sum_{p\in P_w} f_p] \tag{26}$$

s.t.

$$\sum\nolimits_{p\in P_w} f_p = D_w \qquad w \in W \tag{27}$$

$$\sum\nolimits_{p\in Q_a} f_p \leq C_a \qquad a \in A \tag{28}$$

$$C_a \leq C_A \qquad a \in A \tag{29}$$

$$f_p \geq 0 \qquad p \in P_w \qquad w \in W. \tag{30}$$

Our deterministic PCA is a compact arc-path form multicommodity flow problem (MFP) in real variables ([10]). Reliability of the network requires that each origin-destination (O-D) node pair $w \in W$ has a number of node/link disjoint physical paths $p \in P_w$ specified by the network operator. As the problem of finding these paths in a given network is computationally exponential, we precompute the required fixed number of paths to generate the incidence matrices corresponding to OD pair-path incidence set P_w and the arc-path incidence set Q_a. The additional overall link capacity constraints of the form 29 may be set to conform with a suitable transport standard of the optical transmission layer (SDH) with interger multiple transmission rates ([18]).

The PCA problem can be viewed as a hierarchical network planning and traffic management model ([4]). First, GoS requirements are translated into chance-constraints in which the *virtual* capacity of the connection between OD node pair w is a suitable effective bandwidth D for offered stochastic multimedia traffic. For ATM networks this is translated into calculations effective bandwidths for each virtual connection. The further stages involve the solution of a deterministic PCA problem which allocates and routes deterministic equivalent traffic between all specified OD pairs in the network.

The first stage of deterministic optimization involves the routing of traffic over a network with unlimited capacity (uncapacitated PCA) and identification of the bottleneck transmission links. Such a routing usually leads to highly uneven network link load distribution. As a policy, network operators like to spread traffic across the network. This policy is implemented in the PCA model by the constraint 29 on maximum link capacity given by C_A. An optimum value for C_A may be obtained using the *bottleneck PCA model*:

$$\textbf{(BPCA)} \qquad \min[\mathrm{C_A}] \tag{31}$$

s.t.

$$\textstyle\sum_{p\in P_w} f_p = D_w \qquad w \in W \tag{32}$$

$$\textstyle\sum_{p\in Q_a} f_p \leq C_A \qquad a \in A \tag{33}$$

$$f_p \geq 0 \qquad p \in P_w \qquad w \in W. \tag{34}$$

At the next stage the minimized value of the maximum link capacity for a given demand is fixed. The resulting model (PCA+bottleneck) produces optimal traffic routing over all available paths (traffic balancing) with the most stringent possible constraint on maximum capacity installed. (At a lower maximum capacity value all the offered stochastic traffic cannot be carried within the specified GoS requirements.) More detailed design will involve switching and buffering resources and technical decisions at the SDH network layer ([17]). These standards dictate integrality constraints on link capacities. The corresponding modified (PCA+) models are implemented using mixed integer programming solvers.

4 Pricing mechanism

The duals of deterministic PCA problem allow the *pricing* of traffic demands relative to network infrastructure costs.

Let

x_w be the *price for a virtual connection* request of unit capacity

y_a be the *price of unit capacity* on a link, and

z_a be the *peak price of unit capacity* on a link.

With given *costs* b_a of unit capacity per link (usually determined by fibre type and length) and OD pair net *revenue* coefficients r_w, the *dual* PCA model is given by:

$$\textbf{(DPCA)} \qquad \max[\sum_{w \in W} D_w x_w + C_A \sum_{a \in A} z_a] \tag{35}$$

s.t.

$$x_w + \textstyle\sum_{a \in P} y_a \geq -r_w \qquad p \in P_w \qquad w \in W \tag{36}$$

$$y_a - z_a \leq b_a \qquad a \in A. \tag{37}$$

The tariff structure specified by OD pair net revenues $r_w, w \in W$, may be obtained from fixed link costs b_a and in particular the costs b_A of bottleneck links (both based on real costs of infrastructure as noted above) by solving the *dual bottleneck PCA* problem:

$$\textbf{(DBPCA)} \qquad \max[\sum_{w \in W} D_w x_w] \tag{38}$$

s.t.

$$x_w + \textstyle\sum_{a \in P} y_a \geq -r_w \qquad p \in P_w \qquad w \in W \tag{39}$$

$$y_a \leq 1 \qquad a \in A \tag{40}$$

$$y_a \geq 0 \qquad p \in P_w \qquad w \in W \tag{41}$$

to obtain optimal prices x_w^* and y_a^*. Of course these optimal dual variables are a by-product of solving the (primal) bottleneck PCA model using any linear programming technique. From these prices we may set the unit net revenue per virtual connection for the OD pair $w \in W$ as $r_w := (x^* + \sum_{a \in P} y_a^*) b_A$, where b_A is the *maximal* unit cost of bottleneck links.

Next we solve the full (minimal capacity) PCA+Bottleneck problem with these revenue coefficients to obtain the explicit *peak resource tariffs* z_a (and x_w, y_a) for the implementation of *peak capacity* or *marginal cost* pricing. Indeed, from 36, a call of effective bandwidth requirement D between nodes of the OD pair $w \in W$ in the network will be charged:

$$(x_w + \sum_{a \in P} y_a) D \tag{42}$$

with $y_a = b_a$ for non-bottleneck ($C_a < C_A$ and $z_a = 0$) links and $y_a = b_a + z_a$ for bottleneck ($C_a = C_A$ and $z_a > 0$) links on the path p between OD pairs in the network. The optimal traffic spreading across the paths $p \in P_w$ effected by the suggested procedure will keep these individual call unit tariffs maximally homogeneous with respect to actual infrastructure costs (or bulk charges from a backbone network carrier).

Tariff structure at a much more detailed level, including buffering and delay considerations, may be obtained from stochastic traffic resource models which are based on cell flow effective bandwidth charging ([14]). In this context, it is worth noting that the traffic allocations discussed in this paper are also *proportionally fair* in the sense of game theory. For, if we define the network objective as *net revenue*

$$g(f_1, \dots, f_p) := \sum_{w \in W} r_w \sum_{p \in P_w} f_p \tag{43}$$

to be maximized, then by optimality the PCA+bottleneck traffic allocation f^* is *proportionally fair* in that

$$\sum_{w \in W} \sum_{p \in P_w} \frac{\partial g(f^*)}{\partial f_p}(f_p^* - f_p) = \sum_{w \in W} \sum_{p \in P_w} r_w (f_p^* - f_p) > 0 \tag{44}$$

relative to any other feasible allocation f if, and only if, the optimum allocation f^* is unique. Otherwise, it is *weakly proportionally fair* in the sense that the inequality in 44 is non-strict.

5 Implementation

We found in practice that the PCA model described in Section 2 needed to be extended to include ATM and SMD traffic types, and to take into account the cost of switch provisioning. Also, as mentioned above, we considered the effect of imposing integrality constraints on the capacity of links. The full deterministic PCA model is then as follows:

$$\begin{aligned} \min[\sum_{\mathrm{a} \in \mathrm{A}} \mathrm{b_a C_a} + \sum_{\mathrm{n} \in \mathrm{N}} \mathrm{k_n^{ATM} S_n^{ATM} + k_n^{SMD} S_n^{SMD}} \\ - \sum_{w \in W} r_w \sum_{p \in P_w} f_p^{ATM} + f_p^{SMD}] \end{aligned} \tag{45}$$

s.t.

$$\sum_{p\in P_w} f_p^{ATM} = D_w^{ATM} \qquad w \in W \qquad (46)$$

$$\sum_{p\in P_w} f_p^{SMD} = D_w^{SMD} \qquad w \in W \qquad (47)$$

$$\sum_{p\in Q_a} f_p^{ATM} + f_p^{SMD} \leq C_a \qquad a \in A \qquad (48)$$

$$\sum_{p\in R_n} f_p^{ATM} \leq S_n^{ATM} \qquad n \in N \qquad (49)$$

$$\sum_{p\in R_n} f_p^{SMD} \leq S_n^{SMD} \qquad n \in N \qquad (50)$$

$$C_a = kP \qquad k \in \mathbb{N} \qquad (51)$$

$$C_a \leq C_A \qquad a \in A \qquad (52)$$

$$f_p \geq 0 \qquad p \in P_w \qquad w \in W. \qquad (53)$$

where

k_n^{ATM} – ATM switch provisioning cost

k_n^{SMD} – SMD switch provisioning cost

S_n^{ATM} – ATM switch capacity

, S_n^{SMD} – SMD switch capacity

P – STM-1 unit link capacity (155 Mb/s).

Solution of the full PCA model with two traffic types, switch provisioning and a bound on link capacity using a simplex-based LP solver is very fast (see Table 1). The integrality constraint on link size however leads to much longer soution times (using a branch and bond algorithm). However, we found that a good approximation to the optimal integer solution can be found reasonably quickly. The difficulty of finding an integer solution though is heavily dependent on model data. We used algebraic modelling languages (initially MODLER ([8]) and later XPRESS-MP ([2])) to formulate the linear and integer models, and then wrote a grahical user interface and visualization toolkit to provide network operators with the ability to view and manipulate the model data and solutions without the need for knowledge of the algebraic formulation or any particular programming skills.

The user interface is written in Java and connects over a network to a server which runs the effective bandwidth calculator and LP/IP solvers and maintains a database of problem files. This allows us to deploy the user interface on virtually any computing platform, including the world-wide-web. This is important when we consider that there are likely to be multiple classes of end user for such a system. We allow the user high-level access to the model, such as which constraints are imposed, and which traffic types are used, as well as access to model data such as cost and revenue coefficients, the traffic mix and demand profiles.

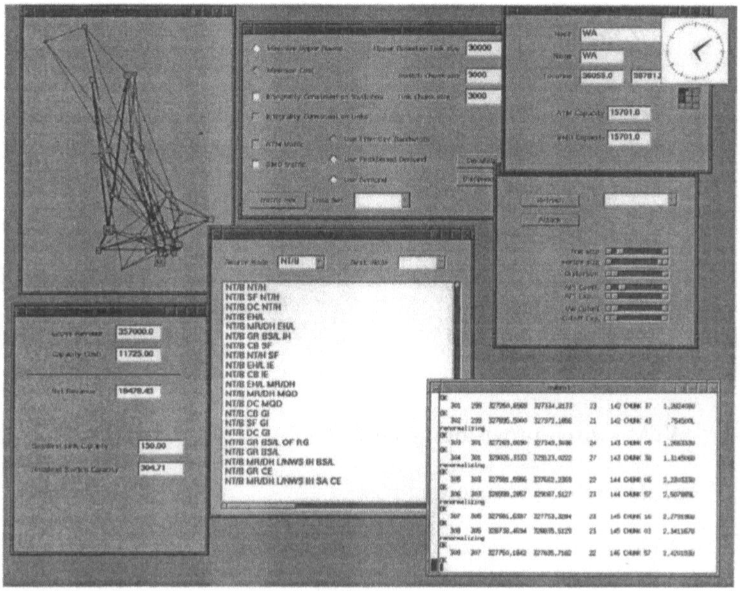

Figure 1: **GUI and Visualization tools for INDS.** These include a *fisheye* network browser, which displays the network topology and a representation of the current capacity allocation, dialog box based control of the solution algorithm, and table based display and manipulation of the model data, such as the current solution, and routing information. The low-level dialog between the solver and the GUI is also shown.

PCA+switch revenue maximization with effective bandwidth calculation					
Network		**LP Dimensions**		**Solution time (seconds)**	
Name	BT	Rows	635	Effective b/w	5.23
Nodes	31	Columns	1376	LP Solution time	0.75
Links	70	Non-zeros	9954		
OD-Pairs	216	Density	1.14%		

Table 1: CPU times for an IBM RS6000/590 running Dash Associates' XPRESS-MP under AIX 4.2.

As the solution times for the effective bandwidth calculation and the LP solution are reasonably short, the user can change the model and see the effects on the solution immediately. A dependency resolution system is employed to ensure that only the minimal amount of recalculation is done for given changes to the model data, so most types of change lead to very fast update of the solution (for more details, see [17]). The addition of integrality constraints on the capacity of links leads to much longer solution times. Keeping interactivity high under such circumstances is difficult unless some approximation to the integer solution is accepted.

6 Conclusions

The results reported in this paper are a summary of research initiated by British Telecom and allowing the modelling of multimedia traffic, optimzation of capacity usage, routing of traffic over the existing physical network and tariffing based on the infrastructure costs. Although our stochastic analysis has been done specifically for an ATM network, an interpretation of quality of service for connections as a chance-constraint provides a general framework for modelling different networks and traffic types. The complex issue of integrating the different transport technologies is accomplished using the Integrated Network Design System which combines optimization modelling tools with an effective graphical user interface.

Acknowledgements

This research has been partially supported by British Telecom Laboratories, the UK EPSRC and the EU Esprit programme. We would like to thank our co-workers M.A.H. Dempster and R.T. Thompson for their collaboration and continued support.

References

[1] M. Bonatti and F. Fantauzzi and A.A. Gaivoronski (1996) *Stochastic programming approach to dynamic virtual path capacity allocation in ATM networks.* In *Proceedings of the 4th International Conference on Telecommunication Systems*, pages 270-275. Vanderbilt University, Nashville.

[2] Dash Associates Limited (1998) *XPRESS-MP User Guide.* Blisworth, Northants, UK. 10th edition.

[3] G. de Veciana, G. Kesidis and J. Walrand (1995). Resource management in wide-area ATM networks using effective bandwidths. *IEEE Journal on Selected Areas in Communications*, 13: 1081-1090.

[4] M.A.H. Dempster (1994). Hierarchical approximation of telecommunications networks. *BT Technology Journal*, 12, pages 40-49.

[5] M. Dempster, E. Medova, H. Azmoodeh, P. Key and S. Sargood(1996). *Design and control of ATM/SDH Networks.* In *Proceedings of the 4th International Conference on Telecommunication Systems*, pages 259-270. Vanderbilt University, Nashville.

[6] A. Elwalid and D. Mitra (1993). Effective bandwidth of general Markovian traffic sources and admission control of high speed networks. In *IEEE/ACM Transactions on Networking*, volume 1, pages 329-343.

[7] L. Fratta, M. Gerla and L. Kleinrock(1973) The flow deviation method: An application to store-and-forward communication network design. *Networks*, 3: 97-133

[8] H. Greenberg (1993). *Modeling by Object-Driven Linear Elemental Relations: A User's Guide for MODLER.* Kluwer Academic Publishers.

[9] R. Guerin, H. Ahmadi and M. Naghshineh (1991). Equivalent Capacity and its Application in High-Speed Networks. *IEEE Journal on Selected Areas in Communication*, 9: 968-981.

[10] J. Hu (1970). *Integer Programming and Network Flows.* Addison-Wesley Publishing Company, Inc.

[11] J.Y. Hui (1988). Resource allocation for broadband networks. *IEEE Journal on Selected Areas in Communications*, 6: 1598-1610.

[12] J.Y. Hui, M.B. Gursoy, N. Moayeri and R.D. Yates (1991). A Layered broadband switching architecture with physical or virtual path configurations *IEEE Journal on Selected Areas in Communications*, 9: 1416-1426.

[13] J.S. Kaufman (1981) Blocking in a shared resource environment. *IEEE Transactions on Communications*, 29, pages 1474-1481.

[14] F.P. Kelly, A.K. Maulloo and D.H.K. Tan (1998). Rate control in communication networks: Shadow prices, proportional fairness and stability. *Journal of the Operational Research Society*, 49, pages 237-252.

[15] F. P. Kelly(1995). Modelling communication networks, present and future. *Proceedings of the Royal Society of London A*, 444, pages 1-20.

[16] E. Medova (1998). Chance-constrained stochastic programming for integrated services network management. *Annals of Operations Research*, 81, pages 213-229.

[17] E.A. Medova and J.E. Scott (2000). Evolving system architectures for multimedia network design. *Annals of Operations Research* (forthcoming).

[18] Nortel(1991). *Synchronous Transmission Systems.* Northern Telecom Europe Ltd, London.

[19] S. Sen, R. Doverspike and S. Cosares(1994). Network planning with random demand. *Telecommunication Systems*, 3, pages 11-30.

[20] A. Weiss (1995). An introduction to large deviations for communication networks. *IEEE Journal on Selected Areas in Communications*, 6, pages 938-953.

Solution of a Product Substitution Problem Using Stochastic Programming

Michael R. Murr (mmurr@lucent.com)
Lucent Technologies
Princeton, New Jersey

András Prékopa (prekopa@rutcor.rutgers.edu)
Rutgers University
New Brunswick, New Jersey

Abstract

Stochastic programming models of optical fiber production planning are presented. The purpose is to set the optimal fiber manufacturing goals while accounting for the uncertainty primarily in the yield and secondly in the demand. The model is solved for the case when the data follows a multivariate discrete distribution, and also for the case of a multivariate normal distribution, which is used to approximate the discrete data.

1 Introduction

The process of manufacturing optical fibers can be divided into two major parts. The first is preform manufacturing. One process to make preforms is called modified chemical vapor deposition (MCVD). In the MCVD process, glass is deposited on the inside of a quartz tube. When the deposition is complete, the tube is collapsed into a solid rod called a preform (Flegal, Haney, Elliott, Kamino, and Ernst).

The second part is fiber draw. In this process the end of the preform is heated in a furnace and fiber is drawn from it. The fiber has the same cross-sectional structure as the preform except that the fiber is much thinner and much longer (Jablonowski, Paek, and Watkins).

S.P. Uryasev (ed.), Probabilistic Constrained Optimization, 252–271.
© 2000 *Kluwer Academic Publishers.*

The fiber draw process produces a variety of lengths of fiber. The fiber lengths that are produced depend on the size of the preform and on the capacity of the fiber spool. When the length reaches the spool capacity, the fiber is cut and a new spool is started. Naturally, the lengths produced also depend on the rate of unplanned fiber breakage.

We assume that the preforms are used to produce the longest fibers, that is, fibers are not cut in the course of the drawing process. They may break, though. The fibers obtained are the primary products. After some cutting has been done to satisfy demands, remnants are produced. Some of these are just thrown away because their lengths are very short but some of them are used to satisfy future demands. These can be called secondary products. It is not necessary, however, to distinguish between the two kinds of products in the model.

In addition to length, fibers have other characteristics, too. For the sake of simplicity we will speak about one additional characteristic that we term "performance" but note that performance is in reality determined by more than one additional measurement.

Our goal is to calculate a recommended production level to meet the demand. Typical results of the calculation are expected to be, for example, 80%, 90%, or 100% of the manufacturing capacity. To reach a sufficient level of accuracy, the model needs to include several features outlined below. Leaving out any of these features will lead to inaccuracies amounting to at least 10% of the manufacturing capacity and will reduce the usefulness of the results of the calculation. The necessary features are:

- substitution of longer length fibers to meet shorter length demand,
- substitution of higher performance fibers to meet adequate performance demand,
- the inventory on hand at each length and performance level,
- the probability distribution of the yield at each length and performance level,
- the probability distribution of the demand at each length and performance level, and
- the opportunity to produce fiber in the current period to meet a surge in demand in a later period.

Stochastic programming is a methodology that is suitable to account for these aspects. The available data makes it possible to formulate a probabilistic constrained stochastic programming problem where the main feature is to ensure the working of the system, in this case the satisfaction of the demands, with a given probability (reliability) level which is near 1 and chosen by ourselves.

The organization of the paper is the following. In Section 1 we present a short description of the probabilistic constrained stochastic programming model construction.

Following this, in Section 2 we present the base problem (or underlying deterministic problem), i.e., the problem (an LP in our case) which is the starting point of the stochastic programming model formulation. The latter is presented in two variants in Sections 3 and 4. Our data are presented in Section 5. The solution methods of the stochastic programming problems are described in Section 6 and the computational results are contained in Section 7. Conclusions are presented in Section 8.

2 Programming under probabilistic constraint

The stochastic programming model mentioned in the title of this section is one of the most powerful model constructions of stochastic programming. In general, if we formulate a stochastic programming problem, we start from an underlying deterministic problem, observe that some of its parameters are random, and then reformulate it by the use of some statistical decison principle. Let the underlying deterministic problem be the following:

Minimize

$$c^T x \tag{1}$$

subject to

$$\begin{aligned} Tx &\geq \xi \\ Ax &= b \\ x &\geq 0, \end{aligned}$$

where we assume that only T and ξ are random.

The reformulated problem, that we call the problem of programming under probabilistic constraint, is the following:

Minimize

$$c^T x \tag{2}$$

subject to

$$\begin{aligned} P(Tx \geq \xi) &\geq p \\ Ax &= b \\ x &\geq 0, \end{aligned}$$

where p is a fixed probability chosen by ourselves. Typical values for p are 0.8, 0.9, 0.95, 0.99. Any feasible solution to this problem guarantees the functioning of the system with the reliability level p.

The model can be extended into a more complex one involving penalties for violating the constraints $T_i x \geq \xi_i$, $i = 1, \ldots, r$, where T_i is the ith row of T and ξ_i is the ith component of ξ. If the penalty for $T_i x < \xi_i$ is $q_i(\xi_i - T_i x)$, $i = 1, \ldots, r$, where $q_1, \ldots, q_r$ are nonnegative constants, the reformulated problem may be the following:

Minimize

$$c^T x + \sum_{i=1}^{r} q_i E([\xi_i - T_i x]_+) \tag{3}$$

subject to

$$\begin{aligned} P(Tx \geq \xi) &\geq p \\ Ax &= b \\ x &\geq 0. \end{aligned}$$

The penalties $q_1, \ldots, q_r$ are, however, frequently unknown and in these cases we restrict ourselves to problem (2). Without the probabilistic constraint problem (3) is the "simple recourse problem" first introduced and studied by Dantzig (1955) and Beale (1955).

For the case of a random vector ξ, with stochastically independent components, problem (2) was first formulated by Miller and Wagner (1965). The general case was first presented by Prékopa (1970). Charnes, Cooper, and Symonds (1958) published the first paper in this area but they use the individual "chance constraints" $P(T_i x \geq \xi_i) \geq p_i$, $i = 1, \ldots, r$ rather than the joint probabilistic constraint $P(Tx \geq \xi) \geq p$ in problems (2) and (3) which renders the problem very simple, from the mathematical point of view.

If ξ has discrete distribution with support $Z = \{z_1, z_2, \ldots\}$, then the probabilistic constraint can be written in an equivalent form by the use of the p-level efficient points (pLEP's), first introduced in Prekopa (1990). A point $z \in Z$ is said to be a pLEP if

$$P(z \geq \xi) \geq p$$

and there is no $y \in Z$, $y \leq z$, $y \neq z$ such that

$$P(y \geq \xi) \geq p.$$

If $\{z^1, \ldots, z^N\}$ is the set of all pLEP's, then problem (2) can be written as:
Minimize

$$c^T x \tag{4}$$

subject to

$$\begin{aligned} Tx &\geq z^k, \text{holds for at least one } k = 1, \ldots, N \\ Ax &= b \\ x &\geq 0. \end{aligned}$$

Problem (4) can be solved by solving N linear programming problems or applying more sophisticated techniques such as the ones mentioned at the end of Section 3 and in Section 11.

For more information about the theory and solution methods of probabilistic constrained stochastic programming see Prékopa (1995).

3 The deterministic problem

Let r be the number of performance levels and assume that these obey a linear ordering, performance level number 1 being the best. Thus, to have a performance level i product satisfy the requirements imposed on a performance level j product, it is necessary to have $i < j$.

The model presented here involves T subsequent periods. The following notation is used:

n: number of different lengths of fibers

l_k: length of fibers of type k

$m_{hk} = \left\lfloor \frac{l_h}{l_k} \right\rfloor$: the number of fibers of length k that can be obtained by cutting one fiber of length h.

T: number of time periods

y^t: overall intended production level in period t, $t = 1, 2, \ldots, T$

a^t_{ih}: expected number of performance level i fibers of length h produced in period t per unit of production, $i = 1, \ldots, r$, $h = 1, \ldots, n$, $t = 1, 2, \ldots, T$

$f^t_{ih}(y^t) = a^t_{ih} y^t + \xi^t_{ih}$: number of performance level i fibers of length h produced in period t, $i = 1, \ldots, r$, $k = 1, \ldots, n$, $t = 1, 2, \ldots, T$. Note that the random components ξ^t_{ih} have an expected value of 0.

ζ^t_{ih}: number of performance level i fibers of length h available at the beginning of period t, $i = 1, \ldots, r$, $h = 1, \ldots, n$, $t = 1, 2, \ldots, T$

c^t_{ih}: cost (per fiber) to produce each performance level i fiber of length h in period t, $i = 1, \ldots, r$, $h = 1, \ldots, n$, $t = 1, 2, \ldots, T$

z^t_{ih}: number of performance level i fibers of length h carried in inventory between period t and period $t+1$, $i = 1, \ldots, r$, $h = 1, \ldots, n$, $t = 1, 2, \ldots, T$

d^t_{jk}: demand for performance level j fibers of length k in period t, $j = 1, \ldots, r$, $k = 1, \ldots, n$, $t = 1, 2, \ldots, T$

x^t_{ijhk}: number of performance level i fibers of length h used to meet demand for performance level j fibers of length k in period t, $1 \le i \le j \le r$, $1 \le h \le k \le n$, $t = 1, 2, \ldots, T$

X^t_{ijhk}: upper bound for x^t_{ijhk}

Summary of notation

Constants: $n,\ l_k,\ m_{hk},\ a_{ih}^t,\ c_{ih}^t,\ X_{ijhk}^t$

Random variables: $\xi_{ik}^t,\ \zeta_{ik}^t,\ d_{ik}^t$

Decision variables: $x_{ijhk}^t,\ y^t,\ z_{ih}^t$

Holding the random variables fixed for now, our downgrading model is the following network flow model:

3.1 Constraints

We need for the inventory of fibers of performance level i and length h at the beginning of the first period plus the number of fibers of performance level i and length h produced during the first period to equal or exceed the number of fibers of performance level i and length h assigned to meet demand during the first period, for each performance level i, $i = 1, \ldots, r$ and each length h, $h = 1, \ldots, n$:

$$\zeta_{ih}^1 + a_{ih}^1 y^1 + \xi_{ih}^1 - z_{ih}^1 \geq \sum_{k=h}^{n} \sum_{j=i}^{r} x_{ijhk}^1. \tag{5}$$

We need for the number of fibers of performance level i and length h carried from the first period to the second period plus the number of fibers of performance level i and length h produced during the second period to equal or exceed the number of fibers of performance level i and length h assigned to meet demand during the second period, for each performance level i, $i = 1, \ldots, r$ and each length h, $h = 1, \ldots, n$:

$$z_{ih}^1 + a_{ih}^2 y^2 + \xi_{ih}^2 - z_{ih}^2 \geq \sum_{k=h}^{n} \sum_{j=i}^{r} x_{ijhk}^2. \tag{6}$$

In general, for each time period t, $t = 2, \ldots, T$, we need for the number of fibers of performance level i and length h carried from period $t-1$ to period t plus the number of fibers of performance level i and length h produced during period t to equal or exceed the number of fibers of performance level i and length h assigned to meet demand during period t, for each performance level i, $i = 1, \ldots, r$ and each length h, $h = 1, \ldots, n$:

$$z_{ih}^{t-1} + a_{ih}^t y^t + \xi_{ih}^t - z_{ih}^t \geq \sum_{k=h}^{n} \sum_{j=i}^{r} x_{ijhk}^t. \tag{7}$$

We need for the sum of the number of fibers assigned to meet the demand to equal or exceed the demand, in each performance level j, $j = 1, \ldots, r$, in each length category k, $k = 1, \ldots, n$, and in each period t, $t = 1, \ldots, T$. The number of fibers assigned needs to be multiplied by the appropriate factor m_{hk} if the length of fibers in length category h is at least twice the length of fibers in category k:

$$\sum_{h=1}^{k} \sum_{i=1}^{j} x_{ijhk}^{t} m_{hk} \geq d_{jk}^{t}. \tag{8}$$

We may want to limit the amount of downgrading and cutting, using upper bounds, for each i and j, $1 \leq i \leq j \leq r$; for each h and k, $1 \leq h \leq k \leq n$; and in each period t, $t = 1, 2, \ldots, T$:

$$0 \leq x_{ijhk}^{t} \leq X_{ijhk}^{t}. \tag{9}$$

3.2 Objective function

The objective function of the underlying deterministic problem is the total production cost equal to

$$\sum_{t=1}^{T} \sum_{h=1}^{n} \sum_{i=1}^{r} c_{ih} f_{ih}^{t}(y) = \sum_{t=1}^{T} \sum_{h=1}^{n} \sum_{i=1}^{r} c_{ih} (a_{ih}^{t} y^{t} + \xi_{ih}^{t}), \tag{10}$$

where the c_{ih} are some positive constants.

The underlying deterministic problem consists of minimizing the objective function (6) subject to the constraints (1) - (5).

4 The stochastic programming problem

The formulation of a multi-period stochastic programming problem, based on the underlying deterministic problem presented above, would lead us to extremely large sizes that we want to avoid. We would like to, however,

- capture the dynamics of the production control process, and
- impose a probabilistic constraint regarding demand satisfiability.

We can take into account the above aspects in a rolling horizon model system where each model encompasses the present and a few future periods. We choose only one period from the future and thus, altogether two periods are included in any model that we formulate and solve. The problem that we have of periods t, $t+1$ contains ζ_{ih}^{t} that we assume to be a known value. In principle the problem contains ζ_{ih}^{t+1} too. However, if the first stochastic constraint in (7) and (8) holds, then $\zeta_{ih}^{t+1} = z_{ih}^{t}$ and therefore we enter z_{ih}^{t} in the second stochastic constraint instead of ζ_{ih}^{t+1}.

We want to ensure that the constraints (1) - (3) are satisfied for periods t, $t+1$ by a prescribed large probability p. Under this condition and the constraints (4), we want to minimize the production cost.

We distinguish two cases:

Case 1: ξ_{ih}^t are random and d_{jk}^t are known. This is a case where we know the future demand.

Case 2: ξ_{ih}^t and d_{jk}^t are random. We may wish to allow for randomness in the future demand.

4.1 Case 1

The optimization problem is the following:

Minimize

$$\sum_{h=1}^{n} \sum_{i=1}^{r} \left[c_{ih}^t a_{ih}^t y^t \ + \ c_{ih}^{t+1} a_{ih}^{t+1} y^{t+1} \right]$$

subject to the probabilistic constraint

$$P \begin{pmatrix} \zeta_{ih}^t + a_{ih}^t y^t + \xi_{ih}^t - z_{ih}^t & \geq & \sum_{k=h}^{n} \sum_{j=i}^{r} x_{ijhk}^t & \text{all } i, h \\ \\ z_{ih}^t + a_{ih}^{t+1} y^{t+1} + \xi_{ih}^{t+1} & \geq & \sum_{k=h}^{n} \sum_{j=i}^{r} x_{ijhk}^{t+1} & \text{all } i, h \end{pmatrix} \geq p \tag{11}$$

and the other constraints

$$\begin{aligned} \sum_{h=1}^{k} \sum_{i=1}^{j} x_{ijhk}^t m_{hk} \ &\geq \ d_{jk}^t && \text{all } j, \ k \\ \sum_{h=1}^{k} \sum_{i=1}^{j} x_{ijhk}^{t+1} m_{hk} \ &\geq \ d_{jk}^{t+1} && \text{all } j, \ k \\ 0 \leq x_{ijhk}^t \ &\leq \ X_{ijhk}^t && \text{all } i, \ j, \ h, \ k, \ t. \end{aligned}$$

4.2 Case 2

In this case, the optimization problem is the following:

Minimize

$$\sum_{h=1}^{n} \sum_{i=1}^{r} \left[c_{ih}^t a_{ih}^t y^t \ + \ c_{ih}^{t+1} a_{ih}^{t+1} y^{t+1} \right]$$

subject to the probabilistic constaint

$$P\left(\begin{array}{lll} \zeta_{ih}^{t} + a_{ih}^{t}y^{t} + \xi_{ih}^{t} - z_{ih}^{t} \geq \sum_{k=h}^{n}\sum_{j=i}^{r} x_{ijhk}^{t} & \text{all } i,\ h \\ z_{ih}^{t} + a_{ih}^{t+1}y^{t+1} + \xi_{ih}^{t+1} \geq \sum_{k=h}^{n}\sum_{j=i}^{r} x_{ijhk}^{t+1} & \text{all } i,\ h \\ \sum_{h=1}^{k}\sum_{i=1}^{j} x_{ijhk}^{t} m_{hk} \geq d_{jk}^{t} & \text{all } j,\ k \\ \sum_{h=1}^{k}\sum_{i=1}^{j} x_{ijhk}^{t+1} m_{hk} \geq d_{jk}^{t+1} & \text{all } j,\ k \end{array}\right) \geq p \qquad (12)$$

and the bounds

$$0 <= x_{ijhk}^{t} <= X_{ijhk}^{t} \qquad \text{all } i,\ j,\ h,\ k,\ t.$$

The solution of Problem (7) (or Problem (8)) yields optimal x_{ijhk}^{t}, y^{t} and x_{ijhk}^{t+1}, y^{t+1} but we accept as final only the x_{ijhk}^{t}, y^{t} whereas the x_{ijhk}^{t+1}, y^{t+1} will be finalized only after the solution of the next problem.

A joint probabilistic constraint is generally given in the form $Tx \geq \xi$, where T is a matrix and x and ξ are vectors. Below is the matrix T for Problem (8). The column headings are the components of the vector x. The right hand side (RHS) is the vector ξ.

x^1_{1111}	x^1_{1112}	x^1_{1122}	x^1_{1211}	x^1_{1212}	x^1_{1222}	x^1_{2211}	x^1_{2212}	x^1_{2222}	y^1	z_{11}	z_{21}	z_{12}	z_{22}	x^2_{1111}	x^2_{1112}	x^2_{1122}	x^2_{1211}	x^2_{1212}	x^2_{1222}	x^2_{2211}	x^2_{2212}	x^2_{2222}	y^2	*RHS*
−1	−1		−1	−1					a^1_{11}	−1														$-\zeta^1_{11}-\xi^1_{11}$
						−1	−1		a^1_{21}		−1													$-\zeta^1_{21}-\xi^1_{21}$
		−1			−1				a^1_{12}			−1												$-\zeta^1_{12}-\xi^1_{12}$
								−1	a^1_{22}				−1											$-\zeta^1_{22}-\xi^1_{22}$
1																								d^1_{11}
			1			1																		d^1_{21}
	2	1																						d^1_{12}
				2	1		2	1																d^1_{22}
										1				−1	−1		−1	−1					a^2_{11}	$-\xi^2_{11}$
											1									−1	−1		a^2_{21}	$-\xi^2_{21}$
												1				−1			−1				a^2_{12}	$-\xi^2_{12}$
													1									−1	a^2_{22}	$-\xi^2_{22}$
														1										d^2_{11}
																	1			1				d^2_{21}
															2	1								d^2_{12}
																		2	1		2	1		d^2_{22}

For the case of Problems (7) and (8) we have, as a special case, the following convexity theorem.

Theorem 1 *If the random variables ξ_{ih}^t, ξ_{ih}^{t+1}, $i = 1, \ldots, r$, $h = 1, \ldots, n$ have continuous joint probability distribution and logconcave joint probability density function then the probability on the left hand side in the probabilistic constraint is a logconcave function of the variables x_{ijhk}^t, x_{ijhk}^{t+1}, (all i, j, h, k) and y^t, y^{t+1}.*

For our problem the data are essentially discrete. The continuous case is an approximation. For the case when the random variables are discrete, Prékopa (1990) proposed a dual type method, Prékopa and Li (1995) presented a more general version of it, and Prékopa, Vizvári, and Badics (1996) proposed a cutting plane method. All these methods allow for the solution of problems which are combinations of probabilistic constrained and simple recourse models and thus, contain as special cases both model constructions. The above methods require the generation of all p-level efficient points (pLEPs). Methods for this have been developed by Murr (1992) and in the above cited paper by Prékopa, Vizvári, and Badics.

T. Szántai (1988) has developed a method for the solution of a probabilistic constrained stochastic programming problem involving multivariate normal, gamma, or Dirichlet distributions. In his method the deterministic constraints as well as the objective function are linear. Successful uses of this method and code are reported in Prékopa and Szántai(1978), Dupaçová, Gaivoronski, Kos, and Szántai (1991), and Murr (1992). Numerical examples for both the discrete case and the continuous approximation will be presented in sections below.

5 Modeling the variability of the production

Let us examine more closely our model for the number of performance level i fibers of length h produced. We are saying that

$$f_{ih}^t(y^t) = a_{ih}^t y^t + \xi_{ih}^t.$$

In other words, the number produced is the decision variable y^t (the intended production level) multiplied by a_{ih}^t (the expected, i. e. mean, capability of the process to produce performance level i, length h fibers) plus a random variable depending on i and h.

We have a problem in that a more valid model would be to say that

$$f_{ih}^t(y^t) = (a_{ih}^t + \xi_{ih}^t)y^t.$$

In other words, the amount of randomness that we anticipate is proportional to y^t rather than independent of y^t.

To modify the model to account for this would introduce random variables on the left hand side of the probabilistic constraint. This would make solution of the model using existing codes difficult or impossible.

Nonetheless we can do almost as well with the model as originally written and with existing codes. Let us consider a model

$$f_{ih}^t(y^t) = a_{ih}^t y^t + \xi_{ih}^t b^t,$$

where b^t is a multiplier of the random variable in time period t.

We recommend solving the model for several values of b^t. A solution is valid only when y^t and b^t are approximately equal.

6 Description of the data

6.1 Data common to continuous case and discrete case

The expected numbers of fibers produced in each category are the following:

a_{11}^t	a_{21}^t	a_{12}^t	a_{22}^t
84	658	126	679

The production cost coefficients for the first time period are:

c_{11}^t	c_{21}^t	c_{12}^t	c_{22}^t
720	720	300	300

The production cost coefficients for the second time period are 5% less:

c_{11}^{t+1}	c_{21}^{t+1}	c_{12}^{t+1}	c_{22}^{t+1}
684	684	285	285

The deterministic demands in the first period are:

d_{11}^t	d_{21}^t	d_{12}^t	d_{22}^t
10	300	30	1000

The deterministic demands in the second period are:

d_{11}^{t+1}	d_{21}^{t+1}	d_{12}^{t+1}	d_{22}^{t+1}
10	300	100	1000

The starting inventories are:

ζ_{11}^t	ζ_{21}^t	ζ_{12}^t	ζ_{22}^t
5	200	10	400

The multipliers for fiber length substitution are:

m_{11}	m_{12}	m_{22}
1	2	1

The upper bounds for the use or downgrading of fiber to meet demand are:

X^t_{1111}	X^t_{1112}	X^t_{1122}	X^t_{1211}	X^t_{1212}	X^t_{1222}	X^t_{2211}	X^t_{2212}	X^t_{2222}
1000	100	1000	40	20	40	1000	1000	1000

The probability level is:

p
0.95

6.2 Data for the discrete case

6.2.1 Case 1

The production variable ξ^t_{11} has 50 possible values, each with probability .02. They are $-25, -24, \ldots, -2, -1, 1, 2, \ldots, 24, 25$.

The production variable ξ^t_{21} also has 50 possible values, each with probability .02. They are $-125, -120, \ldots, -10, -5, 5, 10, \ldots, 120, 125$.

The production variable ξ^t_{12} has 100 possible values, each with probability .01. They are $-50, -49, \ldots, -2, -1, 1, 2, 3, \ldots, 49, 50$.

The production variable ξ^t_{22} has 100 possible values, each with probability .01. They are $-150, -147, -144, \ldots, -6, -3, 3, 6, 9, \ldots, 147, 150$.

The discrete distribution of the production in period $t+1$ is the same as the discrete distribution in period t and is independent of the distribution in period t.

6.2.2 Case 2

The demand variable d^t_{11} has 50 possible values, each with probability .02. They are $0, 1, 2, \ldots, 48, 49$.

The demand variable d^t_{21} has 100 possible values, each with probability .01. They are $251, 252, 253, \ldots, 349, 350$.

The demand variable d^t_{12} has 100 possible values, each with probability .01. They are $21, 22, 23, \ldots, 119, 120$.

The demand variable d^t_{22} has 100 possible values, each with probability .01. They are $902, 904, 906, \ldots, 1098, 1100$.

The discrete distribution of the demand in period $t+1$ is the same as the discrete distribution in period t and is independent of the distribution in period t.

The remainder of the data for Case 2 is identical to the data for Case 1.

6.3 Data for the continuous case

Here we assume that the random variables each have a normal distribution. As stated in the description of the model, the production variables ξ_{11}^t, ξ_{21}^t, ξ_{12}^t, and ξ_{22}^t have mean 0. Their standard deviations are:

ξ_{11}^t	ξ_{21}^t	ξ_{12}^t	ξ_{22}^t
10	45	15	50

The distributions of the production variables within a given time period are correlated. The following correlation matrix will be used:

	ξ_{11}^t	ξ_{21}^t	ξ_{12}^t	ξ_{22}^t
ξ_{11}^t	1			
ξ_{21}^t	0	1		
ξ_{12}^t	0.7	0	1	
ξ_{22}^t	0	0.7	0	1

The correlation of production variables between the two different time periods is assumed to be zero.

The demands are also assumed to have a normal distribution. Their means and standard deviations are:

	d_{11}^t	d_{21}^t	d_{12}^t	d_{22}^t
mean	20	300	70	1000
std. dev.	10	25	25	50

The demands, within and between time periods, are assumed to be uncorrelated.

7 Solution method

An approach to solving discrete probabilistic constrained stochastic programming problems has been developed by Prékopa, Vizvári, and Badics (1996). Let the pLEPs be represented by $z^{(1)}$, $z^{(2)}, \ldots, z^{(N)}$. Recall that the matrix version of the stochastic constraint may be written as:

$$Tx \geq \xi.$$

The probabilistic constraint $P(Tx \geq \xi) \geq p$ can be written in the form: $Tx \geq z^{(i)}$ holds for at least one $i = 1, \ldots, N$. If we also have deterministic constraints $Ax = b$, then the problem to be solved becomes:

Minimize

$$c^T x \tag{13}$$

subject to

$$\begin{aligned} Ax &= b \\ Tx &\geq z^{(i)}, \text{for at least one } i = 1, \ldots, N \\ x &\geq 0. \end{aligned}$$

In the above cited paper the second constraint of problem (9) is approximated by the constraint:

$$Tx \geq \sum_{i=1}^{N} \lambda_i z^{(i)},$$

where

$$\begin{aligned} \sum_{i=1}^{N} \lambda_i &= 1 \\ \lambda_i &\geq 0, \quad i = 1, \ldots, N. \end{aligned}$$

Then the approximate problem to be solved becomes:

Minimize

$$c^T x \tag{14}$$

subject to

$$\begin{aligned} Ax &= b \\ Tx - u - \sum_{i=1}^{N} \lambda_i z^{(i)} &= 0 \\ \sum_{i=1}^{N} \lambda_i &= 1 \\ \lambda_i &\geq 0, \quad i = 1, \ldots, N \\ x &\geq 0 \\ u &\geq 0. \end{aligned}$$

The solution method works in such a way that first we drop the constraint involving the pLEP's and then subsequently build them up, by the use of a cutting plane method.

If the number of pLEP's is small or the sizes of the matrix T are small, then problem (9) can be solved exactly by the solution of N linear programming problems, where the ith one has the constraint $Tx \geq z^{(i)}$. If $x^{(i)}$ is the optimal solution of the ith problem, and $c^T x = \min_{1 \leq i \leq N} c^T x^{(i)}$, then $x^{(i)}$ is the optimal solution of problem (9).

Bounds for the optimum have been developed by Sen (1992), by the use of disjunctive programming.

Systems to exactly solve discrete probabilistic constrained stochastic programming problems have been developed by Maros and Prékopa (1990) and Murr (1992). The former one is based on the linear programming system MILP, developed by Maros (1990), the second one is based on the linear programming subroutine DLPRS of IMSL (1987) available on the Convex C220 machine.

Here we report about results of the use of the second system. It consists of the programs *translator*, *plep*, and *optimizer*. The relationships among them are diagrammed below. Names in boxes represent executable programs. Names associated with arrows represent input and/or output files.

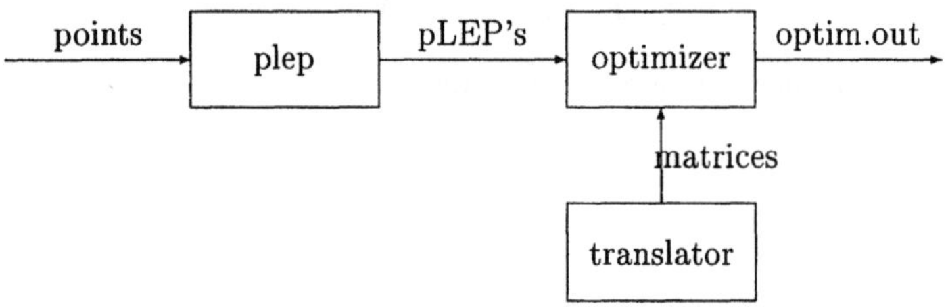

Translator: Contains the matrices and vectors in a human (or, more accurately, programmer) readable format. Writes out these data in a format readable by the stochastic optimization programs.

Plep: Computes p-Level Efficient Points for a multivariate discrete distribution given the discrete distribution of each variable.

Optimizer: It has two input files. One is the output of translator, giving the matrices, vectors, and other data about the optimization problem. The second input file is the list of pLEPs. The optimal solution of the discretized stochastic programming problem is at the end of its output.

On a Convex C220 it took us 6 seconds to run *plep*, and it took us 202 seconds to run *optimizer* to solve the linear program 3532 times and obtain the optimal solution.

For the continuous case, we used *PCSP*, which is described in Szántai (1988). On a Convex C220, the optimal solution was obtained in 408 minutes.

8 Computational results for production and demand both random (Case 2)

8.1 Decision variables

For the first period, the intended production level computed in each case is:

	Discrete Case	Continuous Case
y_1	0.997	1.013

The plan for assigning/downgrading the inventory and production to meet the demand in the first period is:

		Discrete Case	Continuous Case
x^1_{1111}	Use high performance, long length for high performance, long length	48.0	44.6
x^1_{1112}	Cut high performance, long length to high performance, short length	15.7	15.2
x^1_{1122}	Use high performance, short length for high performance, short length	88.6	100.0
x^1_{1211}	Downgrade high performance, long length to adequate performance, long length	0.	0.
x^1_{1212}	Downgrade and cut high performance, long length to adequate performance, short length	0.	0.
x^1_{1222}	Downgrade high performance, short length to adequate performance, short length	0.	0.
x^1_{2211}	Use adequate performance, long length for adequate performance, long length	350.	390.8
x^1_{2212}	Cut adequate performance, long length to adequate performance, short length	86.7	133.4
x^1_{2222}	Use adequate performance, short length for adequate performance, short length	926.7	905.4

8.2 Other Information about the solution

The worst case production levels (values of the random variables) that the model is planning for are as follows. The value p is the (univariate) probability of the variable taking a value equal to or worse than the value shown.

	Discrete Case		Continuous Case	
	ξ	p	ξ	p
ξ_{11}^1	−25.	.00	−24.2	.008
ξ_{21}^1	−125.	.00	−144.1	.001
ξ_{12}^1	−47.	.03	−36.0	.006
ξ_{22}^1	−150.	.00	−182.4	.000

The maximum demand levels in the first period that the model guarantees to satisfy are as follows. Here the value p is the (univariate) probability of the random variable taking a value equal to or greater than the value shown.

	Discrete Case		Continuous Case	
	d	p	d	p
d_{11}^1	48.	.02	44.6	.007
d_{21}^1	350.	.00	390.8	.000
d_{12}^1	120.	.00	130.4	.008
d_{22}^1	1100.	.00	1172.2	.000

9 Conclusion

We have formulated stochastic programming models for the fiber production planning problem, where randomness occurs primarily in the yield but sometimes also in the demand. We have chosen probabilistic constrained models which guarantee safety at a level previously set by ourselves, so that we should be able to meet the deterministic or random demands. Randomness of the yield means that the performance of the produced fiber may be different than originally planned and the lengths may also be different from the required ones due to unpredictable breakage. The solutions of the problems provide us with optimal fiber manufacturing goals to set the production volume parameters to satisfy the mentioned requirements. The practical problem is best described if both the random and the decision variables are discrete. Approximate methods for the solution of problems of this type have been developed by Prékopa, Vizvári, and Badics (1996). Small scale problems can be solved exactly by the solution of as many LP's as the number of p-level efficient points. Maros and Prékopa (1990) and Murr (1992) created codes to solve the problem this way. Here we have used Murr's code which is based on the linear programming subroutine DLPRS of IMSL (1987) available on the Convex C220 machine. The given multivariate discrete probability distribution is also approximated by a multivariate normal distribution. The relevant problem is solved by Szántai's (1988) PCSP code on the Convex C220 machine. The results corresponding to the discrete and continuous cases are close to each other. The developed methods help production planning in determining which orders should be accepted, if they are known prior to the start of the manufacturing process, and set the production goals in both the deterministic and random demand cases. The models can be used in other areas of manufacturing.

References

[1] Beale, E. M. L. 1955. On Minimizing a Convex Function Subject to Linear Inequalities. *J. Royal Statist. Soc., Ser. B* 17, 173-184.

[2] Bitran. G. R. and T-Y. Leong. 1989. Deterministic Approximations to Co-Production Problems with Service Constraints. Working Paper #3071-89-MS, MIT Sloan School of Management.

[3] Charnes, A., W. W. Cooper, and G. H. Symonds. 1958. Cost Horizons and Certainty Equivalents: An Approach to Stochastic Programming of Heating Oil Production. *Management Science* 4, 235-263.

[4] Dantzig, G. B. 1955. Linear Programming under Uncertainty. *Management Science* 1, 197-206.

[5] Deák, I. 1988. Multidimensional Integration and Stochastic Programming. In *Numerical Techniques for Stochastic Optimization*, Yu. Ermoliev and R. J-B Wets (eds.). Springer-Verlag, New York.

[6] Dupačová, J., A. Gaivoronski, Z. Kos, and T. Szántai. 1991. Stochastic Programming in Water Management: A Case Study and a Comparison of Solution Techniques. *European Journal of Operational Research* 52, 28-44.

[7] Flegal, W. M., E. A. Haney, R. S. Elliott, J. T. Kamino, and D. N. Ernst. 1986. Making Single-Mode Preforms by the MCVD Process. *AT&T Technical Journal* 65, 56-61.

[8] Gassmann, H. 1988. Conditional Probability and Conditional Expectation of a Random Vector. In *Numerical Techniques for Stochastic Optimization*, Yu. Ermoliev and R. J-B Wets (eds.). Springer-Verlag, New York.

[9] IMSL. 1987. IMSL Math/Library User's Manual, IMSL, Houston, Texas.

[10] Jablonowski, D. P., U. C. Paek, and L. S. Watkins. 1987. Optical Fiber Manufacturing Techniques. *AT&T Technical Journal* 66, 33-44.

[11] Maros, I. 1990. MILP Linear Programming Optimizer for Personal Computers under DOS. Institut für Angewandte Mathematik, Technische Universität Braunschweig.

[12] Maros, I. and A. Prékopa. 1990. MIPROB, A Computer Code to Solve Probabilistic Constrained Stochastic Programming Problems with Discrete Random Variables. Manuscript.

[13] Miller, B. L. and H. M. Wagner. 1965. Chance Constrained Programming with Joint Constraints. *Operations Research* 13, 930-945.

[14] Murr, M. R. 1992. Some Stochastic Problems in Fiber Production. Ph.D. Dissertation, Rutgers University, New Brunswick, New Jersey.

[15] Prékopa, A. 1970. On Probabilistic Constrained Programming. In *Proceedings of the Princeton Symposium on Mathematical Programming (1967)*, H. Kuhn (ed.). Princeton University Press, Princeton, 113-138.

[16] Prékopa, A. 1971. Logarithmic Concave Measures with Application to Stochastic Programming. *Acta Sci. Math. (Szeged)* 32, 301-316.

[17] Prékopa, A. 1973. On Logarithmic Concave Measures and Functions. *Acta Sci. Math. (Szeged)* 34, 335-343.

[18] Prékopa, A. 1980. Logarithmically Concave Measures and Related Topics. In *Stochastic Programming. Proceedings of the 1974 Oxford International Conference*, M. Dempster (ed.). Academic Press, London, 63-82.

[19] Prékopa, A. 1990. Dual Method for the Solution of a One-Stage Stochastic Programming Problem with Random RHS Obeying a Discrete Probability Distribution. *ZOR* 34, 441-461.

[20] Prékopa, A. 1995. Stochastic Programming. Kluwer Scientific Publishers. Dordrecht, The Netherlands.

[21] Prékopa, A. and W. Li. 1995. Solution of and Bounding in a Linearly Constrained Optimization Problem with Convex, Polyhedral Objective Function. *Mathematical Programming* 70, 1-16.

[22] Prékopa, A. and T. Szántai. 1978. Flood Control Reservoir System Design Using Stochastic Programming. *Mathematical Programming Study* 9, 138-151.

[23] Prékopa, A., B. Vizvári, and T. Badics. 1996. Programming under Probabilistic Constraints with Discrete Random Variables. RUTCOR Research Report 10-96.

[24] Sen, S. 1992. Relaxations for Probabilistically Constrained Programs with Discrete Random Variables. *Operations Research Letters* 11, 81-86.

[25] Szántai, T. 1988. A Computer Code for Solution of Probabilistic-constrained Stochastic Programming Problems. In *Numerical Techniques for Stochastic Optimization*, Yu. Ermoliev and R. J-B Wets (eds.). Springer-Verlag, New York.

[26] Wets, R. J-B. 1983. Solving Stochastic Programs with Simple Recourse. *Stochastics* 10, 219-242.

Some Remarks on the Value-at-Risk and the Conditional Value-at-Risk

Georg Ch. Pflug (georg.pflug@univie.ac.at)
Department of Statistics and Decision Support Systems
University of Vienna

Abstract

The value-at-risk (VaR) and the conditional value-at-risk (CVaR) are two commonly used risk measures. We state some of their properties and make a comparison. Moreover, the structure of the portfolio optimization problem using the VaR and CVaR objective is studied.

Keywords: Risk measures, Value-at-Risk, Conditional Value-at-Risk, Portfolio optimization

1 Introduction

Let Y be a random cost variable and let F_Y be its distribution function, i.e. $F_Y(u) = \mathbb{P}\{Y \leq u\}$. Let $F_Y^{-1}(v)$ be its right continuous inverse, i.e. $F_Y^{-1}(v) = \inf\{u : F_Y(u)) > v\}$. When no confusion may occur, we write simply F instead of F_Y.

For a fixed level α, we define (as usual) the *value-at-risk* VaR_α as the α-quantile, i.e.

$$\mathrm{VaR}_\alpha(Y) = F^{-1}(\alpha). \tag{1}$$

The *conditional value-at-risk* $CVaR_\alpha$ is defined as the solution of an optimization problem

$$\mathrm{CVaR}_\alpha(Y) := \inf\{a + \frac{1}{1-\alpha}\mathbb{E}[Y-a]^+ : a \in \mathbb{R}\}. \tag{2}$$

Here $[z]^+ = \max(z, 0)$. Uryasev and Rockafellar (1999) have shown that CVaR equals the conditional expectation of Y, given that $Y \geq \mathrm{VaR}_\alpha$, i.e.

$$\mathrm{CVaR}_\alpha(Y) = \mathbb{E}(Y | Y \geq \mathrm{VaR}_\alpha(Y)). \tag{3}$$

S.P. Uryasev (ed.), Probabilistic Constrained Optimization, 272–281.
© 2000 *Kluwer Academic Publishers.*

In fact, (3) is the usual definition of CVaR_α.

We will prove some properties of CVaR and VaR and study the relation between these two measures of risk. To begin with, we show that the minimizer in (2) is VaR_α, even if F is not differentiable.

Proposition 1. Suppose that $F(b) \geq \alpha$ and $F(b-) \leq \alpha$. Then

$$b + \frac{1}{1-\alpha}\mathbb{E}[Y-b]^+ \leq a + \frac{1}{1-\alpha}\mathbb{E}[Y-a]^+$$

for all a.

Proof. Suppose first that $b \leq a$. Then

$$\begin{aligned} \mathbb{E}[Y\mathbb{1}_{\{b<Y\}}] - \mathbb{E}[Y\mathbb{1}_{\{a<Y\}}] &= \mathbb{E}[Y\mathbb{1}_{\{b<Y\leq a\}}] \\ &\leq a[F(a) - F(b)] \leq a[F(a) - \alpha] - b[F(b) - \alpha]. \end{aligned}$$

Therefore

$$\begin{aligned} b[1-\alpha] - b[1-F(b)] + \mathbb{E}[Y\mathbb{1}_{\{b<Y\}}] &\leq a[1-\alpha] - a[1-F(a)] + \mathbb{E}[Y\mathbb{1}_{\{a<Y\}}] \\ b[1-\alpha] + \mathbb{E}[(Y-b)\mathbb{1}_{\{b<Y\}}] &\leq a[1-\alpha] + \mathbb{E}[(Y-a)\mathbb{1}_{\{a<Y\}}] \\ b + \frac{1}{1-\alpha}\mathbb{E}[(Y-b)\mathbb{1}_{\{b<Y\}}] &\leq a + \frac{1}{1-\alpha}\mathbb{E}[(Y-a)\mathbb{1}_{\{a<Y\}}]. \end{aligned}$$

Let now $a \leq b$. Then

$$\begin{aligned} \mathbb{E}[Y\mathbb{1}_{\{a<Y\}}] - \mathbb{E}[Y\mathbb{1}_{\{b\leq Y\}}] &= \mathbb{E}[Y\mathbb{1}_{\{a<Y<b\}}] \\ &\geq a[F(b-) - F(a)] \geq b[F(b-) - \alpha] - a[F(a) - \alpha]. \end{aligned}$$

Therefore

$$\begin{aligned} a[1-\alpha] - a[1-F(a)] + \mathbb{E}[Y\mathbb{1}_{\{a<Y\}}] &\geq b[1-\alpha] - b[1-F(b-)] + \mathbb{E}[Y\mathbb{1}_{\{b\leq Y\}}] \\ a[1-\alpha] + \mathbb{E}[(Y-a)\mathbb{1}_{\{a<Y\}}] &\geq b[1-\alpha] + \mathbb{E}[(Y-b)\mathbb{1}_{\{b\leq Y\}}] - b[1-F(b-)] \\ b + \frac{1}{1-\alpha}\mathbb{E}[(Y-b)\mathbb{1}_{\{b<Y\}}] &\leq a + \frac{1}{1-\alpha}\mathbb{E}[(Y-a)\mathbb{1}_{\{a<Y\}}]. \end{aligned}$$

□

As a consequence, one sees that

$$\mathrm{VaR}_\alpha(Y) = F^{-1}(\alpha) \in \text{ argmin } \{a + \frac{1}{1-\alpha}\mathbb{E}[Y-a]^+\}.$$

Alternative, equivalent representations of CVaR are therefore

$$\begin{aligned} \mathrm{CVaR}_\alpha(Y) &= \mathbb{E}[Y|Y \geq F^{-1}(\alpha)] \\ &= \frac{1}{1-\alpha}\int_\alpha^1 F^{-1}(v)\, dv \\ &= \frac{1}{1-\alpha}\int_{F^{-1}(\alpha)}^\infty u\, dF(u). \end{aligned}$$

2 Properties of VaR and CVaR

Properties of risk measures can be formulated in terms of preference structures induced by dominance relations (see Fishburn (1980)).

Let Y_1 and Y_2 be two random variables.

- *Stochastic dominance of order* 1: We say that the relation

  $$Y_1 \prec_{SD(1)} Y_2$$

  holds iff

  $$\mathbb{E}[\psi(Y_1)] \leq \mathbb{E}[\psi(Y_2)]$$

  for all (integrable) monotonic functions ψ.

- *Stochastic dominance of order* 2: We say that the relation

  $$Y_1 \prec_{SD(2)} Y_2$$

  holds iff

  $$\mathbb{E}[\psi(Y_1)] \leq \mathbb{E}[\psi(Y_2)]$$

  for all (integrable) concave, monotonic functions ψ.

- *Monotonic dominance of order* 2: We say that the relation

  $$Y_1 \prec_{MD(1)} Y_2$$

  holds iff

  $$\mathbb{E}[\psi(Y_1)] \leq \mathbb{E}[\psi(Y_2)]$$

  for all (integrable) concave functions ψ .

We have the following trivial consequences

$$Y_1 \prec_{SD(1)} Y_2 \quad \text{implies that} \quad Y_1 \prec_{SD(2)} Y_2$$

$$Y_1 \prec_{SD(2)} Y_2 \quad \text{holds iff} \quad Y_1 \prec_{SD(1)} Y_2 \text{ and } Y_1 \prec_{MD(1)} Y_2.$$

$Y_1 \prec_{SD(2)} Y_2$ is equivalent to $\int_{-\infty}^{x} F_{Y_1}(u)\, du \leq \int_{-\infty}^{x} F_{Y_2}(u)\, du$ for all x. Let us prove the latter statement.

Since $\int_{-\infty}^{x} F(u)\, du = \int_{-\infty}^{\infty} [x-u]^+\, dF(u)$, one sees that $Y_1 \prec_{SD(2)} Y_2$ is equivalent to $\int_{-\infty}^{\infty} \psi(u)\, dF_{Y_1}(u) \leq \int_{-\infty}^{\infty} \psi(u)\, dF_{Y_2}(u)$ for all functions of the form $\psi(u) = \sum_k (-\alpha_k)[x_k - u]^+ + \beta_k$, with $\alpha_k \geq 0$. These functions are dense in the set of all concave, monotonic functions.

We are now ready to state the properties of CVaR_α.

Proposition 2. CVaR_α exhibits the following properties:

(i) CVaR_α is *translation-equivariant*, i.e.

$$\text{CVaR}_\alpha(Y + c) = \text{CVaR}_\alpha(Y) + c.$$

(ii) CVaR_α is *positively homogeneous*, i.e.

$$\text{CVaR}_\alpha(cY) = c\ \text{CVaR}_\alpha(Y),$$

if $c > 0$.

(iii) If Y has a density,

$$\mathbb{E}(Y) = (1 - \alpha)\ \text{CVaR}_\alpha(Y) - \alpha\ \text{CVaR}_{(1-\alpha)}(-Y).$$

(iv) CVaR_α is *convex* in the following sense: For arbitrary (possibly dependent) random variables Y_1 and Y_2 and $0 < \lambda < 1$,

$$\text{CVaR}_\alpha(\lambda Y_1 + (1 - \lambda)Y_2) \leq \lambda\ \text{CVaR}_\alpha(Y_1) + (1 - \lambda)\ \text{CVaR}_\alpha(Y_2).$$

(v) CVaR_α is monotonic w.r.t. SD(2) (and a fortiori w.r.t. SD(1)), i.e. if

$$Y_1 \prec_{SD(2)} Y_2$$

then

$$\text{CVaR}_\alpha(Y_1) \leq \text{CVaR}_\alpha(Y_2).$$

(vi) CVaR_α is monotonic w.r.t. MD(2), i.e. if

$$Y_1 \prec_{MD(2)} Y_2$$

then

$$\text{CVaR}_\alpha(Y_1) \leq \text{CVaR}_\alpha(Y_2).$$

Proof. (i) and (ii) are obvious from the definition of $\text{CVaR}_\alpha(Y)$. Let us prove (iii). Since

$$\begin{aligned}
\text{CVaR}_{(1-\alpha)}(-Y) &= \mathbb{E}(-Y \mid -Y \geq \text{VaR}_{(1-\alpha)}(-Y)) \\
&= \mathbb{E}(-Y \mid -Y \geq -\text{VaR}_\alpha(Y)) \\
&= -\mathbb{E}(Y \mid Y \leq \text{VaR}_\alpha(Y))
\end{aligned}$$

one sees that

$$\begin{aligned}
\mathbb{E}(Y) &= \alpha\mathbb{E}(Y \mid Y \leq \text{VaR}_\alpha(Y)) + (1 - \alpha)\mathbb{E}(Y \mid Y \geq \text{VaR}_\alpha(Y)) \\
&= -\alpha\ \text{CVaR}_{(1-\alpha)}(-Y) + (1 - \alpha)\ \text{CVaR}_\alpha(Y).
\end{aligned}$$

Now we prove (iv). Let a_i be such that $\text{CVaR}_\alpha(Y_i) = a_i + \frac{1}{1-\alpha}\mathbb{E}[Y_i - a_i]^+$. Since $y \mapsto [y-a]^+$ is convex, we have

$$\begin{aligned}
& \text{CVaR}_\alpha(\lambda Y_1 + (1-\lambda)Y_2) \\
\leq\ & \lambda a_1 + (1-\lambda)a_2 + \frac{1}{1-\alpha}\mathbb{E}[\lambda Y_1 + (1-\lambda)Y_2 - \lambda a_1 + (1-\lambda)a_2)]^+ \\
\leq\ & \lambda a_1 + (1-\lambda)a_2 + \frac{\lambda}{1-\alpha}\mathbb{E}[Y_1 - a_1]^+ + \frac{1-\lambda}{1-\alpha}\mathbb{E}[Y_2 - a_2]^+ \\
\leq\ & \lambda\ \text{CVaR}_\alpha(Y_1) + (1-\lambda)\ \text{CVaR}_\alpha(Y_2).
\end{aligned}$$

(v) and (vi) follow from the fact, that $y \mapsto [y-a]^+$ is monotone and convex. □

Artzner, Delbaen, Eber and Heath call a risk measure *coherent*, if it is translation-invariant, convex, positively homogeneous and monotonic w.r.t. $\prec_{SD(1)}$. One sees that CVaR_α is coherent in this sense.

In contrast, VaR_α is not coherent, since it is not convex. On the other hand, it is comonotone additive as is shown below.

Definition. Two random variables Y_1 and Y_2 defined on the same probability space $(\Omega, \mathcal{A}, \mathbb{P})$ are said to be *comonotone*, if for all $\omega, \omega' \in \Omega$,

$$[Y_1(\omega) - Y_2(\omega)][Y_1(\omega') - Y_2(\omega')] \geq 0 \qquad a.s.$$

Equivalently, Y_1 and Y_2 are comonotone, if there is a representation $Y_1 = f(U); Y_2 = g(U)$, with f, g monotonically increasing and U uniform in [0,1] (see Shaun Wang).

Proposition 3. VaR_α exhibits the following properties:

(i) VaR_α is *translation-equivariant*, i.e.

$$\text{VaR}_\alpha(Y + c) = \text{VaR}_\alpha(Y) + c.$$

(ii) VaR_α is *positively homogeneous*, i.e.

$$\text{VaR}_\alpha(cY) = c\ \text{VaR}_\alpha(Y),$$

if $c > 0$.

(iii) $\text{VaR}_\alpha(Y) = -\ \text{VaR}_{(1-\alpha)}(-Y)$.

(iv) VaR_α is monotonic w.r.t. SD(1) i.e. if

$$Y_1 \prec_{SD(1)} Y_2$$

then

$$\text{VaR}_\alpha(Y_1) \leq \text{VaR}_\alpha(Y_2).$$

(v) VaR_α is *comonotone additive*, i.e. if Y_1 and Y_2 are comonotone, then

$$\text{VaR}_\alpha(Y_1 + Y_2) = \text{VaR}_\alpha(Y_1) + \text{VaR}_\alpha(Y_2).$$

Proof. (i) - (iv) are nearly obvious. Only (v) has to be proved in detail. If $Y_1 = f(U)$ with U uniform[0,1] and f monotonically increasing, then $\text{VaR}_\alpha(Y_1) = f(\alpha)$. Similarly $\text{VaR}_\alpha(Y_2) = g(\alpha)$ and therefore $\text{VaR}_\alpha(Y_1 + Y_2) = f(\alpha) + g(\alpha) = \text{VaR}_\alpha(Y_1) + \text{VaR}_\alpha(Y_2)$. □

3 Relations between VaR and CVaR

In principle, VaR and CVaR measure different properties of the distribution. VaR is a quantile and CVaR is a conditional tail expectation. The two values coincide only if the tail is cut off.

Let $[Y]^c$ be the right censored cost variable $[Y]^c = \min(Y, c)$. If we set $c = \text{VaR}_\alpha(Y)$, then $\text{CVaR}_\alpha([Y]^c) = \text{VaR}_\alpha(Y)$.

Proposition 4.

(i) $\text{CVaR}_\alpha(Y) \geq \text{VaR}_\alpha(Y)$

(ii) $\text{VaR}_\alpha(Y) = \sup\{v : \text{CVaR}_\alpha([Y]^v) = v\}$.

(iii) If Y is nonnegative, then

$$\left[\frac{\mathbb{E}(Y^n) - (1-\alpha)\,\text{CVaR}_\alpha(Y^n)}{\alpha}\right]^{1/n} \to \text{VaR}_\alpha(Y)$$

as $n \to \infty$.

Proof. (i) is obvious. To prove (ii) notice that from the representation $\text{CVaR}(Y) = \frac{1}{1-\alpha}\int_\alpha^1 F^{-1}(u)\,du$ one sees that $\text{CVaR}([Y]^c) = \frac{1}{1-\alpha}\int_\alpha^1 \min(F^{-1}(u), c)\,du$, which implies (ii).

The property (iii) is based on the fact that for every nonnegative random variable Z, $[\mathbb{E}(Z^n)]^{1/n} \to \inf\{u : \mathbb{P}\{Z > u\} = 0\}$ as $n \to \infty$. From Proposition 2 (iii) we have

$$-\text{CVaR}_{(1-\alpha)}(-Y) = \frac{1}{\alpha}[\mathbb{E}(Y) - (1-\alpha)\,\text{CVaR}_\alpha(Y)].$$

On the other hand,

$$-\text{CVaR}_{(1-\alpha)}(-Y) = \mathbb{E}(Y | Y \leq F_Y^{-1}(\alpha)) = \sup\{a - \frac{1}{\alpha}\mathbb{E}[Y - a]^- : a \in \mathbb{R}\},$$

which may be proved in analogy to the proof of proposition 1. Since $[\mathbb{E}(Y^n | Y^n \leq F_{Y^n}^{-1}(\alpha))]^{1/n} = [\mathbb{E}(Y^n | Y \leq F_Y^{-1}(\alpha))]^{1/n} \to \text{VaR}_\alpha(Y)$, the result follows. □

4 VaR- and CVaR-optimal portfolios

Let $\xi = (\xi_1, \ldots, \xi_k)$ be a vector of random returns of asset categories $1, \ldots, k$. Let $x = (x_1, \ldots, x_k)$ be the investments in these categories respectively. Without loss of generality, we may assume that the total budget is 1. The return of the total portfolio is $-Y = x^T\xi$ and we aim at minimizing the VaR of Y under the constraint that the expected return exceeds some prespecified level μ.

We consider the following portfolio optimization problems: The VaR-optimization problem

$$\left\| \begin{array}{ll} \text{Minimize (in } x) & \mathrm{VaR}_\alpha(-x^T\xi) \\ \text{subject to} & \\ x^T\mathbb{E}(\xi) \geq \mu & \\ x^T\mathbb{1} = 1 & \\ x \geq 0 & \end{array} \right. \tag{4}$$

and the CVaR-optimization problem

$$\left\| \begin{array}{ll} \text{Minimize (in } x) & \mathrm{CVaR}_\alpha(-x^T\xi) \\ \text{subject to} & \\ x^T\mathbb{E}(\xi) \geq \mu & \\ x^T\mathbb{1} = 1 & \\ x \geq 0 & \end{array} \right. \tag{5}$$

In view of Propositions 2 (iii) and 3 (iii) one could equivalently maximize $\mathrm{VaR}_{(1-\alpha)}(x^T\xi)$ and $\mathrm{CVaR}_{(1-\alpha)}(x^T\xi)$.

Problem (4) is nonconvex in general and may have several local minima. In contrast, (5) can be written as the following (infinite dimensional) linear program:

$$\left\| \begin{array}{ll} \text{Minimize (in } x \text{ and } a) & a + \frac{1}{1-\alpha}\mathbb{E}[Z] \\ \text{subject to} & \\ Z \geq -x^T\xi - a & \text{with probability 1} \\ x^T\mathbb{E}(\xi) \geq \mu & \\ x^T\mathbb{1} = 1 & \\ Z \geq 0 & \\ x \geq 0 & \end{array} \right. \tag{6}$$

From the structure of this program it is clear that if there is a solution at all, this solution is either a singleton or a convex polyhedron. Every local optimum is global. This is the big advantage of the CVaR risk measure over the VaR risk measure.

For practical portfolio optimization, Y is a discrete variable which takes the values $-x^T\xi^i$ with equal probability. The vectors $\xi^i, i = 1, \ldots N$ are called the *scenarios*.

Introduce the function $M_{[k:N]}(u^1, \ldots, u^N)$ to denote the k-th largest among $u^1, \ldots, u^N$. Thus $M_{[1:N]}$ denotes the minimum and $M_{[N:N]}$ the maximum. Calculating CVaR and VaR for the discrete distribution, we get

$$\mathrm{VaR}_\alpha(-x^T\xi) = M_{[\lfloor \alpha N \rfloor : N]}(-x^T\xi^1, \ldots, -x^T\xi^N),$$

$$\mathrm{CVaR}_\alpha(-x^T\xi) = \frac{1}{N} \sum_{\{-x^T\xi^i \geq \mathrm{VaR}_\alpha\}} -x^T\xi^i.$$

The discrete portfolio optimization problem (4) is a nonlinear, nonconvex program:

$$\left\| \begin{array}{ll} \text{Minimize (in } x) & M_{[\lfloor \alpha N \rfloor : N]}(-x^T\xi^1, \ldots, -x^T\xi^N) \\ \text{subject to} & \\ x^T e \geq \mu & \\ x^T \mathbb{1} = 1 & \\ x \geq 0 & \end{array} \right. \tag{7}$$

Here $e = \frac{1}{N}\sum_{i=1}^N \xi^i$ denotes the expected return vector. The discrete version of (5) is linear and may be solved using any LP-solver (see Uryasev and Rockafellar (1999)):

$$\left\| \begin{array}{ll} \text{Minimize (in } x, a \text{ and } z) & a + \frac{1}{(1-\alpha)N}\sum_{i=1}^N z^i \\ \text{subject to} & \\ z^i \geq -x^T\xi^i - a & i = 1, \ldots, N \\ x^T e \geq \mu & \\ x^T \mathbb{1} = 1 & \\ z^i \geq 0 & i = 1, \ldots, N \\ x \geq 0 & \end{array} \right. \tag{8}$$

Notice that the optimal a in (8) is $\mathrm{VaR}_\alpha(-x^T\xi) = -\mathrm{VaR}_{(1-\alpha)}(x^T\xi)$. For the reader, who does not like to formulate a portfolio optimization problem as a minimization program for negative returns, we can reformulate it as a maximization program:

$$\left\| \begin{array}{ll} \text{Maximize (in } x, b \text{ and } z) & b - \frac{1}{(1-\alpha)N}\sum_{i=1}^N z^i \\ \text{subject to} & \\ z^i \geq b - x^T\xi^i & i = 1, \ldots, N \\ x^T e \geq \mu & \\ x^T \mathbb{1} = 1 & \\ z^i \geq 0 & i = 1, \ldots, N \\ x \geq 0 & \end{array} \right. \tag{9}$$

The optimal b in in (9) is $\mathrm{VaR}_{(1-\alpha)}(x^T\xi)$.

4.1 A fixpoint formulation of the VaR optimization problem

Since the CVaR optimization is much simpler in structure, it is desirable to solve the VaR optimization by a sequence of CVaR optimization problems. The idea is to represent the solution of the VaR problem as a fixpoint of CVaR problems using the results of section 2.

Consider the following linear program, which is parametrized by the cut-off point c and an index set I.

$$P(c,I)\left\|\begin{array}{ll} \text{Minimize (in } x, a \text{ and } z) & \{a + \frac{1}{(1-\alpha)N}\left[\sum_{i\notin I} z_i + \sum_{i\in I}(c-a)\right]\} \\ \text{subject to} & \\ z_i \geq -x^T\xi - a & i \notin I \\ -x^T\xi^i \geq c & i \in I \\ c \geq a & \\ x^T e \geq \mu & \\ x^T \mathbb{1} = 1 & \\ z^i \geq 0 & i \notin I \\ x \geq 0 & \end{array}\right. \qquad (10)$$

For a vector x and a value a let $I(x,a) \subseteq \{1,\ldots,N\}$ be the following index set:

$$I(x,a) = \{i : -x^T\xi_i > a\}.$$

Proposition 5. Suppose that x^* is the minimizer and a^* is the minimal value of the VaR optimization problem (7). Then x^* and a^* are the solutions of the linear program $P(a^*, I(x^*, a^*))$. Conversely, every fixpoint i.e. a vector x and a value a such that x and a are the solutions of $P(a, I(x,a))$ is a local minimizer of (7).

Proof. In a neighborhood of (x^*, a^*) we have that

$$[-x^T\xi^i]^a = \begin{cases} a & \text{if} \quad i \in I(x^*, a^*) \\ -x^T\xi^i & \text{if} \quad i \notin I(x^*, a^*) \end{cases}$$

and therefore $\mathrm{VaR}(-x^T\xi) = \mathrm{CVaR}([-x^T\xi]^a)$ there. Since one cannot find a better portfolio w.r.t. VaR within this neighborhood, the local and hence global solution of $P(a, I(x,a))$ must coincide with the solution of the VaR optimization problem. Conversely, let $\hat{x}$, $\hat{a}$ be the solution of $P(\hat{a}, I(\hat{x}, \hat{a}))$. As before, there is a neighborhood of $(\hat{x}, \hat{a})$, in which $\mathrm{VaR}(-x^T\xi)$ and $\mathrm{CVaR}([-x^T\xi]^a)$ coincide, hence $\hat{x}$ is at least a local solution of the VaR minimization problem. □

We are ready to state the fixpoint property of the local minimizers:

The fixpoint property: x^* is a local minimizer of (7), if and only if there is a value a^* and an index set I^*, such that x^*, a^* and I^* are fixpoints in the following sense: The solution of $P(a^*, I^*)$ is x^* and a^* and, in addition, $I(x^*, a^*) = I^*$. Thus, the VaR optimization problem can be reformulated as a fixpoint problem of solutions of linear optimization problems. This leads immediately to a solution strategy.

References

[1] Artzner Ph., Delbaen F., Eber J.-M., Heath D. (1999): Coherent measures of risk. Mathematical Finance 9, 203 – 228.

[2] Fishburn P.C. (1980): Stochastic Dominance and Moments of Distributions. Mathematics of Operations Research 5, 94 – 100

[3] Uryasev S., Rockafellar, R.T. (1999). Optimization of Conditional Value-at-Risk. Research Report 99-4, ISE Dept. Univeristy of Florida.

[4] Uryasev S. (2000). Conditional Value-at-Risk: Optimization Algorithms and Applications. Financial Engineering News 14, February 2000.

[5] Wang Shaun (1997). Axiomatic characterization of insurance prices. Insurance Math. and Economics 21 (2), 173 – 183

Statistical Inference of Stochastic Optimization Problems

Alexander Shapiro (ashapiro@isye.gatech.edu)
Georgia Institute of Technology,
Atlanta, GA 30332-0205, USA

Abstract

We discuss in this paper statistical inference of Monte Carlo simulation based approximations of stochastic optimization problems, where the "true" objective function, and probably some of the constraints, are estimated, typically by averaging a random sample. The classical maximum likelihood estimation can be considered in that framework. Recently statistical analysis of such methods has been motivated by a development of simulation based optimization techniques. We investigate asymptotic properties of the optimal value and an optimal solution of the corresponding Monte Carlo simulation approximations by employing the so-called Delta method, and discuss some examples.

1 Introduction

Consider the optimization problem

$$\operatorname*{Min}_{x \in S} f(x), \tag{1.1}$$

where S is a subset of $\mathbb{R}^m$ and $f : S \to \mathbb{R}$. Suppose that the above optimization problem is approximated by a sequence of problems

$$\operatorname*{Min}_{x \in S} \hat{f}_N(x), \tag{1.2}$$

where $\hat{f}_N(x)$ are random functions converging, as $N \to \infty$, in some probabilistic sense to $f(x)$. We refer to (1.1) and (1.2) as the *true* and *approximating* problems,

S.P. Uryasev (ed.), Probabilistic Constrained Optimization, 282–307.
© 2000 *Kluwer Academic Publishers.*

respectively. Typically the objective function $f(x)$ is given as the expected value function

$$f(x) := \mathbb{E}_P\{g(x,\omega)\} = \int_\Omega g(x,\omega)P(d\omega), \tag{1.3}$$

where $(\Omega, \mathcal{F}, P)$ is a probability space, and the approximating functions $\hat{f}_N(x)$ are constructed by averaging a random sample.

Let v_0, $\hat{v}_N$ and x_0, $\hat{x}_N$ be the optimal values and optimal solutions of the problems (1.1) and (1.2), respectively. In this paper we discuss asymptotic statistical inference of $\hat{v}_N$ and $\hat{x}_N$, as N tends to infinity. We also consider the cases where the feasible set S is subject to perturbations and is given by random constraints. Let us discuss some examples.

Example 1.1 Our first example is motivated by the classical maximum likelihood method of estimation. That is, let $g(y,\theta)$ be a family of probability density functions (pdf), parameterized by the parameter vector $\theta \in \Theta \subset \mathbb{R}^m$, and let $Y_1, ..., Y_N$ be an i.i.d. random sample with a probability distribution P. Define

$$\hat{f}_N(\theta) := -N^{-1}\sum_{i=1}^{N} \ln g(Y_i,\theta).$$

By the Law of Large Numbers we have that, for any fixed value of θ, $\hat{f}_N(\theta)$ converges to

$$f(\theta) := -\mathbb{E}_P\{\ln g(Y,\theta)\} = -\int \ln g(y,\theta)P(dy),$$

with probability one, as $N \to \infty$, provided of course that the above expectation exists. This leads to the "true" and "approximating" optimization problems of minimizing $f(\theta)$ and $\hat{f}_N(\theta)$, respectively, over the parameter set Θ.

In particular, suppose that the distribution P is given by a pdf $g(y,\theta_0)$, $\theta_0 \in \Theta$, from the above parametric family, i.e., the parametric model is correctly specified. Then θ_0 is an unconstrained minimizer of $f(\theta)$, and hence is an optimal solution of the "true" problem. Indeed, by using concavity of the logarithm function, we obtain

$$f(\theta_0) - f(\theta) - \int \ln\left[\frac{g(y,\theta)}{g(y,\theta_0)}\right] g(y,\theta_0)dy \le \int \left[\frac{g(y,\theta)}{g(y,\theta_0)} - 1\right] g(y,\theta_0)dy = 0.$$

There is a large literature on the maximum likelihood method, and the above derivation of optimality of θ_0 is known of course. We will come back to this example later. Let us note at this point that the corresponding random sample usually represents available data and the associated minimizer $\hat{\theta}_N$ of $\hat{f}_N(\theta)$, over Θ, is viewed as the maximum likelihood estimator of the "true" value θ_0 of the parameter vector. There are also various extensions of the maximum likelihood method, in particular the method of M-estimators introduced by Huber [13, 15].

Somewhat different type of examples is motivated by a Monte Carlo simulation approach to numerical solutions of stochastic programming problems. A goal of a such stochastic programming problem is to solve an optimization problem of the form (1.1) with the objective function $f(x)$ given as the expected value in the form (1.3). The probability distribution P is supposed to be known, although may be not given explicitly. However, the corresponding integral (expected value) cannot be calculated in a closed form and has to be approximated. Monte Carlo simulation techniques provide such an approximation by averaging a generated a random sample with an appropriate probability distribution. Let us discuss the following two examples of stochastic programming with recourse and a $GI/G/1$ queue.

Example 1.2 Consider the optimization problem

$$\operatorname*{Min}_{x\in S} c^T x + \mathbb{E}\{Q(x, h(\omega))\}, \tag{1.4}$$

where $c \in \mathbb{R}^m$ is a given vector, $Q(x,h)$ is the optimal value of the optimization problem

$$\operatorname*{Min}_{y\geq 0} q^T y \text{ subject to } Wy = h - Ax, \tag{1.5}$$

and $h = h(\omega)$ is a random vector with a known probability distribution. (For the sake of simplicity we assume that only vector h is random while other parameters in the linear programming problem (1.5) are deterministic.) This is the so-called two-stage stochastic programming problem with recourse, which originated in works of Beale [2] and Dantzig [8]. If the random vector h has a discrete distribution, then the expected value function $\mathbb{E}\{Q(x,h)\}$ is given in a form of summation and problem (1.4) can be written as a large linear programming problem. Over the years this approach was developed and various techniques were suggested in order to make it numerically efficient. The interested reader is referred to recent books by Kall and Wallace [16] and Birge and Louveaux [4], and references therein, for an extensive discussion of these methods.

However, the number of realizations of h (the number of discretization points in case the distribution of h is continuous) typically grows exponentially with the dimensionality of h. Consequently, this number can quickly become so large that even modern computers cannot cope with the required calculations. Monte Carlo simulation techniques suggest an approach to deal with this problem. That is, a random sample $h_1, ..., h_N$ of N independent realizations of the random vector h are generated, and the expected value function $\mathbb{E}\{Q(x,h)\}$ is estimated by the average function $\hat{Q}_N(x) := N^{-1}\sum_{i=1}^N Q(x, h_i)$. Consequently the "true" problem (1.4) is approximated by the problem

$$\operatorname*{Min}_{x\in S} c^T x + \hat{Q}_N(x). \tag{1.6}$$

By calculating an optimal solution $\hat{x}_N$ of the above approximating problem, one obtains an estimator of an optimal solution of the true problem.

By the Law of Large Numbers we have that the average function $\hat{Q}_N(x)$ converges, pointwise, to $\mathbb{E}\{Q(x,h)\}$ with probability one, as $N \to \infty$. The function $Q(\cdot, h)$, and hence the function $\hat{Q}_N(\cdot)$, are piecewise linear and convex. The function $Q(\cdot, h)$ is not given explicitly and in itself is an output of an optimization procedure. Nevertheless, its value and a corresponding subgradient can be calculated, at any given point x, by solving the linear program (1.5). This allows to apply, reasonably efficient, deterministic algorithms in order to solve the approximating problem (1.6). For a discussion of such algorithms and a numerical experience in solving two-stage stochastic programming problems by such methods we refer to Shapiro and Homem-de-Mello [33].

Let us make the following observations. The above example is different from the maximum likelihood example in several respects. In the above example the corresponding random sample is generated in the computer and can be controlled to some extend. The only limitation on the number N of generated points is the computational time and computer's memory capacity. It is also possible to implement various variance reduction techniques which in some cases considerably enhance the numerical performance of the algorithm. Usually the feasible set S is defined by constraints. In this respect inequality type constraints appear naturally in optimization problems. In the maximum likelihood example the optimal solution of the "true" problem is actually an unconstrained minimizer of the objective function. There is no reason for such behavior of an optimal solution of the optimization problem (1.4). As we shall see later this introduces an additional term in the asymptotic expansion of $\hat{x}_N$, associated with a curvature of the set S. Let us finally note that the average function $\hat{Q}_N(x)$ is not everywhere differentiable. If the distribution of h is discrete, this is carried over to the expected value function. On the other hand, if the distribution of h is continuous, then the expected value function is smooth (differentiable). This makes the asymptotics of $\hat{x}_N$ quite different in cases of discrete and continuous distributions of h. We shall discuss that later.

Example 1.3 As our last example we consider a $GI/G/1$ queue whose service times depend on a parameter vector x. Let Y_i be the time between arrivals of the $(i-1)^{th}$ and i^{th} customers, and for a given value of x, let $Z_i(x)$ be the service time of the ith customer, $i = 1, 2, \ldots$. Let $G_i(x)$ denote the ith sojourn time, i.e., the total time spent by the ith customer in the queue. It is assumed that the interarrival and service times are random i.i.d., that the first customer arrives at an empty queue and that for every $x \in S$ the queue is regenerative with the expected number of customers served in one busy period (regenerative cycle) being finite. A recursive relation between the sojourn times is given by Lindley equation

$$G_i(x) = Z_i(x) + [G_{i-1}(x) - Y_i]_+. \tag{1.7}$$

Under standard regularity conditions (e.g., [36]), the long-run average functions

$$\hat{f}_N(x) := N^{-1} \sum_{i=1}^{N} G_i(x)$$

converge pointwise, with probability one, to the expected value (mean) steady state sojourn time $f(x)$. Consider the optimization problem

$$\operatorname*{Min}_{x \in S} f(x) + \psi(x), \tag{1.8}$$

where $\psi(x)$ is a (deterministic) cost function. The above "true" problem can be approximated by generating the i.i.d. sequences of the interarrival and service times and then calculating the sojourn times, by using Lindley equation (1.7), and replacing $f(x)$ with its average estimate $\hat{f}_N(x)$. Let us observe that the sojourn times, used in the averaging procedure, are not independent. The approximating functions $\hat{f}_N(x)$ are piecewise smooth. It is possible to extend the above example to more complex queueing systems. It is somewhat surprising that there are examples of simple queues with deterministic service times, depending on a parameter x belonging to an interval of the real line, such that the corresponding expected value steady state sojourn time is not differentiable at a dense set of points on that interval (Shapiro and Wardi [32]).

2 The Delta method

In order to investigate asymptotic properties of the estimators $\hat{v}_N$ and $\hat{x}_N$ it will be convenient to use the Delta method, which we discuss in this section. Let Y_N be a sequence of random vectors, converging in probability to a vector μ. Suppose that there exists a sequence τ_N of positive numbers, tending to infinity, such that $\tau_N(Y_N - \mu)$ converges in distribution to a random vector Y, denoted $\tau_N(Y_N - \mu) \Rightarrow Y$. Let $G(y)$ be a vector valued function, differentiable at μ. That is

$$G(y) - G(\mu) = M(y - \mu) + r(y), \tag{2.1}$$

where $M = \nabla G(\mu)$ is the Jacobian matrix (of first order partial derivatives) of G at μ, and the remainder $r(y)$ is of order $o(\|y - \mu\|)$, i.e., $r(y)/\|y - \mu\| \to 0$ as $y \to \mu$. It follows from (2.1) that

$$\tau_N[G(Y_N) - G(\mu)] = M[\tau_N(Y_N - \mu)] + \tau_N r(Y_N). \tag{2.2}$$

Since $\tau_N(Y_N - \mu)$ converges in distribution, it is bounded in probability, and hence $\|Y_N - \mu\|$ is of stochastic order $O_p(\tau_N^{-1})$. It follows that

$$r(Y_N) = o(\|Y_N - \mu\|) = o_p(\tau_N^{-1}),$$

and hence $\tau_N r(Y_N)$ converges in probability to zero. Consequently we obtain by (2.2) that

$$\tau_N[G(Y_N) - G(\mu)] \Rightarrow MY. \tag{2.3}$$

This formula is routinely employed in multivariate analysis and is known as the (finite dimensional) Delta Theorem (e.g., [24]).

We need to extend this method in several directions. The random functions $\hat{f}_N$ can be viewed as random elements in an appropriate functional space and the corresponding estimators $\hat{v}_N$ and $\hat{x}_N$ as functions of these random elements. This motivates us to extend formula (2.3) to a Banach space setting. Let B_1 and B_2 be two Banach (i.e., linear normed, complete) spaces, and $G : B_1 \to B_2$ be a mapping. Suppose that G is directionally differentiable at a considered point $\mu \in B_1$, i.e., the limit

$$G'_\mu(d) := \lim_{t \downarrow 0} \frac{G(\mu + td) - G(\mu)}{t} \tag{2.4}$$

exists for all $d \in B_1$. If, in addition, the directional derivative $G'_\mu : B_1 \to B_2$ is linear and continuous, then it is said that G is Gâteaux differentiable at μ. Note that, in any case, the directional derivative $G'_\mu(\cdot)$ is positively homogeneous, that is $G'_\mu(\alpha d) = \alpha G'_\mu(d)$ for any $\alpha \geq 0$ and $d \in B_1$.

It follows from (2.4) that

$$G(\mu + d) - G(\mu) = G'_\mu(d) + r(d)$$

with the remainder $r(d)$ being "small" along any fixed direction d, i.e., $r(td)/t \to 0$ as $t \downarrow 0$. This property is not sufficient, however, to neglect the remainder term in the corresponding asymptotic expansion and we need a stronger notion of directional differentiability. It is said that G is directionally differentiable at μ in the sense of Hadamard if the directional derivative $G'_\mu(d)$ exists for all $d \in B_1$ and, moreover,

$$G'_\mu(d) = \lim_{\substack{t \downarrow 0 \\ d' \to d}} \frac{G(\mu + td') - G(\mu)}{t}. \tag{2.5}$$

It is possible to show that if G is Hadamard directionally differentiable at μ, then the directional derivative $G'_\mu(\cdot)$ is continuous, although possibly is not linear. For a discussion of various concepts of directional differentiability see, e.g., [29].

Now let B_1 and B_2 be equipped with their Borel σ-algebras $\mathcal{B}_1$ and $\mathcal{B}_2$, respectively. (Recall that Borel σ-algebra of a normed space is the σ-algebra generated by the family of its open sets.) An $\mathcal{F}$-measurable mapping from a probability space $(\Omega, \mathcal{F}, P)$ into B_1 is called a random element of B_1. Consider a sequence X_N of random elements of B_1. It is said that X_N converges in distribution (weakly) to a random element Y of B_1, and denoted $X_N \Rightarrow Y$, if the expected values $\mathbb{E}\{f(X_N)\}$ converge to $\mathbb{E}\{f(Y)\}$, as $N \to \infty$, for any bounded and continuous function $f : B_1 \to \mathbb{R}$ (see, e.g., Billingsley [3] for a discussion of weak convergence). Let us formulate now the first version of the Delta Theorem.

Theorem 2.1 *Let B_1 and B_2 be Banach spaces, equipped with their Borel σ-algebras, Y_N be a sequence of random elements of B_1, $G : B_1 \to B_2$ be a mapping, and τ_N be a sequence of positive numbers tending to infinity as $N \to \infty$. Suppose that the space B_1 is separable, the mapping G is Hadamard directionally differentiable at a point $\mu \in B_1$, and the sequence $X_N := \tau_N(Y_N - \mu)$ converges in distribution to a random element Y of B_1. Then*

$$\tau_N[G(Y_N) - G(\mu)] \Rightarrow G'_\mu(Y), \tag{2.6}$$

and

$$\tau_N[G(Y_N) - G(\mu)] = G'_\mu(X_N) + o_p(1). \tag{2.7}$$

Note that, because of the Hadamard directional differentiability of G, the mapping $G'_\mu : B_1 \to B_2$ is continuous, and hence is measurable with respect to the Borel σ-algebras of B_1 and B_2. The above infinite dimensional version of the Delta Theorem appeared in works of Gill [11], Grübel [12] and King [17, 18]. It can be proved easily by using the following Skorohod-Dudley almost sure representation theorem, e.g., [23, p.71]).

Representation Theorem. Suppose that a sequence of random elements X_N, of a separable Banach space B, converges in distribution to a random element Y. Then there exists a sequence X'_N, Y', defined on a single probability space, such that $X'_N \stackrel{\mathcal{D}}{=} X_N$, for all N, $Y' \stackrel{\mathcal{D}}{=} Y$ and $X'_N \to Y'$ with probability one.

Here $Y' \stackrel{\mathcal{D}}{=} Y$ means that the probability measures induced by Y' and Y coincide. We give now a proof of theorem 2.1 for the sake of completeness.

Proof of theorem 2.1. Consider the sequence $X_N := \tau_N(Y_N - \mu)$ of random elements of B_1. By the Representation Theorem, there exists a sequence X'_N, Y', defined on a single probability space, such that $X'_N \stackrel{\mathcal{D}}{=} X_N$, $Y' \stackrel{\mathcal{D}}{=} Y$ and $X'_N \to Y'$ w.p.1. Consequently for $Y'_N := \mu + \tau_N^{-1} X'_N$, we have $Y'_N \stackrel{\mathcal{D}}{=} Y_N$. It follows then from Hadamard directional differentiability of G that

$$\tau_N[G(Y'_N) - G(\mu)] \to G'_\mu(Y') \quad \text{w.p.1.} \tag{2.8}$$

Since convergence with probability one implies convergence in distribution and the terms in (2.8) have the same distributions as the corresponding terms in (2.6), the asymptotic result (2.6) follows.

Now since $G'_\mu(\cdot)$ is continuous and $X'_N \to Y'$ w.p.1, we have that

$$G'_\mu(X'_N) \to G'_\mu(Y') \quad \text{w.p.1.} \tag{2.9}$$

Together with (2.8) this implies that the difference between $G'_\mu(X'_N)$ and the left hand side of (2.8) tends w.p.1, and hence in probability, to zero. We obtain that

$$\tau_N[G(Y'_N) - G(\mu)] = G'_\mu[\tau_N(Y'_N - \mu)] + o_p(1),$$

which implies (2.7). ∎

Let us now formulate the second version of the Delta Theorem where the mapping G is restricted to a subset K of the space B_1. We say that G is Hadamard directionally differentiable at a point μ *tangentially* to the set K if for any sequence d_N of the form $d_N := (y_N - \mu)/t_N$, where $y_N \in K$ and $t_N \downarrow 0$, and such that $d_N \to d$, the following limit exists

$$G'_\mu(d) = \lim_{N\to\infty} \frac{G(\mu + t_N d_N) - G(\mu)}{t_N}. \tag{2.10}$$

Equivalently the above condition (2.10) can be written in the form

$$G'_\mu(d) = \lim_{\substack{t\downarrow 0 \\ d' \to_K d}} \frac{G(\mu + td') - G(\mu)}{t}, \tag{2.11}$$

where the notation $d' \to_K d$ means that $d' \to d$ and $\mu + td' \in K$.

Since $y_N \in K$, and hence $\mu + t_N d_N \in K$, the mapping G needs only to be defined on the set K. Recall that the contingent (Bouligand) cone to K at μ, denoted $T_K(\mu)$, is formed by vectors $d \in B$ such that there exist sequences $d_N \to d$ and $t_N \downarrow 0$ such that $\mu + t_N d_N \in K$. Note that $T_K(\mu)$ is nonempty only if μ belongs to the topological closure of K. By the above definitions we have that $G'_\mu(\cdot)$ is defined on the set $T_K(\mu)$. The following "tangential" version of the Delta Theorem can be easily proved in a way similar to the proof of theorem 2.1 (Shapiro [30]).

Theorem 2.2 *Let B_1 and B_2 be Banach spaces, K be a subset of B_1, $G : K \to B_2$ be a mapping, and Y_N be a sequence of random elements of B_1. Suppose that:* (i) *the space B_1 is separable,* (ii) *the mapping G is Hadamard directionally differentiable at a point μ tangentially to the set K,* (iii) *for some sequence τ_N of positive numbers tending to infinity, the sequence $X_N := \tau_N(Y_N - \mu)$ converges in distribution to a random element Y,* (iv) *$Y_N \in K$, with probability one, for all N large enough. Then*

$$\tau_N[G(Y_N) - G(\mu)] \Rightarrow G'_\mu(Y). \tag{2.12}$$

Moreover, if the set K is convex, then equation (2.7) *holds.*

Note that it follows from the assumptions (iii) and (iv) that the distribution of Y is concentrated on the contingent cone $T_K(\mu)$, and hence the distribution of $G'_\mu(Y)$ is well defined.

Our third variant of the Delta Theorem deals with a second order expansion of the mapping G. That is, suppose that G is directionally differentiable at μ and define

$$G''_\mu(d) = \lim_{\substack{t\downarrow 0 \\ d' \to d}} \frac{G(\mu + td') - G(\mu) - tG'_\mu(d')}{\frac{1}{2}t^2}. \tag{2.13}$$

If the mapping G is twice continuously differentiable, then this second order directional derivative $G''_\mu(d)$ coincides with the second order term in the Taylor expansion of

$G(\mu + d)$. The above definition of $G''_\mu(d)$ makes sense for directionally differentiable mappings. However, in interesting applications, where it is possible to calculate $G''_\mu(d)$, the mapping G is actually (Gâteaux) differentiable. We say that G is second order Hadamard directionally differentiable at μ if the second order directional derivative $G''_\mu(d)$, defined in (2.13), exists for all $d \in B_1$. We say that G is second order Hadamard directionally differentiable at μ tangentially to a set $K \subset B_1$ if for all $d \in T_K(\mu)$ the limit

$$G''_\mu(d) = \lim_{\substack{t \downarrow 0 \\ d' \to_K d}} \frac{G(\mu + td') - G(\mu) - tG'_\mu(d')}{\frac{1}{2}t^2} \tag{2.14}$$

exists.

Note that if G is first and second order Hadamard directionally differentiable at μ tangentially to K, then $G'_\mu(\cdot)$ and $G''_\mu(\cdot)$ are continuous on $T_K(\mu)$, and that $G''_\mu(\alpha d) = \alpha^2 G''_\mu(d)$ for any $\alpha \geq 0$ and $d \in T_K(\mu)$.

Theorem 2.3 *Let B_1 and B_2 be Banach spaces, K be a convex subset of B_1, Y_N be a sequence of random elements of B_1, $G : K \to B_2$ be a mapping, and τ_N be a sequence of positive numbers tending to infinity as $N \to \infty$. Suppose that:* (i) *the space B_1 is separable,* (ii) *G is first and second order Hadamard directionally differentiable at μ tangentially to the set K,* (iii) *the sequence $X_N := \tau_N(Y_N - \mu)$ converges in distribution to a random element Y of B_1,* (iv) *$Y_N \in K$ w.p.1 for N large enough. Then*

$$\tau_N^2 \left[G(Y_N) - G(\mu) - G'_\mu(Y_N - \mu)\right] \Rightarrow \tfrac{1}{2}G''_\mu(Y), \tag{2.15}$$

and

$$G(Y_N) = G(\mu) + G'_\mu(Y_N - \mu) + \tfrac{1}{2}G''_\mu(Y_N - \mu) + o_p(\tau_N^{-2}). \tag{2.16}$$

Proof. let X'_N, Y' and Y'_N be elements as in the proof of theorem 2.1. Recall that their existence is guaranteed by the Representation Theorem. Then by the definition of G''_μ we have

$$\tau_N^2 \left[G(Y'_N) - G(\mu) - \tau_N^{-1}G'_\mu(X'_N)\right] \to \tfrac{1}{2}G''_\mu(Y') \quad \text{w.p.1.}$$

Note that $G'_\mu(\cdot)$ is defined on $T_K(\mu)$ and, since K is convex, $X'_N = \tau_N(Y'_N - \mu) \in T_K(\mu)$. Therefore the expression in the left hand side of the above limit is well defined. Since convergence w.p.1 implies convergence in distribution, formula (2.15) follows. Since $G''_\mu(\cdot)$ is continuous on $T_K(\mu)$, and, by convexity of K, $Y'_N - \mu \in T_K(\mu)$ w.p.1, we have that $\tau_N^2 G''_\mu(Y'_N - \mu) \to G''_\mu(Y')$ w.p.1. Since convergence w.p.1 implies convergence in probability, formula (2.16) then follows. ∎

3 First order asymptotics of the optimal value

In this section we discuss asymptotics of the optimal value $\hat{v}_N$, of the approximating problem, based on first order expansions of the optimal value function. We assume that the feasible set S, of the true and approximating problems, is a *compact* subset of $\mathbb{R}^m$. In many interesting applications such assumption cannot be guaranteed and in fact S can be unbounded. Nevertheless, it can be often showed that an optimal solution of the approximating problem stays with probability one in a bounded subset of $\mathbb{R}^m$, and hence we can restrict the optimization procedure to a compact subset of $\mathbb{R}^m$.

Let us consider the Banach space $C(S)$ of continuous functions $y : S \to \mathbb{R}$ equipped with the sup-norm $\|y\| := \sup_{x \in S} |y(x)|$. We assume that the objective function $f(x)$, of the true problem (1.1), is continuous, and hence $f \in C(S)$, and that the approximating functions $\hat{f}_N$ are *random elements* of $C(S)$. Define the optimal value function $\vartheta : C(S) \to \mathbb{R}$ as $\vartheta(y) := \inf_{x \in S} y(x)$. We have then that $v_0 = \vartheta(f)$ and $\hat{v}_N = \vartheta(\hat{f}_N)$.

It is not difficult to see that the optimal value function ϑ is concave and Lipschitz continuous modulus one, i.e., $|\vartheta(y_1) - \vartheta(y_2)| \leq \|y_1 - y_2\|$ for any $y_1, y_2 \in C(S)$. Moreover, it is possible to show (e.g., [30]) that ϑ is Hadamard directionally differentiable at any point $\mu \in C(S)$ and for any $\delta \in C(S)$,

$$\vartheta'_\mu(\delta) = \inf_{x \in S^*(\mu)} \delta(x), \tag{3.1}$$

where $S^*(\mu) := \arg\min_{x \in S} \mu(x)$. Note that the set $S^*(\mu)$ is nonempty since $\mu(x)$ is continuous and S is compact. Together with theorem 2.1 this leads to the following asymptotic result (Shapiro [30]).

Theorem 3.1 *Suppose that, for a sequence τ_N of positive numbers converging to infinity, the sequence $\tau_N(\hat{f}_N - f)$, of random elements of $C(S)$, converges in distribution to a random element Y of $C(S)$. Then*

$$\tau_N(\hat{v}_N - v_0) \Rightarrow \inf_{x \in S^*(f)} Y(x), \tag{3.2}$$

where $S^*(f)$ is the set of optimal solutions of the true problem (1.1). *In particular, if the true problem has unique optimal solution x_0, then*

$$\tau_N(\hat{v}_N - v_0) \Rightarrow Y(x_0). \tag{3.3}$$

Let us specify the above, somewhat abstract, asymptotic result to the case where f is the expected value function, defined in (1.3), and the approximating functions $\hat{f}_N$ are constructed by averaging a random sample. That is,

$$\hat{f}_N(x) := N^{-1} \sum_{i=1}^{N} g(x, \omega_i), \tag{3.4}$$

where $\omega_1, ..., \omega_N$ is an i.i.d. random sample, in $(\Omega, \mathcal{F})$, with the probability distribution P. Let us make the following assumptions.

(A1) For every $x \in S$, the function $g(x, \cdot)$ is $\mathcal{F}$-measurable.

(A2) For some point $x \in S$, the expectation $\mathbb{E}_P\{g(x,\omega)^2\}$ is finite.

(A3) There exists an $\mathcal{F}$-measurable function $\kappa : \Omega \to \mathbb{R}$ such that $\mathbb{E}_P\{\kappa(\omega)^2\}$ is finite and

$$|g(x_1,\omega) - g(x_2,\omega)| \leq \kappa(\omega)\|x_1 - x_2\| \tag{3.5}$$

for all $x_1, x_2 \in S$ and P-almost all $\omega \in \Omega$.

The above assumptions (A1) - (A3) are sufficient for the Central Limit Theorem to hold in $C(S)$. That is, the sequence $N^{1/2}(\hat{f}_N - f)$, of random elements of $C(S)$, converges in distribution to a random element Y (see Araujo and Giné [1], for details). Note that for any fixed point $x_0 \in S$, $Y(x_0)$ is a real valued random variable having normal distribution with zero mean and variance $\sigma^2(x_0)$ equal to the variance of $g(x_0,\omega)$, i.e.,

$$\sigma^2(x_0) = \mathbb{E}_P\{g(x_0,\omega)^2\} - f(x_0)^2. \tag{3.6}$$

We obtain the following results [30].

Theorem 3.2 *Suppose that f and $\hat{f}_N$ are given in the form* (1.3) *and* (3.4), *respectively, with ω_i being an i.i.d. random sample, that the above assumptions* (A1) - (A3) *hold, and that the true problem* (1.1) *has unique optimal solution x_0. Then it follows that $N^{1/2}(\hat{v}_N - v_0)$ converges in distribution to normal $N(0,\sigma^2)$, with variance $\sigma^2 = \sigma^2(x_0)$ given in* (3.6), *and*

$$\hat{v}_N = \hat{f}_N(x_0) + o_p(N^{-1/2}). \tag{3.7}$$

Formula (3.7) shows that, under the assumptions of the above theorem, the optimal value $\hat{v}_N$ of the approximating problem (1.2) is equivalent, up to order $o_p(N^{-1/2})$, to the value of the problem with the same objective function $\hat{f}_N$ and the feasible set S reduced to the single point x_0. This indicates that the above (first order) asymptotics do not depend on the local structure of the set S near the point x_0. Note that since $\hat{f}_N(x_0)$ is an unbiased estimator of $v_0 = f(x_0)$ and that $\hat{v}_N \leq \hat{f}_N(x_0)$, the estimator $\hat{v}_N$ of v_0 typically has a negative bias. We will derive later an approximation of the asymptotic bias of $\hat{v}_N$, of order $O(N^{-1})$, by using a second order expansion of the optimal value function.

Consider the framework of the maximum likelihood example 1.1. Let Θ_0 and Θ_1 be subsets of $\mathbb{R}^m$ and suppose that we wish to test the null hypothesis $H_0 : \theta \in \Theta_0$ against the alternative $H_1 : \theta \in \Theta_1$. Let

$$\ell_N := 2\left[\inf_{\theta\in\Theta_0}\sum_{i=1}^N \ln g(Y_i,\theta) - \inf_{\theta\in\Theta_1}\sum_{i=1}^N \ln g(Y_i,\theta)\right] \tag{3.8}$$

be the corresponding log-likelihood ratio test statistic. Suppose that $f(\theta) := -\mathbb{E}_P\{\ln g(Y,\theta)\}$ has unique minimizers θ_0 and θ_1 over the sets Θ_0 and Θ_1, respectively. Recall that if the distribution P, of the random sample, is given by a pdf $g(x,\theta_0)$, then θ_0 is an unconstrained minimizer of $f(\theta)$. Moreover, if the parameter vector θ is identified at θ_0, then θ_0 is such unique minimizer. We have by (3.7) that

$$N^{-1/2}\ell_N = 2N^{-1/2}\sum_{i=1}^{N}[\ln g(Y_i,\theta_0) - \ln g(Y_i,\theta_1)] + o_p(1), \tag{3.9}$$

provided that the corresponding regularity assumptions (A1) - (A3) hold. It follows that
$N^{-1/2}(\ell_N - \ell_0)$ converges in distribution to normal $N(0,\sigma^2)$, where ℓ_0 and σ^2 are the mean and the variance, respectively, of the random variable $Z := 2\ln[g(Y,\theta_0)/g(Y,\theta_1)]$.

Note that if $\theta_0 = \theta_1$, then this variable Z degenerates into $Z \equiv 0$. Therefore in cases where vectors θ_0 and θ_1 are close to each other (and usually these are the cases we are interested in), the above normal approximation of the distribution of ℓ_N is not accurate. In fact it is possible to obtain a much better approximation of the distribution of ℓ_N by using a second order expansion of the optimal value function. However, in stochastic programming applications the asymptotic result (3.7) is very useful due to its simplicity and generality. The asymptotic variance $\sigma^2(x)$ can be consistently estimated at each iteration point $x = x^\nu$ of a simulation based optimization algorithm. This allows to incorporate t-test type procedures into such algorithms and to construct confidence intervals for the true optimal value v_0 (see [33]).

Let us consider now a situation where the feasible set is defined by constraints which are not given explicitly and should be estimated. That is,

$$S := \{x \in Q : h_i(x) \le 0,\ i = 1, ..., k\}, \tag{3.10}$$

where Q is a closed subset of $\mathbb{R}^m$ and the constraint functions h_i are given as expected values, $h_i(x) := \mathbb{E}\{g_i(x,\omega)\}$, $i = 1, ..., k$. Suppose that each constraint function $h_i(x)$ is real valued (i.e., the corresponding expectation exists), and that $h_i(x)$ can be estimated, say be a sample average, function $\hat{h}_{iN}(x)$. Then the true problem (1.1) can be approximated by the problem

$$\operatorname*{Min}_{x\in S_N} \hat{f}_N(x), \tag{3.11}$$

where

$$S_N := \left\{x \in Q : \hat{h}_{iN}(x) \le 0,\ i = 1, ..., k\right\}. \tag{3.12}$$

It is possible to show that, under mild regularity conditions, the optimal value $\hat{v}_N$ and an optimal solution $\hat{x}_N$ of the above approximating problem (3.12) are consistent estimators of their "true" counterparts. Let us mention recent work of Dupačová and

Wets [10], King and Wets [19] and Robinson [25], where this consistency problem is studied from the point of view of epi-convergence analysis.

Recall that the true problem (1.1) is said to be *convex* if the set Q is convex and the objective function f and the constraint functions h_i, $i = 1, ..., k$, are convex. The Lagrangian function, associated with problem (1.1), is

$$L(x, \lambda) := f(x) + \sum_{i=1}^{k} \lambda_i h_i(x). \tag{3.13}$$

Suppose that the true problem (1.1) is convex and that the Slater condition holds, i.e., there exists a point $\bar{x} \in Q$ such that $h_i(\bar{x}) < 0$, $i = 1, ..., k$. Then with every optimal solution x_0 of (1.1) is associated a nonempty and bounded set $\Lambda(x_0)$ of Lagrange multipliers vectors $\lambda = (\lambda_1, ..., \lambda_k)$ satisfying the optimality conditions:

$$x_0 \in \arg\min_{x \in Q} L(x, \lambda), \quad \lambda_i \geq 0 \text{ and } \lambda_i h_i(x_0) = 0, \; i = 1, ..., k. \tag{3.14}$$

The set $\Lambda(x_0)$ coincides with the set of optimal solutions of the dual of (1.1) problem, and therefore is the same for any optimal solution of (1.1) (see Rockafellar [26]).

Let the set Q be a compact convex subset of $I\!R^m$ and consider the Banach space $B := C(Q) \times ... \times C(Q)$, given by the Cartesian product of $k+1$ replications of the space $C(Q)$. Note that real valued convex functions are continuous and hence $(f, h_1, ..., h_k) \in B$. Denote by K the subset of B formed by $\xi = (\xi_0, ..., \xi_k) \in B$ such that each function $\xi_i(\cdot)$, $i = 0, ..., k$, is convex on Q. Since problem (1.1) is convex, we have that $(f, h_1, ..., h_k) \in K$. Note that the set K is closed and convex in B. Define the optimal value function

$$\vartheta(\xi) := \inf\{\xi_0(x) : x \in Q, \; \xi_i(x) \leq 0, \; i = 1, ..., k\}. \tag{3.15}$$

Clearly, for $\mu := (f, h_1, ..., h_k)$ and $Y_N := (\hat{f}_N, \hat{h}_{1N}, ..., \hat{h}_{kN})$, we have that $\vartheta(\mu) = v_0$ and $\vartheta(Y_N) = \hat{v}_N$.

It is possible to show that the optimal value function $\vartheta(\cdot)$ is Hadamard directionally differentiable at the point $\mu := (f, h_1, ..., h_k)$, tangentially to the set K, provided the Slater condition is satisfied, which together with the Delta Theorem 2.2 imply the following results (Shapiro [30]).

Theorem 3.3 *Suppose that the true problem is convex and that the Slater condition, for the true problem, is satisfied. Then the optimal value function ϑ is Hadamard directionally differentiable at the point $\mu := (f, h_1, ..., h_k)$ tangentially to the set K and for any $\delta \in T_K(\mu)$,*

$$\vartheta'_\mu(\delta) = \inf_{x \in S^*(\mu)} \sup_{\lambda \in \Lambda(\mu)} \left[\delta_0(x) + \sum_{i=1}^{k} \lambda_i \delta_i(x) \right], \tag{3.16}$$

where $S^*(\mu)$ and $\Lambda(\mu)$ are the sets of optimal solutions and Lagrange multipliers, respectively, of the true problem. If, moreover, $Y_N := (\hat{f}_N, \hat{h}_{1N}, ..., \hat{h}_{kN})$ are random *elements of the Banach space B such that with probability one $Y_N \in K$, i.e., the approximating problem* (3.11) *is convex, and $N^{1/2}(Y_N - \mu)$ converges in distribution to a random element $Y = (Y_0, ..., Y_k)$ of B, then*

$$N^{1/2}(\hat{v}_N - v_0) \Rightarrow \inf_{x \in S^*(\mu)} \sup_{\lambda \in \Lambda(\mu)} \left[Y_0(x) + \sum_{i=1}^{k} \lambda_i Y_i(x) \right]. \tag{3.17}$$

The above formula (3.17) indicates that in order to ensure asymptotic normality of $\hat{v}_N$, one needs to assume that the true problem has unique optimal solution x_0 to which corresponds unique Lagrange multipliers vector $\bar{\lambda} = (\bar{\lambda}_1, ..., \bar{\lambda}_k)$. In that case we obtain, assuming that conditions (A1) - (A3) hold for every function $g_i(x, \omega)$, that

$$N^{1/2}(\hat{v}_N - v_0) \Rightarrow N(0, \sigma^2), \tag{3.18}$$

with $\sigma^2 = var\left\{ g(x_0, \omega) + \sum_{i=1}^{k} \bar{\lambda}_i g_i(x_0, \omega) \right\}$.

Without the convexity assumption an asymptotic analysis of stochastic problems like (3.11) is more involved. It is still possible to derive asymptotic normality of the optimal value $\hat{v}_N$, as in (3.18), but under stronger regularity conditions. In particular, one needs to assume Lipschitz continuity of the involved functions and that assumptions like (A1) -(A3) hold for the corresponding Lipschitz constants as well (Shapiro [30]).

4 Second order expansions of the optimal value and asymptotics of optimal solutions

In this section we discuss second order expansions of the optimal value function, which (as we shall see) are closely related to asymptotics of optimal solutions of the approximating problems. We consider the case where the feasible set S is closed (not necessarily convex) and fixed (deterministic) and only the objective function f is subject to perturbations. Unless stated otherwise, we assume throughout this section that the function f is twice continuously differentiable and that the true problem (1.1) has unique optimal solution x_0. By $\nabla f(x)$ and $\nabla^2 f(x)$ we denote the gradient and the Hessian matrix (of second order partial derivatives), respectively, of f at x.

The following first order necessary conditions hold at the point x_0:

$$w^T \nabla f(x_0) \geq 0, \quad \text{for all } w \in T_S(x_0). \tag{4.1}$$

We say that the second order growth condition holds at x_0 if there exist a constant $c > 0$ and a neighborhood $U \subset \mathbb{R}^m$ of x_0 such that

$$f(x) \geq f(x_0) + c\|x - x_0\|^2, \quad \text{for all } x \in S \cap U. \tag{4.2}$$

This condition is closely related to second order optimality conditions. The set

$$C(x_0) := \left\{ w \in T_S(x_0) : w^T \nabla f(x_0) = 0 \right\} \tag{4.3}$$

is called the *critical cone* of the problem (1.1). It represents those directions for which first order conditions (4.1) do not provide information about optimality of x_0. Note that if $\nabla f(x_0) = 0$, then $C(x_0) = T_S(x_0)$. If the distribution P, in the maximum likelihood example 1.1, is given by a pdf $g(y, \theta_0)$, $\theta_0 \in \Theta$, then θ_0 is an unconstrained minimizer of $f(\theta)$ and hence $\nabla f(\theta_0) = 0$. Therefore in that case the critical and tangent cones to the parameter set Θ coincide at the point θ_0.

It turns out that second order optimality conditions, as well as second order expansions of the optimal value function, involve a term related to the curvature of the set S. There are several ways how the curvature of S can be measured. We approach that problem from the following point of view. The set

$$T_S^2(x, d) := \left\{ w \in I\!R^m : \operatorname{dist}\left(x + td + \tfrac{1}{2}t^2 w, S\right) = o(t^2) \right\} \tag{4.4}$$

is called the second order tangent set, to the set S at the point x in the direction d. Here $\operatorname{dist}(x, S) := \inf_{z \in S} \|x - z\|$ denotes the distance from a point x to the set S. Note that $T_S^2(x, d)$ can be nonempty only if $x \in S$ and $d \in T_S(x)$. Yet even if S is convex and $x \in S$ and $d \in T_S(x)$, it can happen that the corresponding second order tangent set is empty.

We also will need the following technical condition. We say that the set S is *second order regular* at the point x_0 if for any vector $d \in T_S(x_0)$ and any sequence $x_N \in S$ of the form $x_N := x_0 + t_N d + \frac{1}{2}t_N^2 w_N$, where $t_N \downarrow 0$ and $t_N w_N \to 0$, the following condition holds

$$\lim_{N \to \infty} \operatorname{dist}\left(w_N, T_S^2(x_0, d)\right) = 0. \tag{4.5}$$

If $w_N \to w$, then $w \in T_S^2(x_0, d)$ by the definition of second order tangent sets, and hence (4.5) holds. The sequence w_N, however, can be unbounded and it is only required that the term $t_N^2 w_N$, in the expansion of x_N, is of order $o(t_N)$. The above second order regularity condition ensures that $T_S^2(x_0, d)$ provides a "sufficiently tight" second order approximation of the set S in the direction d. This condition and a related second order analysis of optimization problems is extensively discussed in the forthcoming book by Bonnans and Shapiro [5]. Note that the second order regularity condition implies that the set $T_S^2(x_0, d)$ is nonempty, and that $\operatorname{dist}(x_0 + td, S) = o(t)$, $t > 0$, for any $d \in T_S(x_0)$.

Under the second order regularity condition, the following second order optimality conditions are *necessary* and *sufficient* for the second order growth condition (4.2) to hold at the point x_0 ([5]):

$$d^T \nabla^2 f(x_0) d + \inf_{w \in T_S^2(x_0, d)} w^T \nabla f(x_0) > 0, \quad \text{for all } d \in C(x_0) \setminus \{0\}. \tag{4.6}$$

Apart from the quadratic term, corresponding to the second order Taylor expansion of the function f, an additional term, associated with the second order tangent set

$T_S^2(x_0, d)$, appears in the left hand side of (4.6). This terms vanishes if $\nabla f(x_0) = 0$. That is what happens in the maximum likelihood example 1.1.

Example 4.1 Suppose that the set S is defined by equality and inequality constraints

$$S := \{x : h_i(x) = 0,\ i = 1, ..., q;\ h_i(x) \le 0,\ i = q + 1, ..., p\}, \tag{4.7}$$

with the constraint functions h_i, $i = 1, ..., p$, being twice continuously differentiable. Let $L(x, \lambda) := f(x) + \sum_{i=1}^p \lambda_i h_i(x)$ be the Lagrangian function of the true problem. Suppose that the following, Mangasarian-Fromovitz [21], constraint qualification holds, at the point x_0:

- the gradient vectors $\nabla h_i(x_0)$, $i = 1, ..., q$, are linearly independent,
- there exists a vector $w \in I\!R^m$ such that $w^T \nabla h_i(x_0) = 0$, $i = 1, ..., q$, and $w^T \nabla h_i(x_0) < 0$, $i \in \mathcal{I}(x_0)$, where

$$\mathcal{I}(x_0) := \{i : h_i(x_0) = 0,\ i = q + 1, ..., p\} \tag{4.8}$$

denotes the set of active at x_0 inequality constraints.

Then

$$T_S(x_0) = \left\{d \in I\!R^m : d^T \nabla h_i(x_0) = 0,\ i = 1, ..., q;\ d^T \nabla h_i(x_0) \le 0,\ i \in \mathcal{I}(x_0)\right\}, \tag{4.9}$$

and first order (Kuhn-Tucker) necessary optimality conditions take the form: there exists a vector $\lambda = (\lambda_1, ..., \lambda_p)$ such that

$$\nabla_x L(x_0, \lambda) = 0,\quad \lambda_i \ge 0,\quad \lambda_i h_i(x_0) = 0,\ i = q + 1, ..., p. \tag{4.10}$$

Under the Mangasarian-Fromovitz constraint qualification, the set $\Lambda(x_0)$ of all Lagrange multipliers vectors λ, satisfying the above conditions (4.10), is nonempty and bounded, and for any $\lambda \in \Lambda(x_0)$ the critical cone can be written as

$$C(x_0) = \left\{d : d^T \nabla h_i(x_0) = 0,\ i \in \{1, ..., q\} \cup \mathcal{I}_+(\lambda),\ d^T \nabla h_i(x_0) \le 0,\ i \in \mathcal{I}_0(\lambda)\right\}, \tag{4.11}$$

where

$$\mathcal{I}_+(\lambda) := \{i \in \mathcal{I}(x_0) : \lambda_i > 0\} \text{ and } \mathcal{I}_0(\lambda) := \{i \in \mathcal{I}(x_0) : \lambda_i = 0\}.$$

Moreover, the set S is second order regular at x_0, and for $d \in T_S(x_0)$,

$$T_S^2(x_0, d) = \left\{w \in I\!R^m : \begin{array}{l} w^T \nabla h_i(x_0) + d^T \nabla^2 h_i(x_0) d = 0,\ i = 1, \cdots, q, \\ w^T \nabla h_i(x_0) + d^T \nabla^2 h_i(x_0) d \le 0,\ i \in \mathcal{I}_1(x_0, d) \end{array}\right\}, \tag{4.12}$$

where

$$\mathcal{I}_1(x_0, d) := \left\{i \in \mathcal{I}(x_0) : d^T \nabla h_i(x_0) = 0\right\}. \tag{4.13}$$

It follows then by duality arguments that the second order conditions (4.6) can be written in the following equivalent form

$$\sup_{\lambda \in \Lambda(x_0)} d^T \nabla^2_{xx} L(x_0, \lambda) d > 0, \text{ for all } d \in C(x_0) \setminus \{0\}. \tag{4.14}$$

We are prepared now to discuss second order expansions of the optimal value function. We assume that the set S is compact and work in the Banach space $W^{1,\infty}(S)$ of Lipschitz continuous functions $y : S \to \mathbb{R}$ equipped with the norm

$$\|y\| := \sup_{x \in S} |y(x)| + \sup \left\{ \frac{|y(x') - y(x)|}{\|x' - x\|} : x, x' \in S,\ x' \neq x \right\}.$$

Since any function $y \in W^{1,\infty}(S)$ is Lipschitz continuous on S, the above norm of y is finite. Consider the optimal value function $\vartheta(y) := \inf_{x \in S} y(x)$, and let $\chi(y)$ be a corresponding optimal solution, i.e., $\chi(y) \in \arg\min_{x \in S} y(x)$. Note that, since it is assumed that the set S is compact, such optimal solution always exists, although possibly is not unique. Let K be the subset of $W^{1,\infty}(S)$ formed by (Fréchet) differentiable at x_0 functions, i.e., $y \in K$ if there exists $\nabla y(x_0) \in \mathbb{R}^m$ such that $y(x) = y(x_0) + (x - x_0)^T \nabla y(x_0) + o(\|x - x_0\|)$ for $x \in S$. Clearly K is a linear subspace of $W^{1,\infty}(S)$. We have then the following second order expansion of $\vartheta(\cdot)$ and first order expansion of $\chi(\cdot)$ in the space $W^{1,\infty}(S)$ tangentially to K (Bonnans and Shapiro [5]).

Theorem 4.2 *Suppose that:* (i) *the true problem has a unique optimal solution* x_0, (ii) *the function* f *is twice continuously differentiable in a neighborhood of the point* x_0, (iii) *the second order growth condition* (4.2) *holds,* (iv) *the set* S *is second order regular at* x_0. *Then the optimal value function* $\vartheta : W^{1,\infty}(S) \to \mathbb{R}$ *is first and second order Hadamard directionally differentiable at* f *tangentially to the space* K, *and for* $\delta \in K$ *it follows that* $\vartheta'_f(\delta) = \delta(x_0)$ *and*

$$\vartheta''_f(\delta) = \inf_{d \in C(x_0)} \left\{ 2d^T \nabla \delta(x_0) + d^T \nabla^2 f(x_0) d + \inf_{w \in T_S^2(x_0, d)} w^T \nabla f(x_0) \right\}. \tag{4.15}$$

Suppose, further, that: (v) *for any* $\delta \in K$ *the optimization problem in the right hand side of* (4.15) *has unique optimal solution* $\bar{d}(\delta)$. *Then the optimal solution function* $\chi(\cdot)$ *is Hadamard directionally differentiable at* f *tangentially to* K *and* $\chi'_f(\delta) = \bar{d}(\delta)$.

Clearly, if $\nabla f(x_0) = 0$, then the last term in the right hand side of (4.15) vanishes. Another situation where this term vanishes is if the set S is polyhedral, i.e., is defined by a finite number of linear constraints. In general this term is related, through the second order tangent set $T_S^2(x_0, d)$, to the curvature of the set S, at the point x_0. For two-stage stochastic programming problems with recourse expansion (4.15) was derived, and extended further to a case with multiple optimal solutions, in Dentcheva and Römisch [9].

In case the set S is defined by smooth constraints, as in (4.7), and the Mangasarian-Fromovitz constraint qualification holds, the set S is second order regular at x_0 and the second order growth condition (4.2) is equivalent to the second order optimality conditions (4.14). Moreover, it is possible to show, by using formula (4.12) and duality arguments, that the second order expansion (4.15) can be written then in the following equivalent form

$$\vartheta''_f(\delta) = \inf_{d \in C(x_0)} \left\{ 2d^T \nabla \delta(x_0) + \sup_{\lambda \in \Lambda(x_0)} d^T \nabla^2_{xx} L(x_0, \lambda) d \right\}. \tag{4.16}$$

Recall that, under the Mangasarian-Fromovitz constraint qualification, the set $\Lambda(x_0)$ of Lagrange multipliers is nonempty and bounded. Note also that the second order sufficient conditions (4.6) (second order sufficient conditions (4.14)) ensure that the infimum in the right hand side of (4.15) (in the right hand side of (4.16)) is attained, although it can be not unique. The optimization problem in the right hand side of (4.16) has a *unique* optimal solution if the function

$$\psi(d) := \sup_{\lambda \in \Lambda(x_0)} d^T \nabla^2_{xx} L(x_0, \lambda) d$$

is strictly convex on the linear space generated by the critical cone $C(x_0)$. In particular, this holds if the Hessian matrix $\nabla^2_{xx} L(x_0, \lambda)$ is positive definite for every $\lambda \in \Lambda(x_0)$.

The above second order expansion of the optimal value function $\vartheta(\cdot)$ and the corresponding first order approximation of the optimal solution mapping $\chi(\cdot)$, together with the Delta method, imply the following asymptotics of the optimal value $\hat{v}_N$ and an optimal solution $\hat{x}_N$ of the approximating problem.

Theorem 4.3 *Suppose that the assumptions* (i)-(iv) *of theorem* 4.2, *for the true problem, are satisfied. Let* τ_N *be a sequence of positive numbers tending to infinity, and* $\hat{f}_N$ *be a sequence of random elements of* $W^{1,\infty}(S)$ *such that the sequence* $\tau_N(\hat{f}_N - f)$ *converges in distribution to a random element* Y *of* $W^{1,\infty}(S)$ *and that* $\hat{f}_N(\cdot)$ *is (Fréchet) differentiable at* x_0 *w.p.1. Then*

$$\tau_N^2 \left[\hat{v}_N - \hat{f}_N(x_0) \right] \Rightarrow \tfrac{1}{2} \vartheta''_f(Y), \tag{4.17}$$

and

$$\hat{v}_N = \hat{f}_N(x_0) + \tfrac{1}{2} \vartheta''_f(\hat{f}_N - f) + o_p(\tau_N^{-2}), \tag{4.18}$$

where $\vartheta''_f(\cdot)$ *is given in* (4.15). *Suppose, further, that the assumption* (v) *of theorem* 4.2 *holds and let* $\bar{d}(\cdot)$ *be the corresponding (unique) optimal solution function associated with the problem* (4.15). *Then*

$$\tau_N (\hat{x}_N - x_0) \Rightarrow \bar{d}(Y). \tag{4.19}$$

Regularity conditions which are required to ensure convergence in distribution of the sequence $\tau_N(\hat{f}_N - f)$ of random elements of the space $W^{1,\infty}(S)$ may be not satisfied in interesting nondifferentiable examples. Nevertheless, even in such cases formulas (4.17) - (4.19) often give correct asymptotics which can be proved by different methods.

Suppose now that the approximating functions $\hat{f}_N$ are constructed by averaging an i.i.d. random sample, as in (3.4). Suppose further that the function $g(\cdot, \omega)$ is Lipschitz continuous on S and (Fréchet) differentiable at x_0 for P-almost every ω. Moreover, suppose that first order derivatives of $f(x)$ can be taken inside the expected value, i.e., formula

$$\nabla f(x) = \mathbb{E}_P\{\nabla_x g(x, \omega)\} \tag{4.20}$$

holds. Then $N^{1/2}\left(\nabla \hat{f}_N(x_0) - \nabla f(x_0)\right)$ converges in distribution to multivariate normal $N(0, \Sigma)$, with the covariance matrix

$$\Sigma = \mathbb{E}_P\left\{[\nabla_x g(x_0, \omega)][\nabla_x g(x_0, \omega)]^T\right\} - \nabla f(x_0)\nabla f(x_0)^T, \tag{4.21}$$

provided that the second order moments of $\nabla_x g(x_0, \omega)$ do exist. We obtain therefore the following results.

Theorem 4.4 *Suppose that the assumptions* (i)-(iv) *of theorem* 4.2, *for the true problem, are satisfied. Let the approximating function $\hat{f}_N$ be constructed by averaging an i.i.d. random sample, and suppose that the function $g(\cdot, \omega)$ is Lipschitz continuous on S and (Fréchet) differentiable at x_0 w.p.1, that the interchangeability formula* (4.20) *holds, and that $N^{1/2}(\hat{f}_N - f)$ are random elements of $W^{1,\infty}(S)$ converging in distribution. Then*

$$N\left[\hat{v}_N - \hat{f}_N(x_0)\right] \Rightarrow \tfrac{1}{2}\varphi(Z) \tag{4.22}$$

and

$$\hat{v}_N = \hat{f}_N(x_0) + \tfrac{1}{2}\varphi(\zeta_N) + o_p(N^{-1}), \tag{4.23}$$

where $Z \sim N(0, \Sigma)$ is a random vector having multivariate normal distribution with the covariance matrix Σ given in (4.21), *$\zeta_N := \nabla \hat{f}_N(x_0) - \nabla f(x_0)$, and*

$$\varphi(\zeta) := \inf_{d \in C(x_0)} \left\{2d^T\zeta + d^T\nabla^2 f(x_0)d + \inf_{w \in T_S^2(x_0,d)} w^T\nabla f(x_0)\right\}. \tag{4.24}$$

Suppose, further, that for any vector $\zeta \in \mathbb{R}^m$ the optimization problem in the right hand side of (4.24) *has a unique optimal solution, denoted $\bar{d}(\zeta)$. Then*

$$N^{1/2}(\hat{x}_N - x_0) \Rightarrow \bar{d}(Z). \tag{4.25}$$

In case the set S is defined by smooth constraints, as in (4.7), and the Mangasarian-Fromovitz constraint qualification holds, the function $\varphi(\cdot)$, defined in (4.24), can be written in the following equivalent from

$$\varphi(\zeta) = \inf_{d \in C(x_0)} \left\{2d^T\zeta + \sup_{\lambda \in \Lambda(x_0)} d^T\nabla^2_{xx}L(x_0, \lambda)d\right\}. \tag{4.26}$$

In that form formulas (4.23) and (4.25) were derived in Shapiro [28] by a different method. Asymptotics of the optimal solution $\hat{x}_N$ were also derived in King and Rockafellar [20] by the Delta method in a framework of variational inequalities (generalized equations).

Note that the "curvature term" (involving the second order tangent set $T_S^2(x_0, d)$) in the expansion (4.24) vanishes in two cases, namely if $\nabla f(x_0) = 0$ or if the set S is polyhedral. In such cases $\nabla_{xx}^2 L(x_0, \lambda) = \nabla^2 f(x_0)$ for any $\lambda \in \Lambda(x_0)$.

Since $\hat{f}_N(x_0)$ is an unbiased estimator of v_0, we can view the term $\frac{1}{2}I\!E\{\varphi(Z)\}$, where $Z \sim N(0, \Sigma)$, as the asymptotic bias of $\hat{v}_N$, of order $O(N^{-1})$. Note that $\varphi(\zeta) \leq 0$ for any $\zeta \in I\!R^m$, and hence this asymptotic bias is negative.

The optimal solution $\bar{d}(\zeta)$ can be a nonlinear function of ζ even if this optimal solution is unique. In that case the distribution of $\bar{d}(Z)$ is not normal and hence $\hat{x}_N$ is not asymptotically normal (this was pointed out by King [17]). For example, let S be defined by constraints, as in (4.7), and suppose that the the gradient vectors $\nabla h_i(x_0)$, $i \in \{1, ..., q\} \cup \mathcal{I}(x_0)$, are linearly independent. Then $\Lambda(x_0) = \{\bar{\lambda}\}$ is a singleton and $\varphi(\zeta)$ and $\bar{d}(\zeta)$ are the optimal value and an optimal solution of the problem

$$\begin{array}{ll} \text{Min}_{d \in I\!R^m} & 2d^T\zeta + d^T\nabla_{xx}^2 L(x_0, \bar{\lambda})d \\ \text{subject to} & d^T\nabla h_i(x_0) = 0,\ i \in \{1, ..., q\} \cup \mathcal{I}_+(\bar{\lambda}),\ d^T\nabla h_i(x_0) \leq 0,\ i \in \mathcal{I}_0(\bar{\lambda}). \end{array} \tag{4.27}$$

This is a quadratic programming problem. The above linear independence condition implies that it has a unique vector $\bar{\alpha}(\zeta)$ of Lagrange multipliers, and that it has a unique optimal solution $\bar{d}(\zeta)$ if the Hessian matrix $\nabla_{xx}^2 L(x_0, \bar{\lambda})$ is positive definite over the linear space defined by the first $q + |\mathcal{I}_+(\bar{\lambda})|$ (equality) linear constraints in (4.27).

If, furthermore, the strict complementarity condition holds, i.e., $\bar{\lambda}_i > 0$ for all $i \in \mathcal{I}(x_0)$, or in other words $\mathcal{I}_+(\bar{\lambda}) = \mathcal{I}(x_0)$ and $\mathcal{I}_0(\bar{\lambda}) = \emptyset$, then $\bar{d}(\zeta)$ and $\bar{\alpha}(\zeta)$ can be obtained as solutions of the following system of linear equations

$$\begin{bmatrix} H & A \\ A^T & 0 \end{bmatrix} \begin{bmatrix} d \\ \alpha \end{bmatrix} = -\begin{bmatrix} \zeta \\ 0 \end{bmatrix}. \tag{4.28}$$

Here $H := \nabla_{xx}^2 L(x_0, \bar{\lambda})$ and A is the $m \times (q + |\mathcal{I}(x_0)|)$ matrix whose columns are formed by vectors $\nabla h_i(x_0)$, $i \in \{1, ..., q\} \cup \mathcal{I}(x_0)$. We obtain in that case, provided the block matrix in the left hand side of (4.28) is nonsingular, that $N^{1/2}(\hat{x}_N - x_0, \hat{\lambda}_N - \bar{\lambda})$ converges in distribution to normal with zero mean and the covariance matrix

$$\begin{bmatrix} H & A \\ A^T & 0 \end{bmatrix}^{-1} \begin{bmatrix} \Sigma & 0 \\ 0 & 0 \end{bmatrix} \begin{bmatrix} H & A \\ A^T & 0 \end{bmatrix}^{-1}. \tag{4.29}$$

It can happen that the critical cone $C(x_0)$ consists of the single point 0, i.e., $C(x_0) = \{0\}$. In that case the functions $\varphi(\cdot)$ and $\bar{d}(\cdot)$ are identically zero and the corresponding asymptotics are different. For example, if the set S is defined by constraints, as in (4.7), and the Mangasarian-Fromovitz constraint qualification holds, then it follows from formula (4.11) that $C(x_0) = \{0\}$ if the gradient vectors $\nabla h_i(x_0)$, $i \in \{1, ..., q\} \cup \mathcal{I}_+(\lambda)$, generate the space $I\!R^m$. In particular this happens if the number of active inequality constraints at x_0 is $m - q$ (i.e., $|\mathcal{I}(x_0)| = m - q$), the gradient vectors $\nabla h_i(x_0)$, $i \in \{1, ..., q\} \cup \mathcal{I}(x_0)$, are linearly independent and all Lagrange multipliers corresponding to the active inequality constraints are positive.

Suppose that $C(x_0) = \{0\}$. In that case there exists a neighborhood U of $\nabla f(x_0)$ such that if $\nabla \hat{f}_N(x_0) \in U$, then the first order optimality conditions for the approximating problem hold at the point x_0, and x_0 is a locally optimal solution of the approximating problem. By the strong Law of Large Numbers, we have that $\nabla \hat{f}_N(x_0)$ converges to $\nabla f(x_0)$ w.p.1. Consequently, w.p.1 for N large enough, $\nabla \hat{f}_N(x_0) \in U$, and hence x_0 is a locally optimal solution of the approximating problem. It follows then that $\hat{x}_N = x_0$ w.p.1 for N large enough. Moreover, by the Large Deviations theory (e.g., [6]) we have, under mild regularity conditions, that the probability of the event $\nabla \hat{f}_N(x_0) \notin U$ tends to zero exponentially fast as $N \to \infty$, and hence the asymptotic bias of $\hat{v}_N$ approaches zero at an exponential rate.

Let us finally remark that it is also possible to derive similar asymptotics, of the optimal value and optimal solutions, in cases where the feasible set is defined by constraints and the constraint functions are estimated by corresponding sample averages (Rubinstein and Shapiro [27, section 6.6]).

5 Examples and a discussion

Consider the framework of the maximum likelihood example 1.1. Suppose that the parameter set Θ is compact and that the distribution P, of the corresponding random sample, is given by a pdf $g(y, \theta_0)$, $\theta_0 \in \Theta$, from the considered parametric family. Suppose also that for P-almost every y, the function $\ln g(y, \cdot)$ is continuously differentiable in a neighborhood of Θ, and that the corresponding assumptions (A1) - (A3) hold for the function $\ln g(y, \theta)$ and its first order partial derivatives $\partial \ln g(y, \theta)/\partial \theta_i$. Then, since θ_0 is an unconstrained minimizer of $f(\theta)$, and $\nabla f(\theta) = -I\!E_P\{\nabla_\theta \ln g(Y, \theta)\}$, we obtain that $\nabla f(\theta_0) = 0$. Suppose, further, that the expected value function $f(\theta)$ is twice continuously differentiable at θ_0 (note that this property does not follow from the above assumptions), that the parameter vector θ is identified at θ_0 (and hence the minimizer θ_0 is unique), that the second order growth condition holds at θ_0 and that the set Θ is "sufficiently regular" near θ_0. Then the corresponding asymptotic expansions given in theorem 4.3 hold.

Since $\nabla f(\theta_0) = 0$, we have here that $C(\theta_0) = T_\Theta(\theta_0)$ and the third term in the right hand side of (4.24) vanishes. The covariance matrix Σ is equal here to $I(\theta_0)$, where

$$I(\theta_0) := I\!E\left\{[\nabla_\theta \ln g(Y, \theta_0)][\nabla_\theta \ln g(Y, \theta_0)]^T\right\}$$

is Fisher's information matrix. As it is well known, under second order smoothness assumptions about the function $\ln g(y,\theta)$, we also have that $\nabla^2 f(\theta_0) = I(\theta_0)$. Consequently, the second order growth condition is ensured here by the condition: $d^T I(\theta_0) d > 0$ for all nonzero $d \in T_\Theta(\theta_0)$. In particular, this holds if $I(\theta_0)$ is nonsingular, and hence is positive definite. By (4.23) we obtain that

$$\sup_{\theta\in\Theta} \sum_{i=1}^{N} \ln g(Y_i,\theta) = \sum_{i=1}^{N} \ln g(Y_i,\theta_0) + \sup_{d\in T_\Theta(\theta_0)} \left\{ d^T Z_N - \tfrac{1}{2} d^T I(\theta_0) d \right\} + o_p(1),$$

where $Z_N := N^{-1/2}\sum_{i=1}^N \nabla_\theta \ln g(Y_i,\theta)$. Note that $Z_N \Rightarrow N(0, I(\theta_0))$.

Consider the log-likelihood ratio statistic ℓ_N, defined in (3.8), for testing $H_0 : \theta \in \Theta_0$ against $H_1 : \theta \in \Theta_1$. Suppose that the true value θ_0 of the parameter vector belongs to both sets Θ_0 and Θ_1, that the information matrix $I(\theta_0)$ is nonsingular, and define $W_N := I(\theta_0)^{-1} Z_N$. Note that $W_N \Rightarrow N(0, I(\theta_0)^{-1})$. We obtain then the following expansion of ℓ_N,

$$\ell_N = \inf_{d\in T_{\Theta_0}(\theta_0)} (W_N - d)^T I(\theta_0)(W_N - d) - \inf_{d\in T_{\Theta_1}(\theta_0)} (W_N - d)^T I(\theta_0)(W_N - d) + o_p(1).$$

It also follows that if $\hat{\theta}_N$ is the maximum likelihood estimator of θ_0 under H_0 (under H_1), then $N^{1/2}(\hat{\theta}_N - \theta_0)$ converges in distribution to $\bar{d}(W)$, where $W \sim N(0, I(\theta_0)^{-1})$ and $\bar{d}(w)$ is the minimizer of $(w-d)^T I(\theta_0)(w-d)$ over $T_{\Theta_0}(\theta_0)$ (over $T_{\Theta_1}(\theta_0)$). This result goes back to Chernoff [7].

The above discussion shows that the example of maximum likelihood is quite specific from the point of view of general stochastic optimization problems. In that example the gradient of the objective function of the true problem is zero at the optimal solution, and consequently the "curvature term" vanishes from the corresponding second order expansions of the optimal value function.

Before proceeding further let us state the following useful proposition. It can be easily proved by using the Lebesgue dominated convergence theorem (e.g., [27, pp. 70,71]).

Proposition 5.1 *Let $f(x)$ be the expected value function defined in (1.3). Suppose that the expectation $\mathbb{E}_P\{g(x,\omega)\}$ exists for all x in a neighborhood of x_0, that for P-almost every ω the function $g(\cdot,\omega)$ is directionally differentiable at x_0, and that there exists a random variable $\kappa(\omega) \geq 0$ such that $\mathbb{E}_P\{\kappa(\omega)\}$ is finite and*

$$|g(x_1,\omega) - g(x_2,\omega)| \leq \kappa(\omega)\|x_1 - x_2\| \tag{5.1}$$

for all x_1, x_2 in a neighborhood of x_0 and P-almost all ω . Then the function $f(x)$ is Lipschitz continuous near x_0, directionally differentiable at x_0 and

$$f'(x_0,d) = \mathbb{E}_P\{g'_\omega(x_0,d)\}, \tag{5.2}$$

where $g'_\omega(x_0,d)$ denotes the directional derivative of $g(\cdot,\omega)$ at x_0 in the direction d. Moreover, if $g(\cdot,\omega)$ is differentiable at x_0 w.p.1, then the interchangeability formula (4.20) holds at $x = x_0$.

Let us discuss now the two-stage stochastic programming example 1.2. Consider the function

$$G(z) := \inf\{q^T y : Wy = z, \ y \geq 0\}.$$

Clearly the function $Q(x, h)$, given as the optimal value of the problem (1.5), can be written as $Q(x, h) = G(h - Ax)$. By duality arguments of linear programming we have that

$$G(z) = \sup\{\xi^T z : W^T \xi \leq q\},$$

provided the set $\{\xi : W^T\xi \leq q\}$ is nonempty. So let us suppose, for the sake of simplicity, that this set is nonempty and bounded. Then the function $G(z)$ is a real valued piecewise linear convex function. Suppose also that the expectation $\mathbb{E}\{Q(x,h)\}$ exists for all x.

It follows that the approximating function $\hat{f}_N(x) := c^T x + N^{-1}\sum_{i=1}^N Q(x, h_i)$ is a piecewise linear convex function, and hence is not everywhere differentiable. Therefore, the involved asymptotics are quite different depending on whether the distribution of the random vector h is continuous or discrete. Suppose first that the random vector h has a continuous distribution with a density function $g(\cdot)$. Let us fix a point $x_0 \in \mathbb{R}^m$. Since the function $G(z)$ is convex, the set of points where it is not differentiable has Lebesgue measure zero. Since h has a density, it follows then that the function $Q(\cdot, h)$ is differentiable at x_0 w.p.1. Together with (5.2) this implies that $f(x)$ is differentiable at x_0 and $\nabla f(x_0) = c + \mathbb{E}\{\nabla_x Q(x_0, h)\}$. If, moreover, the density function $g(\cdot)$ is continuous, then $f(x)$ is twice continuously differentiable (Wang [35]). In that case the asymptotic formulas (4.22), (4.23) and (4.25), of theorem 4.4, with the covariance matrix Σ of $Z \sim N(0, \Sigma)$ defined in (4.21), make sense. Under some mild assumptions about the density function $g(\cdot)$, these formulas can be proved by a different method, which is based on a stochastic mean value theorem due to Huber [14], (Shapiro [31]).

Let us finally mention that in case the random vector h has a discrete distribution the situation is quite different. It is possible to show that in such case and if the true problem has unique optimal solution x_0, then the probability of the event that $\hat{x}_N$ is exactly equal to x_0 approaches one exponentially fast (Shapiro and Homem-de-Mello [34]).

References

[1] A. Araujo and E. Giné, *The Central Limit Theorem for Real and Banach Valued Random Variables.* Wiley, New York, 1980.

[2] E. Beale, "On minimizing a convex function subject to linear inequalities", *J. Roy. Statist. Soc., Ser. B*, 17 (1955), 173-184.

[3] P. Billingsley, *Convergence of Probability Measures.* Wiley, New York, 1968.

[4] J.R. Birge and F. Louveaux, *Introduction to Stochastic Programming.* Springer, New York, 1997.

[5] J.F. Bonnans and A. Shapiro, *Perturbation Analysis of Optimization Problems.* Springer-Verlag, New York, 2000.

[6] J.A. Bucklew, *Large Deviation Techniques in Decision, Simulation, and Estimation.* Wiley, New York, 1990.

[7] H. Chernoff, "On the distribution of the likelihood ratio", *Ann. Math. Statist.*, 25 (1954), 573-578.

[8] G. Dantzig, "Linear programming under uncertainty", *Management Sci.*, 1 (1955), 197-206.

[9] D. Dentcheva and W. Römisch, "Differential stability of two-stage stochastic programs", *SIAM J. Optimization*, to appear.

[10] J. Dupačová and R.J.B. Wets, "Asymptotic behavior of statistical estimators and of optimal solutions of stochastic optimization problems," *The Annals of Statistics* 16 (1988) 1517-1549.

[11] R.D. Gill, "Non-and-semiparametric maximum likelihood estimators and the von Mises method (Part I)", *Scandinavian Journal of Statistics*, 16 (1989), 97-124.

[12] R. Grübel, "The length of the short", *Ann. Statist.*, 16 (1988), 619-628.

[13] P.J. Huber, "Robust estimation of a location parameter", *Ann. Math. Statist.*, 35 (1964), 73-101.

[14] P.J. Huber, "The behavior of maximum likelihood estimates under nonstandard conditions", *Proc. Fifth Berkeley Symp.Math. Statist. Probab.*, 1 (1967), 221-233, Univ. California Press.

[15] P.J. Huber, *Robust Statistics.* Wiley, New York, 1981.

[16] P. Kall and S.W. Wallace, *Stochastic Programming.* Wiley, Chichester, 1994.

[17] A.J. King, "Asymptotic behavior of solutions in stochastic optimization: Nonsmooth analysis and the derivation of non-normal limit distributions", Ph.D. dissertation, Dept. Applied Mathematics, Univ. Washington, 1986.

[18] A.J. King, "Generalized delta theorems for multivalued mappings and measurable selections", *Mathematics of Operations Research*, 14 (1989), 720-736.

[19] A.J. King and R.J.-B. Wets, "Epi-consistency of convex stochastic programs," *Stochastics*, 34 (1991), 83-92.

[20] A.J. King and R.T. Rockafellar, "Asymptotic theory for solutions in statistical estimation and stochastic programming", *Mathematics of Operations Research,* 18 (1993), 148-162.

[21] O.L. Mangasarian and S. Fromovitz, "The Fritz John necessary optimality conditions in the presence of equality and inequality constraints", *Journal of Mathematical Analysis and Applications,* 7 (1967), pp. 37-47.

[22] E.L. Plambeck, B.R. Fu, S.M. Robinson and R. Suri, "Sample-Path Optimization of Convex Stochastic Performance Functions", *Mathematical Programming,* vol. 75 (1996), no. 2, 137-176.

[23] D. Pollard, *Convergence of Stochastic Processes.* Springer-Verlag, New York, 1984.

[24] C.R. Rao, *Linear Statistical Inference and Its Applications.* Wiley, New York, 1973.

[25] S.M. Robinson, "Analysis of sample-path optimization, " *Math. Oper. Res.*, 21 (1996), 513-528.

[26] R.T. Rockafellar, *Convex Analysis.* Princeton University Press, 1970.

[27] R.Y. Rubinstein and A. Shapiro, *Discrete Event Systems: Sensitivity Analysis and Stochastic Optimization by the Score Function Method.* Wiley, New York, NY, 1993.

[28] A. Shapiro, "Asymptotic Properties of Statistical Estimators in Stochastic Programming," *Annals of Statistics,* 17 (1989), 841-858.

[29] A. Shapiro, "On concepts of directional differentiability," *Journal Optim. Theory and Appl.,* 66 (1990), 477-487.

[30] A. Shapiro, "Asymptotic analysis of stochastic programs," *Annals of Operations Research,* 30 (1991), 169-186.

[31] A. Shapiro, "Asymptotic Behavior of Optimal Solutions in Stochastic Programming," *Mathematics of Operations Research,* 18 (1993), 829-845.

[32] A. Shapiro and Y. Wardi, "Nondifferentiability of the steady-state function in Discrete Event Dynamic Systems", *IEEE transactions on Automatic Control,* 39 (1994), 1707-1711.

[33] A. Shapiro and T. Homem-de-Mello, "A simulation-based approach to two-stage stochastic programming with recourse", *Mathematical Programming,* 81 (1998), 301-325.

[34] A. Shapiro and T. Homem-de-Mello, "On rate of convergence of Monte Carlo approximations of stochastic programs", *SIAM J. Optimization*, to appear.

[35] J. Wang, "Distribution sensitivity analysis for stochastic programs with complete recourse", *Mathematical Programming*, 31 (1985), 286-297.

[36] R.W. Wolff, *Stochastic Modeling and the Theory of Queues*. Prentice Hall, Englewood Cliffs, NJ, 1989.

MIX
Papier aus verantwortungsvollen Quellen
Paper from responsible sources
FSC® C105338

If you have any concerns about our products,
you can contact us on
ProductSafety@springernature.com

In case Publisher is established outside the EU,
the EU authorized representative is:
Springer Nature Customer Service Center GmbH
Europaplatz 3, 69115 Heidelberg, Germany

Printed by Libri Plureos GmbH
in Hamburg, Germany